Tea: Genome and Genetics

Tapan Kumar Mondal

Tea: Genome and Genetics

Springer

Tapan Kumar Mondal
ICAR—National Institute for Plant Biotechnology
Indian Council of Agricultural Research
PUSA, Delhi, India

ISBN 978-981-15-8870-9 ISBN 978-981-15-8868-6 (eBook)
https://doi.org/10.1007/978-981-15-8868-6

This Springer imprint is published by the registered company Springer Nature Singapore Pte Ltd.
The registered company address is: 152 Beach Road, #21-01/04 Gateway East, Singapore 189721, Singapore

Dedicated to my beloved father Dr. Sankardas Mondal and mother late Smt. Gauri Mondal

Foreword

Several million people of different cultures around the world start the day by drinking tea in the morning. Tea is an important plantation crop that provides morning favorite warm drinks. The tea industry not only generates employment but is an eco-friendly industry that conserves the nature. The increasing world population and reduction of land demand the vertical growth of tea productivity. With all great effort, the yield achieved has been 16,000 kg green leaf/ha but being an agro-based industry it is also vulnerable to climate change. Thus, the need of the hour of this industry is to develop high yielding, quality climate resilient tea clone. Thus altogether there is a need to apply the scientific endeavor to increase the yield at lower cost of production, which can only be possible with the application of modern genomics tools along with conventional breeding.

Botanically tea is mainly of two types, i.e., *Camellia sinensis* or china type with short leaf as well as smaller bush size and *Camellia assamica* or Assam type with bigger leaf as well as larger bush type. A special high-quality tea is also well known as "Darjeeling tea" with GI tag grown in Darjeeling, West Bengal. However though they are different species but they bred so freely that distinct morphological; classes are missing and some people also consider the third category *Camellia assamica* subspp. Lasiocalyx. Due to several botanical constrains such as cross-pollination with longer gestation period and nonavailability of mutants, conventional breeding is slow which is restricted to selection of elite bushes only. Systematic plant breeding approaches to harvest the genetic gain such as transgressive breeding or targeted backcross breeding have not been applied in tea for varietal improvement. On the other hand, among more than 300 species in the genus *Camellia* only these two have been domesticated for drinking purpose though *Camellia taliensis* in a limited scale is also used for producing tea in some countries such as China. So far pre-breeding or using wild species as pollen donor in tea breeding has not been attempted seriously, which has the potential to develop climate resilient, high yielding quality tea clones. However among many, one reason could be the non-availability or use of the genomics resource for improvement of this genus.

Nevertheless, since 2018, two Chinese groups and one Indian group simultaneously decoded the genomes of both the species of tea which have made a quantum jump for generation of data across the world in various fields such as transcriptomics, proteomics, and metabolomics which resulted in detailed insights about the genomic resources such as various DNA markers which have been displayed by various data bases for end users. This will surely have an impact for a better understanding of tea quality trait and marker assistant breeding for QTL identification through association mapping approaches.

I am pleased and happy to say that Dr. Tapan Kumar Mondal, Principal Scientist, ICAR-National Institute for Plant Biotechnology, New Delhi, has made significant contributions in the field of tea biotechnology research including its genome sequence. His own research and contributions in the area of chromosome research, breeding, tissue culture, physiology, and structural/functional genomics are synthesized in this book. I am sure this book will be very useful to researchers and planners not only working in the tea industry but also in similar woody species in addition to having utility among policy makers and science managers. Thus, I feel humbled and extremely happy to write the foreword for this book.

Distinguished Biotechnological Research
University of Calcutta
Kolkatta, India

Prof. Swapan Kumar Datta

Ex Vice Chancellor
Visva-Bharati University
Santiniketan, India

Ex-Deputy Director General (Crop Science)
Indian Council of Agricultural research
Government of India
New Delhi, India
30 August 2020

Preface

Tea is an important industrial crop that supports the life of several millions of plantation workers globally mostly in disadvantageous areas. At the same time, it is also the morning drink of several million people worldwide. Interestingly, several wild species such as *C. japonica* are also important due to its elegant flower color. Due to perennial nature with a life span more than 100 years, breeding to improve the cultivars is difficult and limited to few aspects only. During the last four decades, initially as a student, later as a teacher and humble science worker, I was and surely will remain fascinated by this beautiful plant whose not only test but also scenic beauty of plantation always refresh the mind. While working with this plant, at various tea research institutes in nearly four decades, I have experienced the present practices, gaps, and scope of varietal improvement works and felt the need of in vitro culture, molecular breeding, and genomics to supplement the conventional breeding works. With the initiation of cell culture technique in 1968, a significant amount of work on various aspects of cytogenetic, breeding, physiology, biochemistry, and functional genomics of tea and its wild relatives has been taken place. Although several topical reviews, scientific articles, and few books have been published on tea and *Camellia* species, I have felt the need to have an update on this area.

In 2014, my first book *Breeding and Biotechnology of Tea and Wild Relatives* was published, which documented all the research findings till that time. After that 3 genomes of tea were decoded by different groups. Thus, while surveying the literatures, two important things I could notice. First, there are several discoveries of new species of Camellia particularly from Vietnam and some parts of Louse and Eastern China, indicating that speciation under this genus is still very active. Second, massive works have been done on functional genomics followed by metabolomics and proteomics of various kinds mainly targeting some useful traits relevant to tea as well as Camellia or to characterize the mutants that are available in the natural germplasm. Gene expression during manufacturing has also been studied. I have tried to capture all these findings in this book. Besides I have also updated the information related to respective chapters of my previous books.

I owe my sincere debt of gratitude towards my teachers who blessed me to learn about this crop and plant biotechnology as a whole. Therefore, I sincerely acknowledge my thanks to my beloved teachers of Assam Agricultural University, late Prof. P.S. Ahuja, Ex-Director General, CSIR, Government of India and other Scientists of the Institute of Himalayan Bioresource Technology, India, Prof. Pradip Kumar Chand, Vice Chancellor, North Orissa University, India, Scientists of UPASI, Tamil Nadu, Tea Board, Kolkata, Tocklai Tea Research Institute, Assam, and Prof. P. C. Deka, Vice Chancellor, Sir Padampat Singhania University, Rajasthan. Few people also inspired me further to work on tea breeding, and they are the late Prof. N. K. Jain, Mr. P. Haridas, Tata Tea Ltd., Munnar, and some of my planter friends of Southern India, Dooars, West Bengal as well as in Assam. I would also like to give my special thanks to Dr. Tilak Raj Sharma, Deputy Director General (Crop Science), Indian Council of Agricultural Research, New Delhi for his guidance as a mentor.

I would also like to thank my wife Dr. Bipasa Sarkar who helped me to improve the manuscript in several ways. Lastly my son Vaibhav, my younger sister Tia and her family, i.e., Priya and Shyamlendu, elder brother Prof. Swapan Kumar Mondal and his family, i.e., Simadi and Joy, Kaku, and his family, little niece Stuti are also gratefully acknowledged. I am also thankful to the noted tea scientists, Prof. C.R. Park of USA, Prof. A.M. Vieitez of Spain, Prof. S Yamaguchi of Japan, Prof. Z. Chen of China, Prof. I.D. Singh of India, late Dr. S.C. Das, and Prof. L.M.S. Palni, India, for my personal interactions with them since my student days. I apologize for those works, if any, which did not appear in this book despite a detailed search worldwide.

I am also grateful to my Ph.D. students Dr. Pranay Bantawa, Dr. Olivia Saha Roy, Dr. Akan Das, Dr. Pratap Subba, Dr. Mainaak Mukhopadhyay, Dr. Showkat Ahmed Ganie, Miss Soni Chowrasia, Miss Jyoti Nishad, Miss Megha Rohila, Alok kumar Panda, Mr. Abhishek Mazumder, Mr. Hukum C Rawal as I was enriched while working with them. It is my sincere belief that this book will serve the requirement of students, scientists, and industries involved in studies, teaching, and research on breeding and biotechnology of tea and other *Camellia* species with an intension of serving science and society.

New Delhi, India — Tapan Kumar Mondal

About the Book

Tea is an important non-alcoholic beverage plant of the world. The cultivation of tea is very important as it earns revenue for the tea growing nations specially developing countries such as India. To bring down the cost of production, cultivation practices have been standardized over decades of research. Although several improved geographically adopted clones have been developed by conventional breeding which has contributed significantly to industries of various counties, for varietal improvement of tea and other Camellia species with ornamental value. However, genetic gain through systematic breeding with control cross has not been explored. It is slow for several decades due to the constraint of botany of the plant, non-availability of mutant and sufficient wild tea but the main reason was non-availability of genetic and genomic information of tea. But in the past few decades several works related to genetics, genomics, and biotechnology have been done including cytogenetic, traditional breeding, tissue culture, as well as DNA-based markers. With the recent introduction of the genome sequencing of tea, the scope of application of marker-assistant breeding and biotechnology has been increased manyfold. Thus in this book, the state of the art on various aspects of breeding and biotechnology work of tea and other wild *Camellia* species has been compiled in eight chapters. They are (1) Introduction that deals with the origin and descriptions of health benefits as well as morphological classification as first chapter, (2) Genetics and breeding that comprise cytogenetic effects, taxonomy of the genus Camellia, and various conventional breeding techniques of tea along with their genetic resources, (3) Micropropagation that deals with an in-depth study of clonal propagation, (4) Somatic embryogenesis along with alternative techniques such as suspension culture, cryo-preservation, etc., (5) Molecular breeding that deals with the application of various DNA-based markers, QTL discovery, population genetics, linkage map, etc., (6) Genetic transformation and associated factors, (7) Stress physiology that compiles various works done in tea along with its wild relatives on quality, abiotic as well as biotic stress, and (8) Functional genomics that describes the various works of molecular cloning and characterizations, differential gene expression, high-throughput transcriptome sequencing, etc., transcriptomics study that describes the application of next generation sequencing to discover various genes that are related to various traits of tea, and noncoding RNA which describes the discovery of various noncoding RNA in tea and related genera. Sufficient work in the last few years also forced me to decide to have three independent topics. They are metabolomics which deals with different metabolomic study, proteomics which describes the discovery of different proteins, genome sequence which also deals with the work related to three published genome sequences, database management, gene family study, and resequencing of tea genome. Importantly, the author has included exclusive tables in most of the chapters that include a summary of the works in a particular topic. In a nutshell, the book compiles the work that has been done, identifies the problems, analyses the gaps in breeding and biotechnological works of tea as well as its wild species, and discusses the future scope as conclusion. Every effort has been made to include all the published works till June, 2020.

The book will be a useful resource for postgraduate, doctoral as well postdoctoral students working in the field of tea as well as other woody plants. This will also be useful for scientists working in the areas of life sciences, genomics, biotechnology, and molecular biology.

Contents

About the Author

Tapan Kumar Mondal Dr. Tapan Kumar Mondal joined the Institute of Himalayan Bioresource Technology (CSIR), Himachal Pradesh, India, for his Ph.D degree which he completed in 1998. After that, he served as Deputy Manager, Research and Development Department, Tata Tetley Ltd, Munnar, Kerala, till March, 2002. Since then up to 2010, he further served as Assistant Professor at North Bengal Agricultural University, Cooch Behar, West Bengal. Later in February 2010, he joined as Senior Scientist (Plant Biotechnology) at National Bureau of Plant Genetic Resource (ICAR), New Delhi. From 2016, he is working as Principal Scientist at ICAR-National Institute for Plant Biotechnology, New Delhi. He carried out his postdoctoral training with Prof. J. K. Zhu of the University of California, Riverside, USA, on "Regulation of small RNA under cold stress of Arabidopsis" and later worked at the University of Illinois, Urbana-Champaign, USA, on "Identification of nitrogen use efficient genes of maize by RNAseq."

Dr. Mondal has significantly contributed to various areas of biotechnology and genetic resource management of tea. His work leads several maiden findings in tea such as in vivo somatic embryogenesis, discovery of miRNA, lncRNA, and cirRNA of tea for the first time, and the development of first transgenic tea plants. His team decoded the mitochondrial genome of tea for the first time. Recently his team did the genome sequence of Indian Tea cultivar TV-1. He has also submitted several gene sequences of tea at NCBI and also published more than 60 research papers in this area. He was PI of various projects of tea biotechnology funded by DBT, DST, ICAR, and Tea Board, India. He is the recipient of university merit scholarship, scholarship from Indian Tea Association, ICAR JRF, DBT fellowship, and CSIR fellowship and is a life member of several professional societies. He has written one book on tea biotechnology and edited one book on wild species of rice. His team has developed several databases on tea and developed software for gene expression study. He also bagged the "Young Scientist Award" by Korean Society of Tea Science in 2003 and Japan Tea Science Society in 2004.

Abbreviations

ABA	Abscisic acid
AS	Acetosyringone
AMT	Ammonium transporters
ANN	Artificial neural network
BAC	Bacterial artificial chromosome
6-BAP	6-Benzylaminopurine
dCAPS	Derived cleaved amplified polymorphism marker
cM	Centimorgan
CTAB	Cetyl trimethyl ammonium bromide
cv	Cultivar
°C	Degree Celsius
4-CL	4-Coumaroyl-CoA
CM	Coconut milk
2,4-D	2,4-Dichlorophenoxy acetic acid
DMSO	Dimethyl sulfoxide
DDRT	Differential display reverse transcriptase
d	Day(s)
EST	Expressed sequence tag
g	Gram(s)
g/L	Gram(s) per liter
GA_3	Gibberellic acid
GC-MS	Gas chromatography mass spectroscopy
gus	β-glucuronidase gene
gusint	β-glucuronidase gene with an intron
h	Hour(s)
ha	Hectare(s)
hpt	Hygromycin phosphotransferase gene
HPLC	High performance liquid chromatography
IAA	Indole-3-acetic acid
IBA	Indole-3-butyric acid
Kn	Kinetin
KPa	Kilopascal
kb	Kilobase pair

lncRNA	Long noncoding RNA
MAS	Marker-assisted selection
M	Molar
min	Minute(s)
m	Meter(s)
7-NMT	7-N-methyltransferase
mL	Milliliter(s)
miRNA	MicroRNA
mM	Micromolar
MPSS	Massive parallel signature sequencing
MS	Murashige and Skoog's (1962) medium
MYA	Million years ago
μL	Microliter
NAA	Naphthalene acetic acid
NCBI	National Centre for Biotechnology Information
NGS	Next generation sequencing
nM	Nano mole
npt-II	Neomycin phosphotransferase gene
O.D.	Optical density
PA	Proanthocyanidin
PCR	Polymerase chain reaction
PVP	Polyvinylpyrrolidone
pM	Pico mole
%	Percent
QTL	Quantitative trait loci
Q-PCR	Quantitative-PCR
rpm	Revolution per minute
RNAseq	RNA sequencing
RACE	Rapid amplification of cDNA end
RAPD	Random amplified polymorphic DNA
RFLP	Restriction fragment length polymorphism
ROS	Reactive oxygen species
sdH_2O	Sterile distilled water
s	Second(s)
SE	Somatic embryogenesis
SSR	Simple sequence repeat
SSH	Suppression subtractive hybridization
SNP	Single nucleotide polymorphism
t	Tonnes
Taq Pol.	*Taq* DNA polymerase
TBA	Tertiary butyl alcohol
TBE	Tris borate EDTA
TES	Tocklai Experimental Station
TDZ	Thidiazuron

TE	Tris-EDTA
UV	Ultra-violet
UPGMA	Unweighted pair group method with arithmetic mean
UPASI	United Planters' Association of Southern India
v/v	Volume by volume
WPM	Woody plant medium of Lloyd and McCown (1982)
w/v	Weight by volume
mg	Microgram
ng	Nanogram
YE	Yeast extracts
YMB	Yeast mannitol broth

Introduction

1

1.1 Tea and *Camellia*: An Overview

Tea [*Camellia sinensis* (L.) O. Kuntze] belongs to the family Theaceae. It is the oldest non-alcoholic caffeine-containing beverage in the world. Chinese were the first to use tea as medicinal drink and later as beverage and have been doing so for the past 3000 years (Eden 1958). The cultivated taxa comprise three main natural hybrids. They are *C. sinensis* (L.) O. Kuntze or China type, *C. assamica* (Masters) or Assam type, and *C. assamica* subspecies *lasiocalyx* (Planchon ex Watt.) or Cambod or Southern type. However, two species *C. irrawadiensis* and *C. taliensis* have great morphological resemble with cultivated taxa and are used to make tea in some part of China though they have very low content of caffeine and though both these species might have contributed in the cultivated gene pool of tea. Tea is an evergreen, perennial, cross-pollinated plant and grows naturally as tall as 15 m. However, under cultivated condition, the bush height of 60–100 cm is maintained (Fig. 1.1) for harvesting the tender leaves to be processed for making the beverages. The flowers are white in color and born singly or in pairs at the axils. The fruits are green in color with 2–3 seeds and start bearing within 5–6 years after planting. Leaf is the main criterion by which three types of tea are classified. Briefly, they are (1) Assam type with the biggest leaves, (2) China type with the smallest leaves, and (3) Cambod type with leaf sizes in between those of Assam and China types.

Tea thrives well within the latitudinal ranges between 45°N and 34°S though extended beyond 60°N to nearly reaching 52°S covering about 61 countries. Tea occupies about 3.94 million ha of cultivable land of the world with an annual production of about 4 million t (Basu Majumder et al. 2010). Despite occupying only 16.4% of the total tea-growing areas of the world, India ranks second as the producer, consumer, and exporter. Hence tea plays a pivotal role in the national economy of India with an annual turnover of 860 million US dollars.

T. K. Mondal, *Tea: Genome and Genetics*,
https://doi.org/10.1007/978-981-15-8868-6_1

Fig. 1.1 Tea plantation of India. (**a**) Palampur, Himachal Pradesh, India, where tea plant is grown in hilly terrain; (**b**) Valparai, Tamil Nadu, South India, where tea plant is grown mainly in hilly terrain; (**c**) Assam, the "tea bowl" of India. Plants are grown under shade trees and grown mostly in plain area

1.2 History

Tea plants are believed to have been discovered accidentally by the Chinese legendary Emperor Shennong around 2737 BC. As soon as medicinal value began to be attributed to tea by Chinese, a demand for supplies of tea sprang up which results in the cultivation of tea plant in Sichuan province about 3000 years ago. Subsequently, the knowledge of tea cultivation was spread everywhere by the fine arts of Buddhism. Though, in India, wild tea plant was discovered by C. A. Bruce in Assam during 1823, seeds were also brought by G. J. Gordon from China in 1836 for establishing a commercial garden in India. Later C. A. Bruce was appointed as the superintendent of tea plantation who took active interest to cultivate the indigenous tea plant. Soon commercial interests moved in, and the world's first privately owned tea company, the Assam Tea Company, Assam, India, was established on 12 February 1839 with the directives from the British Parliament. This was the beginning of the present-day tea industry of India.

1.3 Origin and Distribution

Southeast Asia is the original home for tea. According to Wight (1959), the primary center of origin of tea was considered around the point of intersection of latitude 29°N and longitude 98°E near the source of the river, Irrawaddy, the point of confluence where lands of Assam, North Burma, Southwest China, and Tibet met. Secondary centers of origin were considered to be located in the Southeast China, Indochina, Mizoram, and Meghalaya (Kingdon-Ward 1950). The above areas were, therefore, considered to be the zone of origin and dispersion of the genus *Camellia* as a whole (Sealy 1958).

Tea was introduced to Japan from China in the early part of the eighth century. From Japan, tea cultivation extended to Indonesia during the seventeenth century. In Sri Lanka, tea was first planted in 1839 when seeds were taken from Calcutta, India. In USSR, tea cultivation started when seeds were imported from China towards the end of last century. Later, from USSR, seeds were exported to Turkey in the year 1939–1940. In Europe, tea was introduced in 1740 by the East India Company's Captain Goff, but those plants, which were planted in the Royal Botanic Garden at Kew in England, could not survive (Sealy 1958), and the first successful introduction was achieved by a British merchant cum naturalist John Ellis during 1768 (Aiton 1789; Booth 1830). From there, tea cultivation was extended to the African countries at the end of the nineteenth century.

1.4 Morphological Descriptions

A summary of the morphological characters of the three races of tea plants as described by Wight (1962), Barua (1963), and Bezbaruah (1971) is given below:

1. The China type [*Camellia sinensis* (L) O. Kuntze]: It is a big shrub, 1–2 m tall with many virgate stems arising from the base of the plant near the ground with hard, thick, and leathery leaf, matty surface, and marginal veins that are indistinct and appear sunken in the lamina. Blade is elliptic with obtuse or broadly obtuse apex, base is cuneate, and margin is bluntly serrulate to sinuate-serrulate with more or less incurved teeth. The leaf is glabrous above and villose below when young, becoming sparsely villose as the leaf ages, ultimately becoming glabrous. Young leaves are garnet-brown through ox-blood to purple in color. Petiole is short, 3–7 mm long, and stout, usually giving the leaf an erect pose. Flowers are borne singly or in pairs in the cataphyllary axils. Pedicel is short, 6–10 mm long, clavate, and glabrous with 2–3 sub-opposite scars little below the middle, marking the position of caducous bractioles 2–5 mm long. Sepals are 5–6 in number, imbricate, persistent, leathery, ovate or orbicular, 3–6 mm long, and glabrous green. Petals are 7–8, shallowly cup-shaped, 1.5–2 cm long, broad oval to sub-orbicular, and generally white sometimes with pale pink pigmentation. Stamen is numerous; arranged in two whorls, inner ones shorter and fewer in number and outer longer and more numerous; 8–13 mm long; and united at the

base for a few mm with the corolla lobes. Ovary is white and densely hairy, three locular ovules are present 3–5 in each loculus, and placentation is axial. Style is generally 3, sometimes up to 5, free for the greater part of their length, occasionally free up to the base of the ovary. Stigma is apical. The number of capsules varied from 1, 2, or 3, coccate, containing 1–3 nearly spherical seeds with 10–15 mm in diameter. Based on leaf sizes, Sealy (1958) recognized two forms of *C. sinensis* (a) f. macrophylla (Sieb.) Kitamura, with wild leaves 4–14 cm long and 2–2.5 cm wide, and (b) f. parvifolia (Miq.) Sealy, with leaves 1.5–1.6 cm long and 1–1.2 cm wide.

2. The Assam type [*C. assamica (Masters)*]: It is a small tree, 10–15 m tall with a trunk sometimes up to one third of its height, possessing a robust branching system. In typical plants, leaf is thin and glossy with more or less acuminate apex with distinct marginal veins. Leaf blade is usually broad elliptic and 8–20 cm long and 3.5–7.5 cm wide, base cuniate, and margin obscurely denticulate to bluntly wide-serrulate, glabrous, or persistently hairy on the midrib below. Flowers are single or in pairs on the cataphyllary axils and pedicels with scars of three caducous bracteoles, smooth and green. Sepals are 5–6 unequal, leathery, and persistent. Petals are white 7–8 in number, occasionally with pale yellow pigmentation at the base of the petals. Stamens are numerous as in *C. sinensis*.
3. The Southern form or Cambod type [*C. assamica Sub species*. lasiocalyx (Planch. MS)]: It is a small fastigiate tree, 6–10 m tall, with several upright, almost equally developed branches. Leaf is more or less erect, glossy, and yellowish-green when young and light green at maturity changing to coppery-yellow or pinkish-red from autumn till the end of the season. Petioles are pinkish-red at the base. Leaf size is intermediate between China and Assam type and broadly elliptic; marginal veins are not very prominent. Ovary is 3–4 in number with 5-locular. Styles are 3–5 in number, free nearly up to half the length, and straight with apical or linear stigma. On the other floral characters, it resembles the Assam plant, with the difference that four or more bracteoles are found on the pedicel of flowers.

1.5 Taxonomy and Nomenclature

The taxonomic position of tea is given below (Fig. 1.2). It is noteworthy to mention that the family comprises 11 genus and the genus *Camellia* has more than 325 species. Out of that, only two are commercially cultivated for producing the tea; other two such as *C. irrawadiensis* and *C. taliensis* are occasionally used in China for making tea.

1.6 Economic Importance and Health Benefits

The economic importance of the genus *Camellia* is primarily due to the tea. Though tea is mainly consumed in the form of "fermented tea" or "black tea," "non-fermented" or "green tea" and lesser known "semi-fermented" or "oolong tea" are

Fig. 1.2 Taxonomic position of tea

Kingdom	Plantae - Plants
Subkingdom	Tracheobionta - Vascular plants
Superdivision	Spermatophyta - Seed plants
Division	Magnoliophyta - Flowering plants
Class	Magnoliopsida - Dicotyledons
Subclass	Dilleniidae
Order	Theales
Family	Theaceae
Genus	*Camellia* L.
Species	*Camellia sinensis* (L.) O. Kuntze - tea
Types	Assam, China, Cambod
Cultivar	600 recognized worldwide.

also available. They differ in their method of manufacture, chemical constituent, appearance, and organoleptic taste. While black tea is widely used in India and other European countries, green tea is traditionally consumed in China, Japan, and Taiwan, but its consumption is increasing gradually across the world due to health benefit properties. Oolong tea is mainly consumed in some parts of China as well as Taiwan. Worldwide 80% black tea, 18% green tea, and 2% oolong tea are being produced.

For black tea, the young tender leaves are completely fermented after withering. The fermentation results oxidation and polymerization of polyphenols, changing the nature of the chemical constituents of tea leaves and forming theaflavin and thearubigin. These polyphenols are responsible for the briskness, strength, color, taste, aroma, and pungency associated with black tea. The infusion of black tea has a bright red or copper color, astringent taste, and characteristic aroma. On the other hand, green tea is unfermented and is the least processed among the three types. The plucked leaves are harvested and steamed immediately to inactivate the enzymes to prevent oxidation and polymerization of primary polyphenols which result in retaining of green color in the finish product. Green tea infusion has a smell of fresh vegetables and low caffeine content. In oolong tea, primary polyphenols are allowed to partly oxidize. Oolong tea is not common and is intermediate in characteristic between green and black tea. Immediately after plucking, the tea leaves are partially fermented for about half the time of black tea. It has the color of black tea and flavor of green tea.

Tea was used initially as medicine and later as beverage and has now been proven well as future potential of becoming an important industrial and pharmaceutical raw material. Scientific reports in the last two decades have validated many beneficial claims for tea. The majority of the beneficial effects have been attributed to the polyphenolic constituents. Several studies suggest that phenolics may be of importance in reducing the incidence of degenerative diseases such as cancer and arteriosclerosis. The most relevant compounds in dietary regime are cinnamic acid derivatives and flavonoids. As natural polyphenols remain unchanged in green tea, it can be said that green tea is more beneficial than black tea. The various health benefits in relation to cancer, arthritis, cardiovascular diseases, diabetes, and obesity are described below:-

1. Antioxidant activity: Most beneficial effects of tea catechins were attributed to their antioxidant properties that sequester metal ions and scavenge oxygen species and free radicals (Wiseman et al. 1977). Among the different components of catechin, (−)-epigallocatechin-3-O-gallate (EGCG) was the most potent chemical of the epicatechin derivatives tested for biological activity. It was thought to prevent tumorigenesis by protecting cellular components from oxidative damage through free radical scavenging. Indeed many of the studies had confirmed the free radical scavenging activity of EGCG as well as its antimutagenic, antiangiogenic, antiproliferating, and/or pro-apoptotic effects on mammalian cells both in vitro and in vivo (Allemain 1999). Tea catechins had been found to be better antioxidants than vitamins C and E and carotene. The polyphenols block free radical damage to lipids (found in cell membranes and serum lipids), nucleic acids, and proteins (like those found as cellular enzymes and structural proteins). Damage to these cell components can lead to tumor formation. The oxidative damage by oxygen free radicals of low-density lipoproteins (LDL) in serum led to arteriosclerosis and coronary heart diseases. The oxidation of cell membranes and other cell components led to aging. The antioxidant activity of tea polyphenols was due to their ability not only to scavenge superoxides but also to increase the activity of some detoxifying enzymes such as glutathione peroxidase, glutathione reductase, glutathione-S-transferase, catalase, and quinine reductase in the small intestine, liver, and lungs. However, the antioxidant activity of tea is diminished by the addition of milk to the infusion due to binding of tea polyphenols to milk proteins.
2. Cardiovascular activity: Tea polyphenols and flavonoids had been reported to inhibit either enzymatic or non-enzymatic lipid peroxidation, an oxidative process implicated in several pathological conditions including atherosclerosis. Specifically, it had been suggested that tea polyphenols lower the oxidation of LDL cholesterol, with a consequent decreased risk of heart diseases. It had been observed that green tea polyphenols significantly reduced the levels of serum LDL, very-low-density lipoproteins (VLDL), and triglycerides. At the same time, they increased the levels of high-density lipoproteins (HDL). This observation had been strengthen by the finding that in hypercholesterolemic rats, green tea polyphenols lowered blood cholesterol levels and reduced blood pressure in spontaneously hypertensive animals. Tea polyphenols also inhibited the absorption of dietary fats and cholesterols (Chen et al. 2000).
3. Anticancer activity: It is the most thoroughly studied function of tea polyphenols. It can protect the cells against cancer at several stages of carcinogenesis including cancer prevention, endogenous carcinogenic activation, DNA damage and destabilization, cell proliferation, neoplastic growth, and metastasis. Tea especially green tea reduced the incidence of cancers of the stomach, small intestine, pancreas, lung, breast, skin, urinary bladder, prostrate, esophagus, and mouth (Vasisht et al. 2003). Also it had been shown to reduce tumor size and growth in cancer-bearing animals. Green tea polyphenols directly inhibited the cytochrome P-450 enzyme systems (phase I enzyme) that played a pivotal role in carcinogenic activation. Concurrently, they boosted the activity

of phase II enzyme (e.g., glutathione transferase) that made xenobiotics hydrophilic for clearance of the body. The process was crucial for carcinogenic detoxification. Recently, it was proposed to associate the anticancer activity of EGCG with the inhibition of urokinase, one of the most frequently expressed enzymes in human cancers. Green tea showed a protective effect against damage produced by UV radiation and reactive oxygen species to the dermis through apostasies and inhibiting lipid peroxidation.

4. Antidiabetic effect: Tea drinking was shown to possess antidiabetic activity and was effective in the prevention and treatment of diabetes. Tea polyphenols lowered the serum glucose by inhibiting the activity of the starch-digesting enzyme, amylase. Polyphenol inhibits both salivary and intestinal amylase. As a result, the starch was broken down more slowly, and the sudden rise in serum glucose was minimized. In addition, tea polyphenols also reduced the intestinal absorption of glucose (Gomes et al. 1995).
5. Antiarthritic activity: Tea polyphenol demonstrated an exceptional protection against arthritis. The major polyphenols showing antiarthritic effect include epicatechin, epigallocatechin, epicatechin-3-gallate, and epigallocatechin-3-gallate. It has been postulated that the antioxidant property of polyphenol might be useful in the prevention and severity of arthritis (Tapiero et al. 2002).
6. Antiplaque activity: Tea polyphenols acted in two different ways to inhibit the growth and adherence of oral bacteria. Firstly, it inhibited the growth of periodontal disease-producing bacterium, *Porphyromonas gingivalis*, and decay-causing bacteria such as *Streptococcus mutans*. Therefore, green tea as mouth rinse resulted in less plaque and periodontal diseases. Secondly, it inhibited the enzyme amylase present in the saliva, and the starch in the mouth did not get converted into glucose and maltose. Less nutrition was thus available to decay-causing bacteria (Yu et al. 1995).
7. Antiviral activity: Tea extract had been shown to have virucidal activity against polio, influenza, and herpes simplex virus (Okubo and Juneja 1997).
8. Anti-AIDS activity: Green tea polyphenols are antimutagenic and act as effective adjuvant to drug therapy. It had been discovered that polyphenols from green tea and their oxidation products could inhibit the reverse transcriptase or polymerase of several types of viruses, including HIV-1 and herpes simplex-1 (Hashimoto et al. 1996). However, research in this area is still in its initial stages.
9. Anorectic effect: The tea polyphenols inhibited catechol-O-methyl transferase, and caffeine inhibited transcellular phosphodiesterase, thus stimulated thermogenesis, and helped to manage obesity. The release of glucose was slowed down by tea, and thus harmful spiking of insulin was prevented (Kwanashie et al. 1989).
10. Antimicrobial activity: The crude catechins and theaflavins had been found to have an antibacterial activity. They were believed to damage bacterial cell membranes. Tea had been used in the treatment of diarrheal infections and cholera. Polyphenols killed the spores of *Clostridium botulinum* and thus displayed antibacterial activity against foodborne diseases and were also effective against heat-resistant bacilli like *Bacillus subtilis*, *Bacillus cereus*, and *Vibrio parahaemolyticus*. Green tea also had protozoacidal properties (Hamilton-Miller 1995).

11. Other biological effects: It had also been reported that green tea polyphenols exhibited neuromuscular, antiangiogenic, antihepatotoxic, antiproliferative/apoptotic, and immunomodulatory effects (Sueoka et al. 2001).

1.7 Landmarks of Biotechnological Works

An overview of various landmark works of tea is depicted in Fig. 1.3. Caffeine (1,3,7-trimethylxanthine) is the main alkaloid for which tea is valued. It was purified from tea leaves of field-grown plants during 1821 (Spedding and Wilson 1964). However with the advancement of cell culture techniques, attentions were paid to

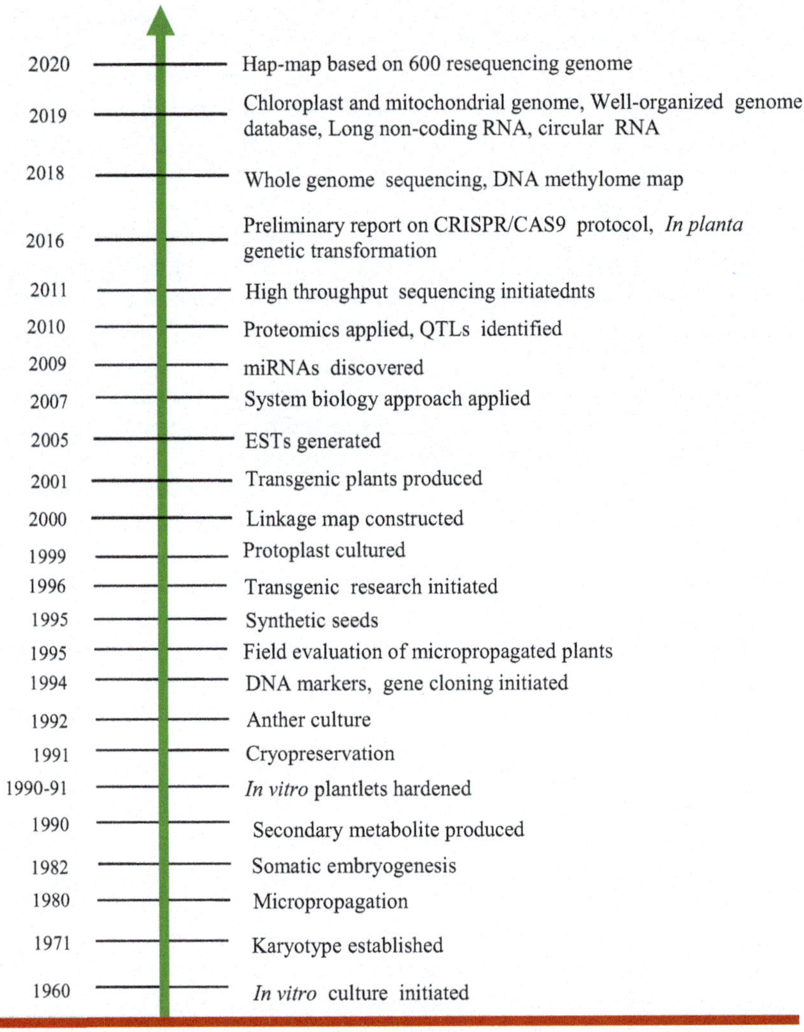

Fig. 1.3 Time-frame flow of landmark discoveries of tea

produce in higher quantity caffeine from the in vitro callus tissue (Ogutuga and Northcote 1970). Simultaneously, cytogenetic works were also initiated. Accordingly chromosome number had been established for the most available taxa of *Camellia* including tea (Bezbaruah 1971) at Tea Experimental Station, Jorhat, India, and elsewhere. However, it was evident from the literature that while Forrest (1969) was pioneer to establish the in vitro culture of tea, Kato (1985) did a systematic study on micropropagation. Since then a significant amount of work has been done in tea and its wild relatives on various aspect of in vitro culture (Fig. 1.4). In tea, somatic embryogenesis had been fully exploited for clonal

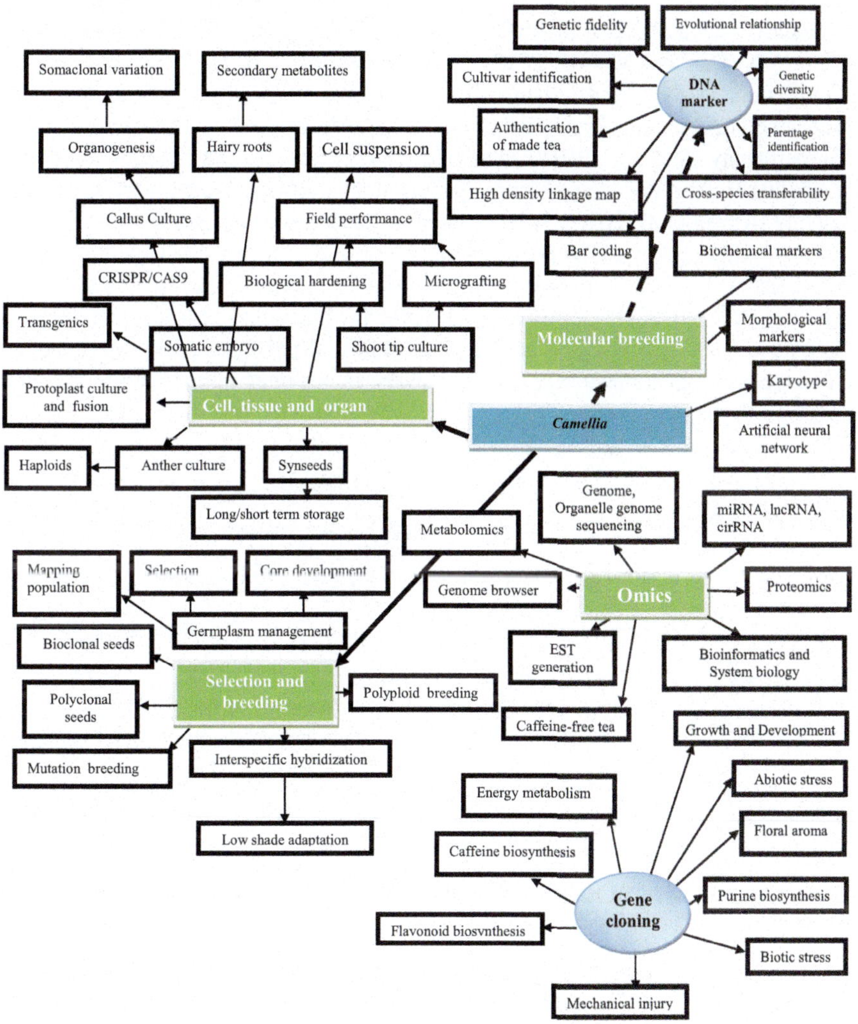

Fig. 1.4 Schematic explanation of *Camellia* improvement. The bold arrows are the major areas of research. Dotted arrows are the sub-areas. Thin arrows are the different applications with a major or sub-major area

propagation (Mondal et al. 2001a), genetic transformation (Mondal et al. 2001b), artificial seed production (Mondal et al. 2000), embryo rescue in some interspecific hybrid crosses of *Camellia* (Nadamitsu et al. 1986), and haploid plant production (Chen and Liao 1982). Randomly Amplified Polymorphic DNA (RAPD) was first DNA marker that was applied by Wachira et al. (1995) to study the genetic diversity of Kenyan tea germplasm; following that, several DNA-based markers such as Inter-Simple Sequence Repeat (ISSR), Simple Sequence Repeats (SSR), Restriction Fragment Length Polymorphism (RFLP), and Amplified Fragment Length Polymorphism (AFLP) were utilized by several workers across the world (Mukhopadhyay et al. 2013, 2016). In tea, functional genomics was initiated with the isolation of chalcone synthase gene by Takeuchi et al. (1994); at Japan, the first cDNA microarray was developed in 2006 (Zhao et al. 2006). Later non-coding RNA such as miRNA (Das and Mondal 2010), lncRNA (Varshney et al. 2019), and cirRNA (Tong et al. 2018) were discovered in tea. In-depth study of transcriptomics using next-generation sequencing was first started with Jiang et al. (2011) and by cDNA microarray by Ma et al. (2012). Although chloroplast genome of *Camellia* (Yang et al. 2013) was decoded during 2013, mitochondrial genome was sequenced recently (Rawal et al. 2020). Advancement of tea genome sequence was also done recently. Although first Chinese tea plant genome sequence was made available in 2017 (Xia et al. 2017), *C. assamica* genome was sequenced for the first time recently (Mondal et al. 2019). Presently several works on functional genomics and proteomics have been initiated and achieved at such a speed that it will speed up the downstream application such as QTL discovery and its integration to improve the tea genotype in the near future.

References

Aiton W (1789) Hortus kewensis, or a catalogue of the plants. Royal Botanical Garden, Kew. Printed for George Nicol, London, pp 48–57

Allemain G (1999) Multiple actions of EGCG, the main component of green tea. Bull Cancer 86:721–724

Barua PK (1963) Classification of tea plant. Two Bud 10:3–11

Basu Majumder A, Bera B, Rajan A (2010) Tea statistics: global scenario. Int J Tea Sci 8:121–124

Bezbaruah HP (1971) Cytological investigation in the family theaceae-I. Chromosome numbers in some *Camellia* species and allied genera. Caryologia 24:421–426

Booth WB (1830) History and description of the species of *Camellia* and Thea. Trans Hort Soc Lond 7:519–562

Chen Z, Liao H (1982) Obtaining plantlet through anther culture of tea plants. Zhongguo Chaye 4:6–7

Chen ZY, Law WI, Yao XQ, Lau CW, Ho WK, Huang Y (2000) Inhibitory effects of purified green tea epicatechins in construction and proliferation of arterial smooth muscle cells. Acta Pharma Sci 21:835–840

Das A, Mondal TK (2010) Computational identification of conserved microRNAs and their targets in tea (*Camellia sinensis*). Am J Plant Sci 1:77–86

Eden T (1958) The development of tea culture. In: Eden T (ed) Tea. Longman, London, pp 1–4

Forrest GI (1969) Studies on the polyphenol metabolism of tissue culture derived from the tea plant (*C. sinensis* L.). Biochem J 113:765–772

Gomes A, Vedasiromoni JR, Das M, Sharma RM, Ganguly DK (1995) Antihyperglycemic effect of black tea (*Camellia sinensis*) in rat. J Ethnopharmacol 45:223–226

Hamilton-Miller JM (1995) Antimicrobial properties of tea (*Camellia sinensis* (L) kuntze). Antimicrob Agents Chemother 39:2375–2377

Hashimoto F, Kashiwada Y, Nonaka GI, Nishioka I, Nohara T, Cosentibno LM, Lee KH (1996) Evaluation of tea polyphenols as anti-HIV agents. Med Chem Lett 6960:695–700

Jiang C, Wen Q, Chen Y, Xu LA, Huang MR (2011) Efficient extraction of RNA from various *Camellia* species rich in secondary metabolites for deep transcriptome sequencing and gene expression analysis. Afr J Biotechnol 10:16769–16773

Kato M (1985) Regeneration of plantlets from tea stem callus. Jap J Breed 35:317–322

Kingdon-Ward F (1950) Does wild tea exist? Nature 165:297–299

Kwanashie HO, Usman H, Nkim SA (1989) Screening of "Kargasok tea": anorexia and obesity. Biochem Soc Trans 17:1132–1133

Ma CL, Chen L, Wang XC, Jin JQ, Ma JQ, Yao MZ, Wang ZL (2012) Differential expression analysis of different albescent stages of 'Anji Baicha' (*Camellia sinensis* (L.) O. Kuntze) using cDNA microarray. Sci Hortic 148:246–254

Mondal TK, Bhattacharya A, Sood A, Ahuja PS (2000) Factor effecting induction and storage of encapsulated tea (*Camellia sinensis* L. O. Kuntze) somatic embryos. Tea 21:92–100

Mondal TK, Bhattacharya A, Ahuja PS (2001a) Induction of synchronous secondary embryogenesis of Tea (*Camellia sinensis*). J Plant Physiol 158:945–951

Mondal TK, Bhattacharya A, Ahuja PS, Chand PK (2001b) Factor effecting *Agrobacterium tumefaciens* mediated transformation of tea (*Camellia sinensis* (L). O. Kuntze). Plant Cell Rep 20:712–720

Mondal TK, Rawal HC, Bera B, Kumar PM, Choubey M, Saha G, Das B, Bandyopadhyay T, Ilango V, Sharma TR, Barua A, Radhakrishnan B, Singh NK (2019) Draft genome sequence of a popular Indian tea genotype TV-1 [*Camellia assamica L.* (*O*). Kuntze]. BioRxiv:762161

Mukhopadhyay M, Sarkar B, Mondal TK (2013) Omics advances in tea (*Camellia sinensis*). In: Bhar D (ed) Omics applications in crop science. CRC Press, Taylor and Francis Group, Boca Raton, FL, pp 347–366. ISBN:978-1-4665-8582

Mukhopadhyay M, Mondal TK, Chand PK (2016) Biotechnological advances in tea (*Camellia sinensis* [L.] O. Kuntze): a review. Plant Cell Rep 35(2):255–287

Nadamitsu S, Andoh Y, Kondo K, Segawa M (1986) Interspecific hybrids between *Camellia vietnamensis* and *C. chrysantha* by cotyledon culture. Jap J Breed 36:309–313

Ogutuga DBA, Northcote DH (1970) Caffeine formation in tea callus tissue. J Exp Bot 21:258–273

Okubo T, Juneja LR (1997) Chemistry and application of green tea. CRC Press, New York, NY, pp 109–121

Rawal HC, Bera B, Mohan Kumar P, Singh NK, Mondal TK (2020) Decoding and analysis of organelle genomes of Indian tea (*Camellia assamica*) for phylogenetic confirmation. Genomics 112:659–668

Sealy JR (1958) A revision of the genus *Camellia*. Royal Horticultural Society, London

Spedding DJ, Wilson AT (1964) Caffeine metabolism. Nature 204:73

Sueoka N, Suganuma M, Sueoka E, Okabe S, Matsuyama S, Imai K, Nakachi K, Fujiki H (2001) A new function of green tea: prevention of lifestyle-related diseases. New York Acad Sci 928:274–280

Takeuchi A, Matsumoto S, Hayatsu M (1994) Chalcone synthase from *Camellia sinensis* isolation of the cDNAs and the organ-specific and sugar-responsive expression of the genes. Plant Cell Physiol 35:1011–1018

Tapiero H, Tew KD, Ba GN, Mathe G (2002) Polyphenols: do they play a role in the prevention of human pathogens? Biomed Pharmacother 56:200–207

Tong W, Yu J, Hou Y et al (2018) Circular RNA architecture and differentiation during leaf bud to young leaf development in tea (*Camellia sinensis*). Planta 248:1417–1429

Varshney D, Rawal HC, Dubey H, Bandyopadhyay T, Bera B, Mohan Kumar P, Singh NK, Mondal TK (2019) Tissue specific long non-coding RNA landscape and their association with aroma formation of tea. Indust Crop Prod 133:79–89

Vasisht K, Sharma PD, Karan M, Rakesh D, Vyas S, Sethi S, Manktala R (2003) Study to promote the industrial exploitation of green tea poly-phenols in India. ICS-UNIDO, Trieste, pp 15–22

Wachira FN, Waugh R, Hackett CA, Powell W (1995) Detection of genetic diversity in tea (*Camellia sinensis*) using RAPD markers. Genome 38:201–210

Wight W (1959) Nomenclature and classification of tea plant. Nature 183:1726–1728

Wight W (1962) Tea classification revised. Curr Sci 31:298–299

Wiseman SA, Balentine DA, Frei B (1977) Antioxidant in tea. Crit Rev Food Sci Nutr 37:705–718

Xia EH, Zhang HB, Sheng J, Li K, Zhang QJ, Kim C, Zhang Y, Liu Y, Zhu T, Li W (2017) The tea tree genome provides insights into tea flavor and independent evolution of caffeine biosynthesis. Mol Plant 882(10):866–877

Yang JB, Yang SX, Li HT, Yang J, Li DZ (2013) Comparative chloroplast genomes of Camellia species. PLoS ONE 8(8):e73053

Yu H, Oho T, Xu LX (1995) Effects of several tea components on acid resistant of human tooth enamel. J Dental Sci 23:101–105

Zhao LP, Gao QK, Chen L, Wang XC, Yao MZ (2006) Development and preliminary application of cDNA microarray of tea plant (*Camellia sinensis*). J Tea Sci 5:3–7

Genetics and Breeding 2

2.1 Introduction

Although applications of breeding in tea are difficult, the entire varietal development in tea and other *Camellia* species has been done through conventional breeding which started way back in 1939. Since then, several developments of genetics and breeding have taken place which are discussed here.

2.2 Genome Size

The genome size of tea plant was initially estimated to be 4.0 G bases (Hanson et al. 2001; Tanaka et al. 2005) though intraspecific and interspecific variations of 2C DNA content were observed in the genus *Camellia*. However, recently the genome size of an Indian tea (cultivar TV 1) was found to be 3 Gb through flow cytometer (Mondal et al. 2019). It had been found that the while intra-population variations of 2C DNA content of tea were 5.87–6.4 pg of DNA, the inter-population variations of 2C DNA (pg) content of different species varied from 2.5 to 25 (Huang et al. 2013). The reason for higher DNA content was due to higher levels of ploidy. For example, *C. oleifera* and *C. sasanqua* had 2C DNA (pg) 17.47 and 18.79 as they were found to be octaploids (Huang et al. 2013). Tea chromosomes were found to be small and had the tendency to clump together due to "stickiness." Tea is diploid ($2n = 30$; basic chromosome number, $x = 15$), and karyotype ranges from 1.28 to 3.44 μm (Bezbaruah 1971). The r value (ratio of long arm to short arm) for all the 15 pairs of chromosomes ranged from 1.00 to 1.91 (Fig. 2.1). This consistency in diploid chromosome number suggested a monophyletic origin for all *Camellia* species. However, few higher ploidy levels, such as triploids ($2n = 45$), tetraploids ($2n = 60$), pentaploids ($2n = 75$), hexaploids ($2n = 90$), octaploids ($2n = 120$), and aneuploids ($2n \pm 1$ to 29), had also been identified (Singh 1980; Zhan et al. 1987; Huang et al. 2013).

T. K. Mondal, *Tea: Genome and Genetics*,
https://doi.org/10.1007/978-981-15-8868-6_2

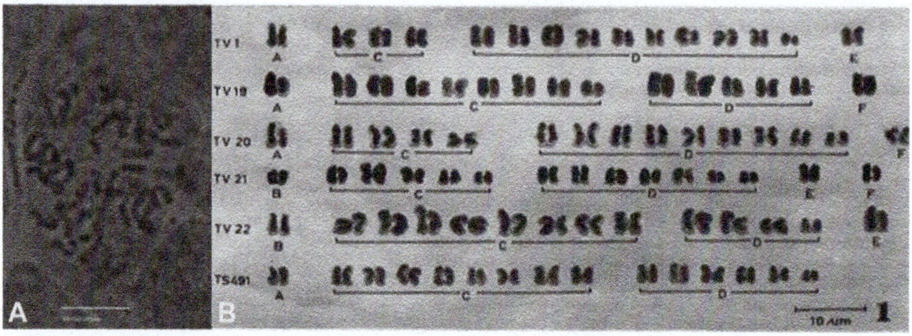

Fig. 2.1 Karyotyping of tea. (**a**) Metaphase chromosome of tea plant. (**b**) Karyotyping of some Indian tea cultivar. (This photo is taken from Datta and Agarwal 1992)

Fig. 2.2 Flower of wild camellia. (**a**) *C. vuquangensis* Luong, Tran, and L. T. Nguyen; (**b**) *C. tuyenquangensis* D. V. Luong, N. N. H. Le, and N. Tran; (**c**) *C. namkadingensis* Soulad and Tagane

2.3 Diversity of the Genus

The genus *Camellia* had 40 species in 1920. The number of species was increased to 87 in 1958 (Sealy 1958), and more than 267 species were registered in 1982 (Chang and Bartholomew 1984). Presently, this genus is believed to comprise more than 300 species (Mondal et al. 2004) with a discovery of *C. cherryana* in 2012 (Orel and Wilson 2012a, b) and more recently *C. vuquangensis* and *C. hatinhensis* (Nguyen et al. 2018), suggesting the genetic instability and high out-breeding nature of the genus. In a conservative estimation, there are more than 30,000 cultivated varieties of ornamental *Camellia* worldwide. The *Camellia* is the largest genus of the family "Theaceae." The genus is valued for tea due to the presence of caffeine, a purine alkaloid, which acts as a stimulus for the central nervous system of the human being. Apart from that, the wild *Camellia* has a very attractive flower which is why they are commercially important (Fig. 2.2). Nagata and Sakai (1984) reported the distribution of caffeine in 23 species of *Camellia*. The caffeine content on a dry weight basis in some of them was as follows: *C. sinensis* var. *sinensis* (3.5%), *C. sinensis* var. *assamica* (4%), *C. taliensis* (2.54%), and *C. kissi* (0.02%). Among these, *C. kissi* belongs to the section Paracamellia, and the other genera belong to the section Thea. The other three genera in the family are Eurya with 140 species, Ternstroemia with 130 species, and Adinandra with 100 species. Apart from having caffeine in the 3 species, around 50 species of the genus *Camellia* are known to produce oil with industrial uses (Mondal 2011).

The classification of the genus *Camellia* is very dynamic in a sense several new species have been reported in recent past. This had been revisited many times by several workers (Sealy 1958; Ming and Bartholomew 2007; Chen et al. 2000), yet Chang and Bartholomew (1984) remained the most popular who divided the whole *Camellia* genus into 4 subgenera and 20 sections totally, which are depicted below with the example of some prominent species in each section. In addition to this, several other sections reported by various workers are also reported here.

I Subgenus	**Protocamellia**
Section I	Archecamellia
	C. albogigas
	C. cattienensis (Orel and Wilson 2012a, b)
	C. granthamiana
	C. pleurocarpa
	C. pukhangensis (Do et al. 2019)
Section II	Stereocarpus
	C. dormoyana
	C. krempfii
	C. liberistyla
	C. liberistyloides
	C. maiana (Orel and Wilson 2010a)
	C. yunnanensis
Section III	Piquetia
	C. dalatensis
	C. dongnaiensis
	C. honbaensis (Truong et al. 2018)
	C. longii (Orel et al. 2013)
	C. piquetiana
	C. sonthaiensi
	C. sonthaiensis (Luu et al. 2015)
	C. yenhoae (Luu et al. 2018)
II Subgenus	**Camellia**
Section IV	Olifera
	C. gauchowensis
	C. hatinhensis
	C. hunanica
	C. oleifera
	C. sasanqua
	C. vietnamensis

(continued)

Section V	Furfuracea
	C. crapnelliana
	C. furfuracea
	C. integerrima
	C. latipetiolata
	C. oblata
	C. parafurfuracea
	C. polypetala
	C. gaudichaudii
Section VI	Paracamellia
	C. brevissima
	C. brevistyla
	C. chinmeii (Lee and Yang 2019)
	C. confusa
	C. fluviatilis
	C. grijsii
	C. kissii
	C. lutescens
	C. maliflora
	C. microphylla
	C. miyagii
	C. obtusifolia
	C. phaeoclada (Vijayan et al. 2009)
	C. puniceiflora
	C. shensiensis
	C. tenii
	C. tenuiflora (Vijayan et al. 2009)
	C. weiningensis
	C. yuhsienensis (Vijayan et al. 2009)
Section VII	Pseudocamellia
	C. chungkingensis
	C. henryana
	C. ilicifolia
	C. szechuanensis
	C. trichocarpa
Section VIII	Tuberculata
	C. acutiperulata
	C. anlungensis (syn: *C. acuticalyx*)
	C. hupehensis
	C. litchii
	C. obovatifolia
	C. parvimuricata
	C. rhytidocarpa
	C. rhytidophylla
	C. rubituberculata
	C. tuberculata

(continued)

Section IX	Luteoflora
	C. luteoflora
Section X	Camellia
	C. albescens
	C. albovillosa
	C. azalea (syn: *C. changii*) (Orel et al. 1985)
	C. bailinshanica
	C. boreali-yunnanica
	C. chekiangoleosa
	C. compressa
	C. cryptoneura
	C. delicata
	C. edithae
	C. hiemalis
	C. hongkongensis
	C. japonica (syn: *Camellia bonnardi* Berl)
	C. jinshajiangica
	C. kweichowensis
	C. lapidea
	C. longicaudata
	C. lucidissima
	C. lungshenensis
	C. magnocarpa
	C. mairei
	C. multiperulata
	C. namkadingensis (Souladeth et al. 2019)
	C. oconoriana (Orel et al. 2013)
	C. omeiensis
	C. oviformis
	C. paucipetala (Vijayan et al. 2009)
	C. phellocapsa
	C. pitardii
	C. polyodonta
	C. reticulata (syn: *C. borealiyunnanica*; *C. brevicolumna*; *C. brevigyna*; *C. brevipetiolata*)
	C. rusticana (Vijayan et al. 2009)
	C. saluenensis
	C. saluensis (Vijayan et al. 2009)
	C. semiserrata
	C. setiperulata
	C. subintegra
	C. trichosperma
	C. tunganica
	C. uraku
	C. villosa

(continued)

	C. xifongensis (Vijayan et al. 2009)
	C. xylocarpa
	C. yokdonensis (Hakoda and Kirino 2007)
III Subgenus	**Thea**
Section XI	Corallina
	C. acutiserrata
	C. atrothea
	C. brachygyna
	C. corallina
	C. coralline
	C. fleuryi
	C. kwangsiensis
	C. lienshanensis
	C. nitidissima
	C. paucipunctata
	C. pentamera
	C. pilosperma
	C. ptilophylla
	C. scariosisepala
	C. tonkinensis
	C. waldeniae
	C. wardii
Section XII	Brachyandra
	C. amplexifolia
	C. brachyandra
	C. crassipetala
	C. gilbertii
	C. muricatula
	C. nematodea
	C. nervosa
	C. pachyandra
	C. parviflora
	C. szemaoensis
	C. xanthochroma
	C. yangkiangensis
Section XIII	Longipedicellata
	C. amplexicaulis (syn: *C. cylindracea*, Zhao 2019)
	C. indochinensis
	C. longipedicellata
	C. petelotii
Section XIV	Chrysantha
	C. aurea
	C. chrysantha
	C. chrysanthoides
	C. euphlebia

(continued)

	C. flava (syn: *C. cucphuongensis*, Zhao 2019)
	C. flavida
	C. hirsute (Hakoda and Kirino 2007)
	C. impressinervis
	C. nitidissima
	C. phanii (Hakoda and Kirino 2007)
	C. pingguoensis
	C. pubipetala
	C. thanxaensa (Hakoda and Kirino 2007)
	C. thuongiana
	C. tunghinensis
	C. velutina (Pham et al. 2019)
Section XV	Calpandria
	C. connata
	C. lanceolata
Section XVI	Thea
	C. angustifolia
	C. assamica
	C. costata
	C. crassicolumna
	C. crispula
	C. fangchensis
	C. formosensis
	C. gymogyna
	C. irrawadiensis
	C. leptophylla
	C. longtousanica
	C. parvisepala
	C. pentastyla
	C. ptilophylla
	C. pubicosta
	C. pubilimba
	C. quinquelocularis
	C. sinensis
	C. tachangensis
	C. taliensis
	C. waldenae
	C. yunkiangensis
Section XVII	Longissima
	C. gracilipes
	C. longissima
Section XVIII	Glaberrima
	C. glaberrima
	C. kwangtungensis

(continued)

IV Subgenus	**Metacamellia**
Section XIX	Theopsis
	C. acutisepala
	C. acutissima
	C. brevipes
	C. buxifolia
	C. callidonta
	C. campanisepala
	C. chekiangensis
	C. costei
	C. crassipes
	C. cuspidata
	C. dubia
	C. elongata
	C. euryoides
	C. forestii
	C. fraterna
	C. grandiflora
	C. handelii
	C. lancicalyx
	C. lancilimba
	C. longicalyx
	C. longicarpa
	C. longicuspis
	C. lutchuensis
	C. macrosepala
	C. membranacea
	C. minutiflora
	C. nokoensis
	C. parvicaudata
	C. parvicuspidata
	C. parvilapidea
	C. parvilimba
	C. parviovata
	C. percuspidata
	C. pubisepala
	C. quangcuongii (Dung et al. 2016)
	C. rosaeflora
	C. rosthorniana
	C. stuartiana
	C. subacutissima
	C. subglabra
	C. synaptica
	C. transarisanensis
	C. transnokoensis

(continued)

	C. triantha
	C. trichandra
	C. trichoclada
	C. tsaii
	C. tsingpiensis
	C. tsofuii
	C. viridicalyx
Section XX	Eriandra
	C. assimilis
	C. assimiloides
	C. candida
	C. caudata
	C. cordifolia
	C. cratera
	C. edentata
	C. lawaii
	C. melliana
	C. punctata
	C. salicifolia
	C. trigonocarpa
	C. villicarpa
	C. wenshanensis
Some new section as described by various workers	
Section	Heterogena (Sealy 1958; Ming and Bartholomew 2007; Chang 1998)
	C. bolovenensis
	C. cherryana
	C. curryana (Orel et al. 2014a)
	C. fleurii
	C. furfuracea
	C. henryana
	C. mileensis
	C. pubifurfuracea
	C. tennii
	C. wardii
Section	Furfuraceae (Chang 1998)
	C. crapnelliana
	C. octopetala
	C. parafurfuracea
Section	Eriandra (Vijayan et al. 2009; Chang 1998)
	C. lawii
Section	Bidoupia (Orel et al. 2012)
	C. inusitata
Section	Dalatia (Orel and Wilson 2010a, b)
	C. bugiamapensis (Orel et al. 2014b)
	C. luteocerata (Orel and Wilson 2010a, b)

(continued)

Section	Oboidea (Tran and Luong 2013)
	C. dilinhensis
Section	Capitatae (Orel et al. 2014a, b)
	C. capitata
Section	Pierrea (Orel et al. 2014a, b)
	C. ligustrina
Section	Heterogeneae (Orel et al. 2014a, b)
	C. duyana
Section	Obvoidae (Nguyen et al. 2018)
	C. vuquangensis
Species for which taxonomic position is yet to confirm	
	C. achrysantha (Chang and Liang 1994)
	C. atuberculata (Chang 1991a)
	C. axillaris
	C. bambusifolia
	C. banksiana
	C. biflora
	C. bugiamapensis (Hu et al. 2019)
	C. capitata (Hu et al. 2019)
	C. campanisepala
	C. changningensis
	C. crassiphylla (Ninh and Hakoda 1998)
	C. cucphuongensis (Ninh and Rosmann 1998)
	C. debaoensis (Hu et al. 2019)
	C. drupifera (Souladeth et al. 2019)
	C. fascicularis (Chang 1991b)
	C. hakodae (Ninh and Ninh 2014)
	C. hamyenensis (Manh et al. 2019)
	C. hengchuensis (Lee and Yang 2019)
	C. buisanensis (Su et al. 2004)
	C. hengchunensis (Vijayan et al. 2009)
	C. hirsute (Ninh and Ninh 2014; Manh et al. 2019)
	C. huana (Manh et al. 2019)
	C. huulungensis (Manh et al. 2019)
	C. lanceisepala (Linag and Fujianica 1988).
	C. langbianensis (Zhao et al. 2017 and syn: *C. vidalii*, Zhao 2019)
	C. laotica (Souladeth et al. 2019)
	C. leptopetala (Manh et al. 2019)
	C. limonia (Manh et al. 2019)
	C. longruiensis (Manh et al. 2019)
	C. longzhouensis (Manh et al. 2019)
	C. luteopallida (Hu et al. 2019)
	C. megasepala (Ninh 2007)
	C. meiocarpa (Vijayan et al. 2009)
	C. micrantha (Manh et al. 2019)

(continued)

	C. mingii (Hu et al. 2019)
	C. multipetala (Manh et al. 2019)
	C. murauchii (Manh et al. 2019)
	C. oconariana (Hu et al. 2019)
	C. parvifolia (Manh et al. 2019)
	C. parvipetala (Manh et al. 2019)
	C. phanii (Ninh and Ninh 2014)
	C. quephongensis (Manh et al. 2019)
	C. quinqueloculosa (Manh et al. 2019)
	C. rosacea (Souladeth et al. 2019)
	C. rosmannii (Manh et al. 2019)
	C. rubriflora (Manh et al. 2019)
	C. tamdaoensis (Duong 2011)
	C. terminalis (Manh et al. 2019)
	C. tianeensis (Manh et al. 2019) and syn: *C. huana* (Ming and Zhang 1993)
	C. tienii (Ninh and Ninh 2014)
	C. tuyenquangensis (Le et al. 2017)

Owing to extensive internal hybridization between different *Camellia* taxa, several intergrades, introgressants, and putative hybrids had been found. They were arranged in a gradient based on morphological characteristics that extended from China types through intermediates to those of Assam types (Mukhopadhyay et al. 2013, 2016). Indeed, because of the high cross-pollination, existence of the extreme pure homozygous archetypes of tea was doubtful (Visser 1969). Numerous hybrids therefore are referred to as China, Assam, or Cambod tea depending on morphological proximity to the main taxon (Banerjee 1992a). Tea breeds well with some of the wild relatives, and thus taxonomists had always been interested to identify such hybrids due to suspected involvements in tea genetic pool. Two particularly interesting taxa were *C. irrawadiensis* and *C. taliensis* whose morphological distributions overlapped with that of tea (Banerjee 1992a) although caffeine contents in these two are 0.02% and 2%, respectively (Nagata and Sakai 1985). It had also been postulated that some desirable traits such as anthocyanin pigmentation or special quality characteristics of Darjeeling tea might had introduced from these two wild species (Wood and Barua 1958).

Other *Camellia* species, which were suspected to have contributed to the tea genetic pool by hybridization, include *C. flava* (Pifard) Sealy, *C. petelotii* (Merrill) Sealy (Wight 1962), and possibly *C. lutescens* Dyer (Sharma and Venkataramani 1974). The role of *C. taliensis* was, however, not clear because the species itself was considered to be a hybrid between *C. sinensis* and *C. irrawadiensis* (Wood and Barua 1958; Visser 1969) though it is being used regularly in China for drinking tea. Therefore, it was generally agreed that at least three taxa, i.e., *C. assamica*; *C. sinensis*; and *C. assamica* subsp. *lasiocalyx*, and to an extent *C. irrawadiensis* had mainly contributed to the genome of tea. The term “tea” should, therefore, cover progenies of these taxa and the hybrids thereof or between them.

Apart from this natural diversity, the different tea research institutes and dedicated planters had further developed a number of varieties with better yield, quality, and traits such as tolerance to drought, diseases, etc. In my estimation, more than 1200 such commercial cultivars of tea have been developed and released for cultivation worldwide, and many of them have special characteristics (Table 2.1).

2.4 Karyotype

Karyotype was considered to be the most important cytological markers for identification of the species. Karyotypes had been established for the most available taxa of *Camellia* including tea (Kondo 1975; Fukusima et al. 1966; Ackerman 1971; Datta and Agarwal 1992). However, karyotype grouping based on chromosome size was not successful in the *Camellia* taxa due to high stickiness of the chromosomes. Furthermore, even in the best preparation, homologous chromosome pairs could not appear identical in *Camellia* (Kondo 1975). Relatively little intraspecific karyotypic variation had been observed for the cultivated species of *Camellia* studied (Kondo 1975). Sat-chromosomes in karyotypes within mass accessions of certain *Camellia* species were morphologically and quantitatively variable. Thus, karyotypes including characteristics of sat-chromosomes were not of taxonomic significance for *Camellia* taxa. Among the diploid species of *Camellia* studied, *C. japonica* L. sensu lato showed the greatest karyotypic variation; many of the studied accessions indicated similar karyotypic patterns to each other (Kondo 1975). Actually, *C. japonica* L. var. macrocarpa Masamune had satellites on four submetacentric chromosomes, and the other accessions had satellites on two submetacentric chromosomes (Kondo and Parks 1980). Later, it was shown by Kondo and Parks (1979) that the C-banding method could be applied to the somatic mid-metaphase chromosomes in *Camellia* taxa. These differentially stained bands in mid-metaphase chromosomes permitted the identification of 238 individual chromosomes and made it possible to match the homologous pairs of chromosomes more precisely. Karyotypic variability and divergence among the seven accessions of *C. japonica* L. sensu lato with aceto-orcein staining were revealed by C-banding method (Kondo and Parks 1980). By this way, the cytological marker was used to sort and classify the genotypes.

The karyotype characteristics of the some *Camellia* species were listed in Table 2.2. The cardinal chromosome number of the 29 species was found to be stable, $2n = 30$. No polyploidy were found, excluding cultivated *C. sinensis* and *C. assamica*. It indicated that the evolution of karyotype of section Thea was mainly through the gene similar to other tropical woody plants, different from other species of genus *Camellia* in the temperate zone. The karyotype of most species in Thea was M (metacentric) and SM (submetacentric) chromosomes; only few species with ST (subtelocentric) chromosomes with the order of the number were M > SM > ST.

Interestingly, *C. reticulata* had a series of polyploid varying from $2n = 2x = 30$, $2n = 4x = 60$, to $2n = 6x = 90$, with a basic chromosome number of $x = 15$. The hypothetic allopolyploid origin and parental genomes of these polyploidy types

Table 2.1 Different tea cultivars with special characteristics[a]

Special characteristics	Clone	Originator	Reference
Biggest leaf	Betjan	Betjan T.E, India	Singh (1980)
Waterlogged tolerant	TV-9	TES, Assam, India	Singh (1980)
Wind tolerance	UPASI-2, UPASI-10	UPASI-TRF, India	Sharma and Satyanarayana (1987)
Drought resistance	UPASI-9	UPASI-TRF, India	Sharma and Satyanarayana (1987)
Blister blight tolerance	TRI-2043, DT-1	TRI, Sri Lanka	Sivapalan et al. (1995)
High pubescence content	TRI/2043	TRI, Sri Lanka	Sivapalan et al. (1995)
High anthocyanin pigmentation	TRI/2025	TRI, Sri Lanka	Sivapalan et al. (1995)
High tolerance to pH	TN-14-3	TRF, Kenya	Anon (1999)
Poor fermenter	12/2	TRF, Kenya	Anon (1999)
Mite tolerance	7/9	TRF, Kenya	Anon (1999)
Scale insect tolerant	TN 14-3	TRF, Kenya	Anon (1999)
High polyphenol content (53.7%)	Luxi white tea	TRI, China	Yu and Xu (1999)
High amino acid content (6.5%)	Anji white tea	TRI, China	Yu and Xu (1999)
Low caffeine content (0.14%),	Guangdong tea	TRI, China	Yu and Xu (1999)
High caffeine content (6.96%)	Wild tea at Yunnan	TRI, China	Yu and Xu (1999)
Gray blight resistant	P11P11P12P12	NIVOT, Japan	Takeda (2002)
Early germination	Tianfu 28	CAS, China	Wang et al. (2003)
Very early budding	Zhingcha 108	CAS, China	Yang et al. (2003)
Loop hopper resistance	EF	CAS, China	Hu et al. (2003)
Trichomeless mutant	Progeny of Surugawase	NIVOT, Japan	Takyu et al. (2003)
Small leaf	CH-1, Vimtal	IHBT and TES, Assam, India	Mondal et al. (2004)
High somatic embryogenesis	Makura-Ck2	NIVOT, Japan	Furukawa and Tanaka (2004)
Frost resistance/tolerance	BS 53	HPKV-TES, India	Deka et al. (2006)
Anthracnose	Abo	NIVOT, Japan	Yoshida and Takeda (2006)
Temperature-sensitive albino tea cultivars with white leaves	White leaf No. 1		Du et al. (2008)
High anthocyanin pigmentation	Cha Chuukanbohon Nou 6	NIVOT, Japan	Nesumi et al. (2012)
Light-sensitive albino tea cultivars with yellow leaves, such as	'Huangjinya'		Wang et al. (2014)

(continued)

Table 2.1 (continued)

Special characteristics	Clone	Originator	Reference
Dongcha11	Tolerant to winter dormancy	South China Normal University	Liu et al. (2017)

UPASI United Planters' Association of Southern India, *HPKV* Himachal Pradesh Krishi Viswavidyalaya, India, *TRI* Tea Research Institute, Sri Lanka, *NIVOT* National Institute of Vegetable, Ornamental Crops and Tea Science, Japan, *TRF* Tea Research Foundation, Kenya, *CAS* Chinese Academy Of Sciences, China

[a]Adapted and modified from Mondal (2009)

remained unknown. Genomic in situ hybridization (GISH) was used to study the genome organization and evolution of *C. reticulata*. Total genomic DNAs from closely related diploid species (*C. pitardii* and *C. saluenensis*), with the chromosome number $2n = 2x = 30$, were labeled and hybridized in the presence of blocking DNA onto metaphase spreads of *C. reticulata*. The *C. pitartii* probe painted part of the tetraploid and hexaploid *C. reticulata* genomes, whereas the *C. saluenensis* probe delineated part of the hexaploid *C. reticulata* genome. The results provided compelling evidence for the allopolyploid origin of *C. reticulata* genomes and demonstrated that (1) the diploid *C. reticulata*, *C. pitardii*, and *C. saluenensis* were the progenitors of polyploid *C. reticulata*, (2) hybridization between diploid *C. reticulata* and diploid *C. pitardii* gave birth to allotetraploid *C. reticulata*, and (3) subsequent hybridization between allotetraploid *C. reticulata* and diploid *C. saluenensis* formed the allohexaploid *C. reticulata* (Gu and Xiao 2003).

2.5 Propagation

Tea and its wild species are mainly propagated by three means, i.e., seeds, vegetative cuttings, and nursery grafting, although budding or grafting at mature plants is also followed but very rarely (Mukhopadhyay et al. 2013).

Seeds Conventionally tea is propagated through seeds. Seeds are generally produced in "seed bari" (seed orchard). A fully matured healthy seed while attaching to the plants or immediately dehisced is collected from the ground of the seed orchards. This is primarily due to the fact that tea seeds being recalcitrant have low viability (Mondal 2008).

After eliminating the very small seeds, the remaining seeds are transferred to a tank or trough filled up with water and allowed to soak for 2–3 h. The sinker seeds are taken out of water and examined for mechanical, insect, and pest damage. The usual practice is to cut open a sample of 50–100 seeds from the batch to examine starred, cheesy, shrunken seeds or otherwise damaged seeds by pests or diseases. Floater seeds are discarded as such seeds are found to have dried cotyledons, which normally fail to germinate. Floaters are frequently the result of punctured made by

Table 2.2 The karyotype of some *Camellia* species (Liang et al. 1994; Chen et al. 2000)

Species	Karyotype
C. assamica	$2n = 30 = 18m + 12sm$
C. sinensis	$2n = 30 = 20m + 8sm + 2st$
C. grandibracteata	$2n = 30 = 24m + 6sm$
C. kwangnaica	$2n = 30 = 22m + 8sm$
C. quinquelocularis	$2n = 30 = 24m + 6sm$
C. tachangensis	$2n = 30 = 22m + 8sm$
C. gymnogynae	$2n = 30 = 22m + 6sm + 2st/20m + 8sm + 2st$
C. ygmnogynoides	$2n + 30 = 22m + 6sm + 2st$
C. jungkiangensis	$2n = 30 = 20m + 8sm + 2st$
C. tetracocca	$2n = 30 = 22m + 8sm$
C. nanchuanica	$2n = 30 = 20m + 8sm + 2st/24m + 6sm$
C. crassicolumma	$2n = 30 = 22m + 8sm/2n = 30 = 18m + 9sm + 3st$
C. atrothea	$2n = 30 = 20m + 6sm + 4st$
C. taliensis	$2n = 30 = 22m + 8sm$
C. taliensis var. *bangweicha*	$2n = 30 = 22m + 6sm + 2st$
C. irrawadiensis	$2n = 30 = 18m + 12sm/22m + 8sm$
C. rotundata	$2n = 30 = 20m + 10sm$
C. makuanaica	$2n = 30 = 22m + 8sm/20m + 10sm$
C. manglaensis	$2n = 30 = 22m + 8sm + 4st$
C. leptophylla	$2n = 30 = 24m + 4sm + 2st$
C. dehungensis	$2n = 30 = 20m + 10sm + 2st$
C. gymnigyna	$2n = 30 = 20m + 8sm + 2st$
C. costata	$2n = 30 = 20m + 8sm + 2st$
C. parvisepaloides	$2n = 30 = 22m + 8sm$
C. gymnagynoides	$2n = 30 = 22m + 6m + 2/22m + 6 + 2/$ $20m + 8 + 2st$
C. purpurea	$2n = 30 = 22m + 4sm + 4st$
C. polyneura	$2n = 30 = 22m + 4sm + 4st$
C. sinensis	$2n = 30 = 18m + 12sm + 2st$
C. sinensis var. *pubilimba*	$2n = 30 = 20m + 10sm/18m + 10 + 2st$
C. sinensis var. *kucha*	$2n = 30 = 22m + 8sm + 2st$
C. ptilophylla	$2n = 30 = 22m + 8sm + 4st$
C. assamica	$2n = 30 = 22m + 8sm$
C. yankiangcha	$2n = 30 = 22m + 8sm + 4st$
C. arborescens	$2n = 30 = 20m + 10sm$
C. tachangensis	$2n = 30 = 23m + 7sm$
C. taliensis	$2n = 30 = 1m + 9sm$
C. crassicolumna	$2n = 30 = 20m + 9sm + 1st$
C. gymnigyna	$2n = 30 = 21m + 8sm + 1st$
C. sinensis	$2n = 30 = 21m + 8sm + 1st$
C. sinensis var. *sinensis*	$2n = 30 = 20m + 9sm + 1st$
C. sinensis var. *assamica*	$2n = 30 = 22m + 7sm + 1st$
C. sinensis var. *pubulimba*	$2n = 30 = 21m + 9sm$

(continued)

Table 2.2 (continued)

Species	Karyotype
C. quinquelocularea	2*n* = 30 = 22m + 8sm
C. trilocularea	2*n* = 30 = 21m + 9sm
C. trilocularea var. *macrophyllaea*	2*n* = 30 = 20m + 9sm
C. trilocularea var. *micromidphyllaea*	2*n* = 30 = 20m + 10sm
C. trilocularea var. *kuiea*	2*n* = 30 = 22m + 8sm
C. cryploneura Chang	2*n* = 90 = 61m + 28sm + 1st
C. oblate Chang ex Chang	2*n* = 30 = 24m + 5sm + 1st
C. meiocarpa Chang	2*n* = 60 = 37m + 18sm + 5st
C. oleifera Abel	2*n* = 90 = 60m + 29sm + 1st
C. grijsii Hance	2*n* = 30 = 25m + 5sm
C. forrestii (Diels) Cohen-Sturt	2*n* = 60 = 39m + 19sm + 2st
C. tsaii	2*n* = 60 = 38m + 21sm + 1st
C. tsingpienensis Hu	2*n* = 30 = 21m + 5sm + 4st
C. yunnanensis (Pitard ex Diels) Cohen-Stuart	2*n* = 30 = 19m + 11sm
C. chrysantha (Hu) Tuyama	2*n* = 30 = 17m + 10sm + 3st
C. chrysantha var. microcarpa Mo	2*n* = 30 = 22m + 8 + sm
C. impressinervis Chang	2*n* = 30 = 21m + 8sm + 1st
C. impressinervis Chang	2*n* = 30 = 20m + 9sm + 1st

tea seed bug (*Poecilocoris latus*). As soon as possible, seeds are packed after grading and sorting. For transport over long distances, packing is done in wooden boxes in units of 20 kg using moist sand, sub-soil, powered charcoal, or ash or mixture of two or more of these as packing material. Moisture content of the packing material varies from 10% to 30%, while in the case of powered charcoal, it may vary from 25% to 30%. Seeds are spread in layers along with some packing materials, and each layer is separated from the one on the top by a thin sheet of tough paper. A kilogram of graded and sorted seed may contain from 300 to 500 seeds depending on the size of the grader used.

After 45 d, the germinated seeds are transferred to the polythene sleeves and kept under the shaded nursery for another 12–18 months (Mukhopadhyay and Mondal 2016).

Vegetative cuttings Seeds were the only commercial method for propagation till the beginning of the nineteenth century. However, due to out-breeding nature of the plant, seedling shows a wide variability for the attributes, such as yield and quality, and so it forced people to find some alternatives. The first attempt for vegetative propagation of tea was done in Indonesia by budding as well as grafting. However, due to slower speed, this method could not serve the purpose of rapid multiplication. Hence, the faster propagation by single-leaf cutting was developed simultaneously in India, Sri Lanka, and Indonesia (Mondal 2011). This was further fine-tuned later to fit for commercial venture that exists now. Cuttings from green and semi-hard wood are usually taken from current-year growth. Cuttings are then immediately subjected to fungicide as well as commercial-grade rooting hormone treatment and inserted in the nursery for root induction for 45–60 d depending on the location, planting material, and so forth. The successful rooted cuttings are then transferred to polythene sleeves filled with a good virgin soil (pH around 4.5) having adequate water-holding capacity and kept for another 8–12 months in the nursery, which by then become ready for field transfer. Meanwhile, propagation by cuttings was attempted in several parts of the tea-growing areas around the world (Tunstall 1931a, b; Tubbs 1932; Wellensiek 1933); however, standardization of the technique of single-leaf internode cuttings, practiced today, took a long time to be successful. Following this technique, TES, Assam, India, released the first lot of three clones in 1949, which revolutionized the tea industry in North-Eastern India, and more clones have since been released from time to time.

Grafting In recent years, nursery grafting as an alternative propagation technique has gained considerable popularity. In this technique, fresh single-leaf internode cuttings of both rootstock and scion are generally taken. Scion commonly a quality cultivar is grafted on rootstock, which is either a drought-tolerant or high-yielding cultivar. On grafting, the scion and stock influence each other, and thus composite plants combine both the characteristics, resulting in 100% increase of yield with better quality than either of the ungrafted cultivar. Recently, a modified improved "second-generation" grafting had been developed, where a tender in vitro-derived shoot was grafted on the young seedlings of tea, which had an additional advantage over conventional grafting due to presence of taproot system (Prakash et al. 1999). Nevertheless, with the increasing demand for clonal tea, vegetative propagation with single-leaf internode cuttings remains the best choice in the tea industry worldwide.

2.6 Floral Biology and Pollination Mechanism

The significant differences in flower description between the China and Assam types of tea were reported by several workers (Wellensiek 1933; Bezbaruah 1975) which had been described in Sect. 1.4. Characteristics, such as the length of the style and style arm, the number and length of the outer stamens, or the size of the inner petals, were able to indicate the difference in floral characteristics among the varieties. Importantly, anatomical study suggested that tea flower should be classified as central placental type instead of parental placental type (Bezbaruah 1975).

Tea plants showed an appreciable degree of self-sterility and invariably set a better crop of seeds with pollen from another bush nearly four times than that of selfed seeds (Wight 1938; Wu 1964). Generally selfed seeds exhibited reduced germination. Progenies of self-pollination were inferior in vigor to those of cross-pollination. Investigations on the mechanism of pollination revealed that tea pollens were heavy and sticky in nature and occurred mostly in clumps, a condition which was not favorable for carrying out by wind; rather only non-viable dried up pollen grains can move long distance by wind. However, insects, such as bees and wasps

(Hymenoptera), were found to carry pollens from bush to bush. Besides, Bezbaruah (1975) observed that syrphid flies (Diptera) were the most common insect for natural cross-pollination. The tea flower secretes considerable amount of honey, but they contain high phenols that caused indigestion. Perhaps that may be the reason for not attracting diverse insects for pollination (Barua 1989). Therefore, for higher seed setting, it may be necessary to take measures to attract insects in tea seed orchards.

After 24–48 h of pollination, the corolla withers off and drops from the pedicel along with the anther lobes leaving the ovary exposed. The persistent calyx lobes close flatly over the ovary and the style as well and the stigma gradually withers off. Though pollination takes place during flowering growth, that is, September to January, in India, the first external sign of development of fruits becomes evident by March, and fruits gain considerable size by May of the same year. By about August, the fruits attain full size with completely developed embryo and cotyledon. The mature embryo with two large cotyledons remains covered by a hard, deep brown testa, formed by the outer integument. The pericarp encloses 1–3 seeds inside and is made of thick, parenchymatous tissue when young but becomes sclerotic on maturity. The ripening of fruit generally takes 12 months from the time of flowering till maturity (Singh 1999).

The mechanism of self-incompatibility in tea and related species remains a challenge. A study of reproduction barrier limiting interspecific hybridization between *C. japonica* and *C. chrysantha* was undertaken in intra- and interspecies crosses. Pre-zygotic barriers were not important; pollen type growth and penetration were good in all combination of crosses. Zygote formation and early embryo development were similar in all the crosses. While intraspecies embryos developed smoothly, interspecies embryos were aborting in various stages. Interestingly full size but empty ovules in mature capsule resulted from embryo abortion (Hwang et al. 1992). It was reported that tea could have been considered a facultative out-breeder but with a homomorphic gametophytic self-incompatibility system. The observation of successful self-pollen tube growth to the ovary and ovule penetration clearly indicated that tea had a late self-incompatibility type of selfing control (Bezbaruah 1975).

Tea flower had also been used for value addition. Importantly, tea companies in China had begun to process fresh tea flowers for blending into specialty teas to make them flavory. However, chemical composition of tea flowers and leaves in terms of catechins and caffeine was comparable. The flowers contained less caffeine but equivalent amount of catechins (Su et al. 2000; Lin et al. 2003). Recently, Joshi et al. (2011) did a detail characterization of various catechins and volatile compounds and found that like leaf, unopened young flower bud contained maximum amount of flavor compounds than the fully open flower of tea.

2.7 Seed Biology

Generally 1–3 seeds of 1.5–2.0 cm diameter are found in each capsule. The seeds have a hard testa outside, and the embryo is covered in between two large cotyledons. Tea seeds are highly recalcitrant and lose viability within a few days after shedding from the plant (Bhattacharjee and Singh 1994). However, their viability can be maintained by surface sterilizing with mercuric chloride solution (0.01%) for 15 min and subsequent cold storage at 4 °C. Though seeds are generally stored in moist charcoal for few days, it is advisable to use the healthy seeds for propagation as early as possible (Singh 1999). For commercial propagation, tea seeds are produced in seed baries (orchard) planted specially for the purpose. After the release of the clonal seeds, commercial seed baries have been established for production of hybrid clonal seeds. After collection, seeds are passed through a rotary-type shifter to eliminate the very small seeds. Tea seeds are normally varied in size from 10 to 20 mm in diameter.

2.8 Breeding Techniques

Tea breeding objectives vary from country to country, depending on the local needs, which is illustrated below (Table 2.3). However, by and large it is aimed to improve the quality and yield. However, generally the breeding works at black tea-producing countries, such as India, Kenya, and Sri Lanka, are biased towards the development of high-yielding and high-quality clones, whereas the tea-producing countries near the equator, such as Japan and China, are focused on the development of cold tolerance and frost tolerance, as these countries primarily produced green tea where the quality of made tea does not have much influence on price. Today's modern tea variety has evolved through the sincere efforts for many years of both the tea breeder and experienced planter through different stages of plant introduction, selection hybridization, and physical as well as chemical breeding. The different techniques are discussed here.

2.8.1 Introduction

Introduction may involve de novo addition of new varieties, wild relatives, or totally new species in a particular area. Often materials are introduced from other countries or continents. In tea the unorganized agents such as Buddhist pilgrimage or colonial soldiers made the primary introduction. However, later the secondary introduction in various tea-growing countries was done by the experienced British planters or scientific communities which are discussed below.

2.8.1.1 Primary Introduction

Introduction of tea into Korea was done by the troops of Emperor Wu Di during his invasion to Korea from China. In Japan tea was introduced during the eighth century

Table 2.3 Breeding objectives of tea[a]

Objectives	Importance	Regions
Improving quality	Directly linked to the profitability	Black tea-producing countries such as India, East Africa, Sri Lanka, Bangladesh, and Indonesia
Increasing yield	Horizontal increase of production by extension planting is limited	Worldwide
Drought tolerance	Reduce productivity and occur all tea-growing regions of the world	Worldwide where tea is grown as a rain-fed crop
Reduce winter dormancy	No leaf production during winter months and occurs in Northeast India, Japan, China, etc.	Tea plantation in near equator
Hail/frost resistance	Causes economic loss as young leaves during rainy season are mostly affected.	Hilly region of the tea-producing countries
Water log tolerance	Reduce productivity during rainy season. Generally occurs in North-Eastern India	Northeast India
Cold hardiness	Reduced productivity during winter due to snow. Generally occurs in China, Japan, and Russia	Mainly in Japan, Russia, and China
Disease resistance, such as blister blight, stem canker, etc.	Blister blight causes severe damage as only young leaves are infected. Generally occurs in Japan, Sri Lanka, South India, and Darjeeling hills of Northeast India	Mainly India, Sri Lanka, Indonesia, and Japan
Pest resistance, such as red spider mite, tea mosquito bug, leaf-sucking pest, etc.	Most important biotic stress as all cause severe damage to the leaves. Generally, occurs in all the tea-growing regions in the world	Worldwide
Suitability to type of manufacturing	For matching the customer's demand as well as better recovery percentage in made tea	Black tea-producing countries such as India, East Africa, Sri Lanka, Bangladesh, and Indonesia
Low input responsive clone	Required for organic tea farming	Organic tea
Purple tea	Required for product diversification due to health benefit	Black tea-producing countries such as India, East Africa, Sri Lanka, Bangladesh, and Indonesia

[a]Adapted from Mondal (2009) with modification

when Buddhist monk Saichō who returned from studies of Guo Xing Si on Tiantai Mountain of Zhejiang provinces brought tea seeds to Japan and planted them at the foot of Mountain Hi Yei in the village of Sakomoto of Omi County. Later on in 1191, A. D. Yei Sai another Buddhist re-introduced tea seeds from China to Japan and planted them on the hilly terrain of the Seburi Mountain, South-West of Castle Fukuoka in Chikuzen province. Yei Sai not only planted and cultivated but also regarded tea plant as the source of a sacred remedy.

In 1690 A.D., the governor of Indonesia, J. Camphuys, brought tea seeds from China and planted the same nearby Djakata region. Between 1828 and 1833, Mr. Jackson of the East India Company went to China and brought tea seeds again which were planted in Indonesia.

Tea was brought to Russia during 1567. In 1735, tea was first carried overland by governmental caravans. From Chinese border, this legendary trail lay North-Westwards across 800 miles of inhospitable Gobi Desert through Ular Bator in Mongolia and into Russia skirting lake Baykal to the town of Irkutsk.

Initially, Sri Lanka began to cultivate coffee but without success. Therefore, they decided to start tea plantation during 1841. M.B. Worms, a German who lived in Ceylon, visited China and brought tea seedlings to Pussellawa. Simultaneously, seeds were also brought from the Botanical Garden of Calcutta, India, during 1839 and planted in the nurseries of Royal Botanical Gardens at Peradeniya near Kandy, Sri Lanka.

2.8.1.2 Secondary Introduction

Information regarding the secondary introduction is rather scanty due to the fact that the presence of stringent rules worldwide prevents trans-border movements of the plant propagule in later years. Additionally, the secondary introduction of tea always happened off the record by the commercial tea growers. Interestingly, the early history of the planting materials clearly indicates that being the oldest tea research institute in the world, TES, Assam, India, contributed significantly towards the secondary introduction of tea worldwide. In a conservative estimation, 60% of the world acreage had received its initial planting materials directly or indirectly from tea genetic resources of TES, Assam, India. Initially the commercial tea estates were established later that had been used as parent materials for developing the new cultivars.

2.8.2 Hybridization

In natural hybridization, based on known desirable characteristics, such as previous performance of yield, quality, or disease tolerance capability, two parents are planted side by side in an isolated place to bear fruits. Subsequently, the seeds (F_1) are harvested, raised, and planted. If average performances of these plants are found to be better than either parent, then seeds are released as hybrids or bi-clonal seeds. However, some of the outstanding performers among these progenies are marked and verified for multi-location trials and still, if found suitable, released as clone.

These clones are geographically specific, and most of the tea research institutes of the world have generated the clones for their own region. Sometime in the above process more than two parents are used and known as polyclonal seeds. The idea is to introduce more variability among the F_1 seeds. Since it is difficult to know about the pedigree of the cultivar (as pollen may come from any male), hence the chance of reproducibility of performance is low, and therefore the process thus least preferred currently.

Hand pollination or control cross, despite being an important approach, has made a limited success in tea breeding. The reasons identified could be (1) low success rate, (2) availability (2–3 months) of tea flowers for a short period, (3) longer time taken for seed maturation (12–18 months), and (4) difference in flowering time for different clones.

2.8.3 Selection

The seeds from a particular "seed orchard" are known by the name of that orchard or locality and are called a *jats* or population, and those *jats* are the main source of planting materials of tea. There are wide variations among the offsprings for morphology, yield, and quality parameters generated from a particular *jats* even though the source of seed was the same. This is primarily due to high out-breeding nature of tea plants. Wight (1939) showed an interesting fact. About 10% of the bushes in a commercial tea garden of North-Eastern India produced only 2% of the total crops, i.e., green leaf, and about 0.5% bushes produced as much as or more than 300% of the average crops. Thus, the planters thought that new areas planted with seeds from those 0.5% bushes will produce more uniform tea plant with better-quality tea. Those selected plants were used for future plantations. Thus, the process of "selection" started. The first scientific attempt to select improved tea in North-Eastern India was made by Stiefelhagen brothers during 1860 by establishing standard sources of tea seeds. Subsequently, scientific improvements of tea by selection were followed in many countries, such as Indonesia (Wellensiek 1934), Java (Cohen Stuart 1929), Russia (Bakhtadze 1935), and North-Eastern India (Wight 1939). Mother bushes were selected based on morphological characteristics followed by anatomical (Wight 1956) as well as organoleptic performance of made tea (Timoshenko 1936). Indigenous Assam tea was improved by following the technique of mass selection. The yield increased considerably, because of line breeding for desirable morphological features that are genetically linked with the characteristics of Assam tea. After that, line breeding techniques were followed to improve further by mainly evolving more uniform tea plant with better quality and adaptability. In the earliest effort, two selected *jats* were hybridized to combine desirable characteristics into their progenies. The "Rajghur" *jats* was developed by combining the high quality of a light leaf local *jats* with the vigor of a dark leaf "Manipuri" *jats*.

However, the seed-grown plants were not uniform as their characteristics were governed by genotypes of their parents, which were diverse in some phenotypic characteristics in relation to environmental and soil conditions. In some cases, the yield and quality were unpredictable. It was, therefore, felt necessary to develop clonal cultivars in tea like many other fruit crops by multiplying the selected bushes vegetatively. Today all tea-growing countries of the world have developed clonal materials as per their requirements.

2.8.4 Polyploidy Breeding

Yield is the major criteria in commercial tea cultivation which depends on the size and density of the leaves in the plucking table, i.e., upper surface area of the tea bush. A positive correlation between size of the leaf and the yield in tea was well-established (Satyanarayan and Sharma 1982). Therefore, the development of tea genotypes with bigger leaves through polyploidy breeding may be useful to increase the yield of tea. Further, being vegetatively propagated plants, polyploidy breeding can be used effectively. In tea, significant amount of works on polyploidy breeding had been done (Gunasekara and Ranatunga 2003), which are reviewed below.

2.8.4.1 Naturally Occurred Polyploids

Although cultivated species of tea was diploid (Morinago et al. 1929; Barua 1989), naturally occurring intra- and interspecies polyploids of tea (Janaki Ammal 1952; Bezbaruah 1971; Jayasuriya and Govindarajulu 1975; Wachira and Kiplangat 1991) and its wild relatives were also reported (Kondo 1977). Interestingly, natural polyploids were more common in tea populations (Table 2.4) of Japan than any other countries (Banerjee 1992b; Simura and Inabe 1952). Bezbarua (1968) reported that in tea, the triploids, tetraploids, pentaploids, and aneuploids, resulting from open-pollinated progenies, occurred naturally but in extremely low frequency.

2.8.4.2 Artificial Induction of Polyploids

Since the discovery of the effect of colchicine in the 1930s on cell division for mitotic doubling of the chromosome number, it was widely used to induce the level of polyploidy in plant (Blakeslee and Avery 1937). Colchicine inhibits mitosis in cells by interfering with the structure of the mitotic spindle, thus resulting in formation of cells with a doubled chromosome number.

Similar to other plants, colchicine had been used to induce artificial polyploids in tea (Table 2.5). In Sri Lanka, Sebasthiampillai (1976) produced five tetraploid plants, namely, TRI 2023, 2024, 2025, and 2026 and DT 95, by treating the meristematic tissues of the terminal bud for 2–7 d with colchicine impregnated in agar. Although he found the differential response of tea genotypes with the colchicine treatment, his ploidy plants were tetraploid as he confirmed through the cytological examination of root tip cells. However, attempts to induce polyploids using ethyl methanesulfonate (EMS) at TES, Assam, India, though tried, were not successful. Nevertheless, more than 170 and 70 polyploids were subsequently generated in the same institute through conventional hybridization and colchicine

Table 2.4 Triploid tea cultivar used under commercial cultivation along with their promising characteristics (Gunasekara and Ranatunga 2003)

Country	Polyploid cultivar	Level of ploidy	Type of polyploidy	Promising characteristic(s)	Reference
India	Sundaram	3*n*	Natural	High yield and quality	Sharma and Ranganathan (1986)
India	UPASI 3	3*n*	Natural	High yield and overall quality	Satyanarayan and Sharma (1986)
India	UPASI 20	3*n*	Natural	Moderate yield, highly tolerant to drought	Satyanarayan and Sharma (1986)
India	TV 29	3*n*	Natural	High quality	Barbora et al. (1996)
Japan	Not known	3*n*	Natural	Hardier and cold resistant	Simura and Inabe (1952)
Kenya	382/1	3*n*	Natural	High yield	Wachira (1994)
Sri Lanka	TRI 3069	4*n*	Artificial	High yield and drought tolerant	Kulasegaram (1980)
Sri Lanka	HS 10A	3*n*	Natural	Cold resistant	Kulasegaram (1980)
Sri Lanka	GF 5/01	3*n*	Natural	High yielding	Anon (1973)

Table 2.5 Details of induction of tea colchiploids (Gunasekara and Ranatunga 2003)

Plant part used	Treatment	% success	Reference
Axillary buds of etiolated shoots	Cotton wool moistened with colchicine (0.2%) and treatment was given in the dark	13.0%	Katsuo (1966)
Terminal buds of active shoots developing from pruned bushes	Agar impregnated with colchicines (0.2–0.5%) for 5–6 d	13.5%	Sebasthiampillai (1976)
Terminal buds	Immersion in aqueous colchicines (1–2%) for 5–7 d	6–17%	Anon (1979)
Flower buds	Colchicines (0.05%) injection and drop application for 2–6 d	30.0%	Osone (1958)

treatment, respectively (Singh 1999), yet in tea, as high as 30% ploidy had been achieved using colchicines as mutagenic agents (Table 2.5).

2.8.4.3 Morphological, Anatomical, and Cytological Markers in Polyploid Teas

Screening of polyploids has not been widely exploited in tea due to lack of reliable markers. Hence, the identification of markers related to morphology, anatomy, or cytology is of great importance in screening for polyploids, whether they are artificially induced or naturally occurred. Generally, the ploidy level of tea is determined by counting chromosome numbers in meristematic tissue, i.e., root tip cells or pollen mother cells.

Chaudhuri (1979) found a wide range of phenotypic and anatomical variations, such as frequency and size of stomata and sclereids, among the progenies of triploid tea (Chaudhuri and Bezbaruah 1985). Similarly, to assess the effects of the level of polyploidy on the morphogenetic attributes of the F_1 seedling population, generated from a cross between diploid and tetraploid cultivars, a clear relationship was shown between ploidy levels and morphogenetic variations (Rashid et al. 1985).

Among the different morphological markers, while leaf area was found to be higher in triploid, its expansion, i.e., growth, was lower in triploids in comparison with diploid tea leaves (Ng'etich and Wachira 1992). The reason for this may be that the attributes considered were more affected by environmental factors than by genotypes.

Anatomical marker such as stomatal density was used to differentiate the triploid than diploid as it was found to be lower in triploid than diploid cultivars (Amma 1974; Chaudhuri and Bezbaruah 1985; Wachira 1994). It was found that triploid plants had a lower stomatal density than the diploid genotypes (Wachira 1994). However, this marker could not always be used as a reliable marker for identification of polyploids in tea. In fact Chaudhuri and Bezbaruah (1985) had indeed reported that there was a lack of correlation between level of ploidy and stomatal density. Similarly guard cell sizes as well as stomata size were larger in tetraploids and triploid teas than its diploid counterparts (Amma 1974). On the contrary, Wachira (1994) found that the length/breadth ratio of the guard cells was not significantly different between diploids and triploids. Later, chloroplast number in the guard cells had been identified as a reliable ploidy marker in tea (Ahmed and Singh 1993; Koskey and Wachira 2000; Ranatunaga and Gunasekare 2002; Chen and Ye 1989). Further, Koskey and Wachira (2000) found that the ratio of the guard cell chloroplast numbers in diploids, triploids, and tetraploids was found to be 2:3:4, which was the same as the ratio of their chromosome numbers (30:45:60) (Ahmed and Singh 1993). Therefore this finding indicated that the ploidy level of tea could be accurately and rapidly identified by the chloroplast count method, rather than by criteria based on the size and density of stomata.

Reproductive organs such as pollen grain in most of the induced tetraploid cultivars are found to be higher than that of its diploid counterparts (Gunasekara 2000). But in vitro germination of pollen grain was poorer in tetraploid cultivars than in diploid cultivars (Thirukkumaran and Gunasekare 2001). Only 2% of the pollen grains of natural triploids were found to be viable (Bezbaruah 1971). It had been reported that pollen viability and fertility of triploid cultivars were unable to set seeds and fruits. In general, these morphological as well as anatomical markers were not consistent which was why they were not accepted by the tea breeder.

Therefore, the alternative cytological markers such as chromosome counts were found to be better reliable to differentiate the triploids from diploids. Chromosome counting in pollen mother cells, root tip cells, and meristematic tissue cells at the shoot tip was subsequently standardized in tea (Gunasekara and Ranatunga 2003). Wachira and Muoki (1997) devised a new cytological technique to assess the activity of nucleoli and nucleolus-organizing regions of polyploids and diploids. Their study revealed that the mode of nucleolar number corresponded to multiples of the somatic cell number and hence was a good marker for ploidy. Therefore, it is assumed that the attributes, related to anatomical features, are much more precise than the morphological characteristics which have been used to screen polyploidy genotypes in tea due to the fact that the latter has larger environmental influence than the former.

2.8.4.4 Use of Polyploids in Tea Breeding

Generally, tea polyploids often lack desirable traits (Bezbarua 1968; Sarmah and Bezbaruah 1984), and polyploidy breeding therefore requires planned hybridization, selection of promising polyploids, and proper evaluation to confirm their performance as potential cultivars. High-yielding polyploids with low quality of made tea had been improved through hybridization with a diploid cultivar of high-quality traits (Sarmah and Bezbaruah 1984). Triploids had been produced by hybridizing tetraploid tea with diploid tea in Japan (Osone 1958), India (Chaudhuri 1979), and Bangladesh (Rashid et al. 1985). It had been shown that it was possible to combine good cup quality, with the superior vigor and hardiness of the polyploids, by crossing tetraploid progeny for commercialization. For example, open-pollinated tetraploid with inferior cup quality but with higher growth vigor (Bezbaruah 1976) was improved to higher cup quality by crossing with high-quality diploid clone as the male parent (Bezbaruah 1991).

The conventional method of producing triploids is through artificial induction of tetraploids, followed by hybridization with diploid cultivars. For example, out of 238 hybrids produced through hybridization between tetraploids and diploids at TES, Assam, India, only 79 hybrids were found to be triploids (Barbora et al. 1996). Since induction of diploid was time-consuming, Osone (1958) used diploidized pollen of immature flowers to pollinate diploid plants for producing triploids. However there is no evidence that this method had been widely practiced in polyploid breeding programs. Recently, to verify the quality of triploid cultivars of tea, Das et al. (2013) profiled caffeine and catechins of 97 F_1 segregating progenies of a common tetraploid and diploid parents. Catechins and caffeine level of the triploid progenies were compared against their diploid parents. Some of the progenies were found to be better-quality clones than their diploid parents. Most of the progenies of the diploid *C. sinensis* crossed with tetraploid showed heterosis for caffeine and catechins. The genomic contributions of the diploid parents seem to be the main factor in the variation between the two populations. They demonstrated quantitative enhancement of some of the quality-related parameters in tea, providing a platform to refocus on this classical breeding approach for developing quality cultivars in tea.

2.8.4.5 Commercial Exploitation of Polyploids

Although emphasis was given to identify natural polyploids and to develop artificial polyploids, reports on their performance and trait evaluation are scare. After the discovery of natural polyploids of tea (Karasawa 1932; Bezbaruah 1971; Amma 1974; Katsuo 1966; Sebasthiampillai 1976), natural polyploids were included in cultivar selection programs to identify desirable agronomic traits. Certain studies indicated that natural polyploids found in Southern India possess attributes for high yields as well as quality (Sharma and Ranganathan 1986). On the other hand, Banerjee (1992b) had reported that though polyploids showed high vigor and tolerance to environmental stresses, they did not always contribute towards high yields and even sometimes produced low quality of tea (Bezbarua 1968; Sarmah and Bezbaruah 1984). The prolific growth in polyploids may be attributed to increase photosynthesis owing to the increase of chloroplast number in the guard cells.

The effects of ploidy on yield and its components had been studied in tea (Amma 1974; Banerjee 1992b; Wachira 1994; Wachira and Ng'etich 1999). It was found that triploid cultivars produced larger and heavier but fewer harvestable shoots per unit area, compared to diploids due to which triploids yield less than diploids, despite higher shoot weight (Wachira 1994). In another study by Singh (1980), it was found that out of the different types of polyploids produced in India, the dry weights of five fully formed leaves in triploids and tetraploids were higher than that of diploid leaves by 14% and 109%, respectively. Other pentaploids and aneuploids, however, had relatively low leaf dry weights.

Although it was shown that polyploidy in tea enhanced the yield (Jayasuriya and Govindarajulu 1975; Kulasegaram 1980; Sharma and Ranganathan 1986), this was not always, the case, as there were instances where increased polyploidy was demonstrated to depress the productivity (Banerjee 1992b; Wachira 1994). However, it was clear from those studies that, though increased ploidy depressed yields, a significant difference in production could also be observed among the genotypes at the same ploidy level. In certain cases, triploids out-yielded diploids, which indicated the potential for selecting or developing high-yielding polyploid cultivars. Nevertheless, rooting ability, leaf size, and leaf dry weights of triploids and tetraploids were higher than that of diploids but lower in pentaploids and aneuploids (Banerjee 1992b).

Two triploid cultivars were commercially successful in the tea industry of Sri Lanka. The first one, TRI 3069, which was an induced tetraploid of TRI 2025, had been accepted commercially and possesses many improved traits. The second cultivar, HS 10A, which was a natural triploid selected from a seedling tea population on Hethersett Estate of Sri Lanka, was found to be better adapted to high elevations than diploid cultivars (Kulasegaram 1980). It had been reported that triploid forms of tea were hardier and more resistant to cold conditions than diploids, and a clone which was widely recommended for planting in Southern India was reported to be a natural triploid (Jayasuriya and Govindarajulu 1975). Commercially acceptable polyploid tea cultivars developed in some tea-growing countries are detailed in Table 2.4 which show that only eight polyploids have been found their way into cultivation, including one artificially induced polyploid.

Although extensive works have been done to identify precise markers for ploidy level in tea, it is clear from the above discussion that the results obtained are not consistent. However, among the criteria studied, the number of chloroplasts in the guard cells and the stomatal density can be used with some reliability for ploidy-level analysis. These markers may be used for the screening of polyploids from a larger number of tea genotypes, although chromosome counts remain the most reliable and this procedure could minimize the time and resources needed for subsequent cytological studies.

2.8.5 Mutation Breeding

The work on mutation breeding in tea was initiated during 1967–1968 at TES, India, with the objectives of increasing genetic variability for possible use in evaluation of superior planting materials. However, except a preliminary report on irradiation with ϒ rays on cuttings, no progress had been achieved till now (Singh 1984). Studies done elsewhere had shown that a wide range of variations can be created by irradiating various plant parts, such as seeds, leaf cuttings, and auxiliary and apical buds, of tea to induce mutations (Tavadgiridze 1979).

2.8.6 Pre-breeding and Distance Hybridization

Tea breeds freely among the two cultivated species, i.e., *C. sinensis* and *C. assamica*, and up to a limited extent with few wild relatives. Earlier, Wight and Barua (1957) hybridized *C. irrawadiensis* with *C. sinensis*. Though the progenies were resembled later but failed to attack tester tong due to inferior quality. Later Bezbaruah and Gogoi (1972) made a successful hybridization between *C. japonica* and *C. sinensis*. Morphologically progenies were found to be intermediate but produced low yield as well as quality. However, a commercial high-yielding clone TV-24 was produced at TES, Assam, India, from the cross between F_1 hybrids from *C. irrawadiensis* and TV-2 an Assam–China hybrids. Six interspecific backcross progenies were generated by crossing hybrids of *C. saluensis* X *C. japonica* (*C. X williamsii*) back to *C. japonica*. Segregation data were obtained within these six families for five traits, three involving flavonoid constituents and two pubescence characteristics. A single major gene seemed to be primarily involved in each cases, and two of the traits exhibited linkage behavior. The taxonomic value of such marker trait was also discussed (Parks and Kondo 1974). Li et al. (2005) made a cross between high-quality *C. sinensis* and *C. ptilophylla*. Out of the 62 progenies, 3 identified to be commercially viable on the basis of the biochemical parameters or organoleptic test. At the same time, these three progenies were cold tolerant, a character which came from *C. ptilophylla*, thus making them suitable for cultivation in subtropical region of China.

Interspecific cross-compatibility between tea and its allied 26 species in the genus *Camellia* was examined. The interspecific crossing abilities varied among the cross combinations, and the fruit-bearing rates were in the range from 0% to 42.6%. The interspecific hybrids obtained from the crossings of *C. sinensis* with *C. japonica*, *C. pitardii*, *C. assimilis*, *C. caudata*, *C. salicifolia*, *C. irrawadiensis*, and *C. taliensis* showed very low pollen fertilities. Morphological characteristics of those hybrids, including size and shape of leaves, flowers, and tree performance, were generally intermediate of their parental species. The hybrids between *C. sinensis* and *C. japonica* showed a high level of tolerance to diseases such as tea gray bright and tea anthracnose and to cold damage during winter as well. It was presumed that the F_1 hybrids which were obtained from the crossing of *C. sinensis* with *C. sasanqua*, *C. brevistyla*, and *C. oleifera* used as a male parent might be developed through parthenogenesis of a reduced gamete. The cross-compatibilities between *C. sinensis* and subgeneric sections of genus *Camellia* were found to be as follows: Thea>Camelliopsis>Paracamellia=Camellia=Theopsis>Heterogenea=Corallina (Takeda 1990).

In a natural cross between *C. taliensis* and *C. sinensis*, two low-caffeine but high-theobromine containing plants were produced. Genetic analysis indicated that caffeine-less character might be controlled by one recessive locus. Thus this caffeine-less plant might be used as genetic resource for introducing caffeine-less trait in cultivated tea through breeding (Ogino et al. 2009).

2.9 Genetic Resources of Tea

Progress and achievements of tea breeding works in certain tea-producing courtiers had been well-reviewed (Singh 1999; Ghosh Hajra 2001; Deka et al. 2006). Those literatures indicated that the initial emphasis was to collect and evaluate the either indigenous or exotic germplasm for befitting the local environment. However, with the increase of the region-specific need of the industry, all most all tea-producing countries have developed their past specific clones or seed stocks which are reviewed here.

As mentioned earlier, the breeding works at TES, Assam, India, which started since 1949, developed a total 33 clones (Table 2.6), 16 bioclonal seed stocks (Table 2.7), and 134 TRA/garden series clones (Table 2.8). In Southern India, the breeding works started at UPASI, Tamil Nadu, during the early 1960s, which resulted in the release of 33 clones (Table 2.9) and 5 bioclonal stocks, i.e., BSS-1 (UPASI-10 X TRI-2025), BSS-2 (UPASI-2 X TRI-2025), BSS-3 (UPASI-9 X TRI-2025), BSS-4 (UPASI-15 X TRI-2025), and BSS-5 (CR6017 X UPASI-8). Further to widen the genetic base, clones developed in Sri Lanka were introduced and experimented, and it was found that TRI-2024 and TRI-2025 were suitable which gained popularity in the industry (Sharma and Satyanarayana 1987). Efforts to generate and conserve the tea genetic resources by private funding were also made. For example, the Research and Development Department of Tata Tea Ltd, Kerala, India, had developed 7 promising cultivars and several bioclonal seed stock such as TTSS-1 and TTSS-2; besides, they maintained a tea germplasm collection more than 100 popular tea genotypes and wild species (Haridas et al. 2006) (Table 2.10).

The plant improvement programs in Sri Lanka started during the 1930s when seeds of Betjan, Manipur, and Rajghur from Assam were planted in Peradeniya Botanical Garden in Sri Lanka to establish the first tea garden. However, scientific works started only during 1937 when Dr. F. R. Tubbs brought few seeds of ST 4/10 from TES, Assam, India, and seedlings were raised at the Tea Research Institute of

Table 2.6 Descriptions of different TV clones (Deka et al. 2006)

Name	Year of released	Preference for manufacture		Remarks
		First preference	Second preference	Assam China hybrid
TV-1	1949	CTC	Orthodox	Assam type
TV-2	1949	Orthodox	CTC	Assam type
TV-3	1949	Orthodox	CTC	Assam type
TV-4	1959	Orthodox	CTC	Assam type
TV-5	1959	Orthodox	CTC	Assam type
TV-6	1959	Orthodox	CTC	Assam type
TV-7	1959	Orthodox	CTC	China hybrid
TV-8	1959	Suitable for both		Assam type
TV-9	1959	CTC	Orthodox	Cambod type
TV-10	1963	CTC	Orthodox	Assam type
TV-11	1963	Orthodox	CTC	Assam type
TV-12	1963	Orthodox	CTC	Assam type
TV-13	1965	Orthodox	CTC	Assam type
TV-14	1967	CTC	Orthodox	Assam hybrid
TV-15	1967	Orthodox	CTC	Assam type
TV-16	1968	CTC	Orthodox	Assam hybrid
TV-17	1968	CTC	Orthodox	Assam hybrid
TV-18	1970	CTC	Orthodox	Cambod type
TV-19	1973	CTC	Orthodox	Cambod type
TV-20	1974	CTC	Orthodox	Cambod type
TV-21	1976	Orthodox	CTC	Assam type
TV-22	1976	CTC	Orthodox	Cambod type
TV-23	1976	CTC	Orthodox	Cambod type
TV-24	1979	CTC	Orthodox	Cambod and species hybrid
TV-25	1982	CTC	Orthodox	Cambod type
TV-26	1982	CTC	Orthodox	Cambod type
TV-27	1985	CTC	Orthodox	Cambod type
TV-28	1985	CTC	Orthodox	Cambod type
TV-29	1990	CTC	Orthodox	Cambod, triploid
TV-30	1993	CTC	Orthodox	Cambod type
TV-31	2006	CTC	Orthodox	Assam hybrid
TTRI I	2014	CTC	Orthodox	Assam hybrid
TTRI II	2014	CTC	Orthodox	Assam hybrid

Sri Lanka, and subsequently eight clones were released. Since then, several clones were released which were popularly known as 20 series, 30 series, and 40 series clones. At present, Sri Lanka has more than 57% clonal tea area. Out of this, around

Table 2.7 Descriptions of bioclonal seedling developed by TES, Assam, India (Deka et al. 2006)

Cultivar	Parent combination	Year of release	Suitable for area
TS 378	14.5.35 X 14.6.28	1968	Hills (Darjeeling)
TS 379	14.5.35 X 14.12.16	1989	Hills (Darjeeling)
TS 397	TV-1 X 19.35.2	1976	Plains
TS 449	TV-1 X 270.2.14	1970	Plains
TS 450	TV-2 X 270.2.13	1970	Hills (Darjeeling)
TS 462	TV-1 X 124.48.8	1980	Plains
TS 463	TV-1 X TV-19	1984	Plains
TS 464	TV-1 X 19.29.2	1984	Plains
TS 491	TV-1 X S_3A_3	1989	Plains
TS 506	TV-1 X 19.22.4	1994	Plains
TS 520	TV-19 X TV-20	1992	Plains
TS 557	AV-2 X Teen Ali 17	1996	Hills (Darjeeling)
TS 569	AV-2 X Tukdah-78	1996	Hills (Darjeeling)
TS 589	TV-20 X Heeleakah 22/14	1996	Plains
TSS1	TV13 X TV 17	2015	Plains
TS560	TRA/AV2 X TRA/P312	2017	Hills (Darjeeling)

Table 2.8 Descriptions of different region-specific garden series clones (Deka et al. 2006)

Area	Clone
Darjeeling	Phoobsering 312, Phoobsering 1404, Phoobsering 1258, Kopati 1/1, Happy Valley 39, Bannockburn 157, Tukdah 145, AV2, Tukdah 253, Tukdah 246, Bannockburn 777, Rungli Rungliot 4/5, Bannockburn 688, Tukdah 78, Tukdah 383, Rungli Rungliot 17/144, CP-1, Teesta Valley 1, Badamtam 15/263, Balasun 7/1A/76, Balasun 9/3/76, Thurbo 3, Thurbo 9, and Lingia 12
South India	ATK-1 (drought-tolerant clone), C-17, D-12/A2, C-1, CR-6017 (quality clone), SMP-1 (resistant to blister blight), W-35, SA-6, TTL-1, TTL-2, TTl-4, and TTL-5
Kangra Valley	Kangra and Jawala
Tripura	Huplongcherra 18, Huplongcherra 26, Meghlibundh 11, Meghlibundh 20, and Meghlibundh 25
Barak Valley	Narinpore 4, Narinpore 18, Narinpore 22, Chandighat 9, Longai 17, Longai 26, Poloi 23, and Lalamookh 7
Dooars and Terai	Hantapara 12, Huldibari 19, Leesh River 9/34, Sukna 7, Sanyasithan 8, Kamalpur 6, Mohargung, and Gulma 25

80% is composed by only three popular clones, i.e., TRI-2023, TRI-2025, and TRI-2026.

Bangladesh Tea Research Institute, Maulvibazar, Bangladesh, had developed 13 clones and 2 bioclonal seed stocks so far. These clones are known as BT-1 to BT-13. Between the two bioclonal seed stocks, i.e., BST1, BST-2, former one is more popular which is a cross between BT-1 and popular Indian tea cultivar, TV-1 (Deka et al. 2006).

Table 2.9 Descriptions of different UPASI clones (Sharma and Satyanarayana 1987)

Name	Description
UPASI-1	Resistant to drought and tolerant to wind, very upright and vigorous growth, and forming as compact bush. Suitable for mid elevations
UPASI-2	Excellent spread with a dense plucking table fairly hardy tolerant to drought and wind. Suitable for all elevation
UPASI-3	A triploid clone with an excellent spread and dense plucking table flourishing at all elevation, bright infusion and liquor
UPASI-4	With profuse branching. Suitable for mid and high elevations
UPASI-5	Compact bush and dense plucking table suitable for mid and high elevation
UPASI-6	With good spread and high plucking density, fairly tolerant to drought, suitable for mid and higher altitude
UPASI-7	With profuse branching, good spread with a dense plucking table. Tolerant to drought and wind damage. Suitable for mid and high elevation
UPASI-8	Vigorous growth, with a good spread and dense plucking table. Can grow at all elevation particularly in warm humid zones
UPASI-9	Most popular standard clone in South India, fairly tolerant to drought and suitable to all altitude, and can withstand the soil pH near neutral (6.8)
UPASI-10	Forms a broad dense plucking table, hardy clone, resistant to drought, fairly tolerant to wind. Thrives even in soils with pH near neutral (6.8)
UPASI-11	Vigorous orthotropic grower, tolerant to drought, suitable for mid altitude
UPASI-12	Semi-orthotropic, good grower, suitable for mid elevation
UPASI-13	Profuse branching with dense plucking table. Suitable for high altitude
UPASI-14	Spready bush with high plucking density. Suitable for high altitude and produced flavory tea
UPASI-15	Resistant to drought and fairly tolerant to wind and mild frost. Suitable to mid altitude and produce quality tea throughout the year
UPASI-16	Tolerant to drought and mild frost. Suitable for all elevation. Flushes during winter
UPASI-17	Good spread and dense plucking table. Suitable for mid and high elevations. Produced very bright liquor
UPASI-18	Semi-orthotropic and vigorous grower with good branching, fairly tolerant to drought. Suitable mid and high altitude
UPASI-19	Hardy, resistant to drought, and tolerant to mild frost. Flushes throughout the year. Tolerance to both drought and winter. Suitable for high altitude. Produced flavory tea
UPASI-20	Semi-orthotropic grower, resistant to drought, suitable for mid altitude. Produce bright liquor
UPASI-21	Excellent spread and dense plucking table. Suitable for mid and high altitude. Produce flavor, bright liquor
UPASI-22	Vigorous grower but with sparse branching. Makes good, tippy, orthodox tea with good flavor and quality
UPASI-23	Good rooter, excellent spread, tolerant to drought fairly well. Suitable for mid and high elevation

(continued)

Table 2.9 (continued)

Name	Description
UPASI-24	Flushes throughout the year. Suitable for mid and high elevation. Produce scented flavor
UPASI-25	Recovery from pruning is quick; dense plucking point. Suitable for high altitude
UPASI-26	Recovery from pruning is quick. Resistant to drought; flushes throughout the year. Suitable for high altitude
UPASI-27	Compact plucking table with moderate spread. Tolerant to drought
UPASI-28	It is high yielder and fairly tolerant to drought
TRF-1	High-yielding clone with profuse branching
TRF-2	High-quality clone suitable for hilly terrain of South India
TRF-3	High-yielding clone suitable for Karnataka state of South India
TRF-4	Quality clone; fairly drought tolerant
TRF-5	It is a quality clone, with high leaf production and suitable for manual as well as mechanical harvesting

Table 2.10 Clones developed by Tata Tea Ltd (Haridas et al. 2006)

Clones	Parentage	Salient features
TTL-1	UPASI 9X TRI-2025	High yielder and moderate quality, drought tolerant
TTL-2	Estate selection	Average yielder and excellent quality
TTL-3	Estate selection	Average yielder, moderately prone to drought
TTL-4	UPASI-10 X TRI-2025	High yielder, easy rooter, good quality
TTL-5	UPASI-10 X TRI-2025	High yielder, easy rooter, good quality, fairly tolerant to drought
TTL-6	UPASI-9 X TRI-2025	High yielder, broad leaved with larger shoots
TTL-7	UPASI-9 X TRI-2025	High yielder, moderate quality, drought tolerant

Initially, the Tea Research Institute of Vietnam, Vĩnh Phúc, had started the breeding of tea by collecting the planting materials from India. Many elite clones and *jats* such as PH-1, PH-3, and IA had been evolved through selection and hybridization (Tien 1993). Later, two clones, namely, LDP1 and LDP2, had been released with a yield potential of 17,500 kg green leaf/ha and 16,900 kg green leaf/ha, respectively. Two quality clones 276 and 215 (cross between PH1 and Shan tea) had also been developed (Toan and Tao 2005).

In Indonesia, the breeding of tea which was initiated during the 1980s was confined to selection only. Later ten clones, namely, GMB-1 to GMB-10, had been released which were developed by hybridization at the Research Institute for Tea and Cinchona, Gambung, Indonesia. While GMB-1 to GMB-5 had the potential to produce 3500 kg/ha/year of made tea, GMB-6 to GMB-10 had a productivity potential of 5000 kg/ha/year of made tea. Interestingly, few clones were also found to be tolerant to blister blight (Arfin and Semangun 1999).

Tea improvements in Kenya started with the introduction of seeds from Assam, India, which were used to establish the first tea plantation. Since these progenies had not been particularly selected for high yield and quality, the resultant seedling populations of mixed genotypes were genetically inferior, though diverse. With these populations, organized tea improvement started with the formation of the *Tea Research Institute of East Africa* in 1961 and later the Tea Research Foundation of Kenya in 1980 with a mandate for research on all aspects of tea. Thus in the *first phase of the tea improvement, mass selection among introduced seedling jats based on morphological characteristics was done.* As a result, several cultivars such as TRFK 6/8, TRFK 7/3, TRFK 7/9, TRFK 7/14, TRFK 11/4, TRFK 12/12, TRFK 12/19, TRFK 31/8, TRFK 31/11, TRFK 31/27, TRFK 31/28, TRFK 31/29, TRFK 54/40, TRFK 55/55, TRFK 55/56, TRFK 56/89, TRFK 100/5, and TRFK 108/82 were released for the industry. However, being heterogeneous genotypes, they formed a good breeding material for the second phase of mass selection. It was basically done by hybridization of selected parental stocks, superior in certain attributes that they were selected for. Several cultivars such as TRFK 303/35, TRFK 303/152, TRFK 303/156, TRFK 303/179, TRFK 303/186, TRFK 303/199, TRFK 303/216, TRFK 303/231, TRFK 303/259, TRFK 303/248, TRFK 303/352, TRFK 303/366, TRFK 303/388, TRFK 303/577, TRFK 303/745, TRFK 303/791, TRFK 303/978, TRFK 303/999, TRFK 303/1199, TRFK 347/314, TRFK 347/326, TRFK 347/336, and TRFK 347/573 were developed. The third phase involving selections from bi-clonal full-sib progeny resulted in the release of clones TRFK 337/3, TRFK 337/138, and TRFK 338/13. Later, the limitations of phenotypic selection encouraged and spurred the search and development of superior but genetically uniform tea clones. Presently 45 clones had been developed, and out of which, 24 were selected from seedling populations including the most popular clone 6/8. Thus in total, 27 clones (60%) shared the genetic pedigree of clone 6/8. Among the cultivars, few were credited with some special characteristics such as S 15/10; a high-yielding clone recently developed and registered for 10,000 kg made tea/ha/annum. Similarly SFS 150 and 303/577 were accounted for drought tolerant and TN 14-3 was accounted for tolerant to high soil pH, SFS 150 and TN 14-3 were accounted for cold tolerant, 12/2 was accounted for poor fermenter, 311/287 was accounted for tetraploids, and 7/9, 57/15, SC 31/27, and S 15/10 were accounted for mite tolerant (Seurei 1996).

About 2665 accessions of tea including few species of *Camellia* were maintained at the China National Germplasm Tea Repository which had contributed to develop more than 200 improved cultivars (Wang et al. 2011). Out of which, 97 were nationally registered cultivars, among them 17 were *jats*, 80 were clones, 30 were landraces, and 67 were improved clones. They were bred by 23 different institutions including the national and local tea research institutes, agricultural universities, local agricultural departments, tea experimental stations, etc. There were also about 130 registered cultivars, among them 16 were *jats*, 114 were clones, 29 were landraces, and 101 were improved clones, respectively. However among them, only 54 were popular clones suitable for green tea, 32 cultivars for black tea, and 33 cultivars were for oolong tea. Few cultivars such as Zhuyeqi and Fuding Dabaicha processed stronger prune-shock. The plucking surface of Zhuyeqi and Fuding Dabaicha contained more dense leaves, branches, and shoots than that of popular clone 'Xianggbolu'; hence, the former two were recommended for mechanical plucking (Yongming 1999; Chen and Zhou 2005).

Apart from the Chinese main land, commendable works on tea breeding had also been done at Taiwan Tea Experimental Station, Taoyuan, Taiwan. Among the 66 cultivars, Chinsin Oolong, Chinsin Dapan, TTES No 12, TTES No 13, and Shy jih Chuen were very popular. Nevertheless, Chinsin Oolong was the most important tea cultivars in Taiwan tea industry occupying half of total tea acreage. Besides, Chinsin Dapan had a wide manufacturing adaptability, suitable for making green tea, paochung tea, white-tip oolong tea, and black tea. TTES No 12 (Kinshen) was known for its light milky flavor and hence very popular. On the other hand, TTES No 13 had satisfactory tolerance levels of environmental stress such as drought and dieback diseases as compared to Chinese Oolong tea (Toan and Tao 2005).

In Korea, tea breeding began at Tea Experimental Station, Boseong, which was established in 1992 and developed seven tea cultivars till recently. Those cultivars were being propagated as cuttings for distribution to tea farmers. Some of the popular cultivars were Bohyang, Myungseon, Chamnok, Seonhyang, Mihyang, Jinhyang, and Oseon. Since then, massive breeding works such as introduction of green tea cultivars and hybridization using cv. Yabukita as parent had been undertaken (Jeong et al. 2005).

Tea breeding was started way back in 1920 at Tea Experimental Station, Shizuoka, Japan. Additionally, several private tea breeders started the varietal improvement of tea which resulted in the development of many clones. Among them, Hikosaburo Sugiyama (1857–1941), a noted tea breeder popularly known as "Burbank of tea" after US plant scientist, Luther Burbank, developed the popular clones of Japan 'Yabukita', 'Koyanishi', and 'Rokuro'. At present, 'Yabukita' alone planted 76% of all Japanese tea plantations. However, clonal selection program was more intensified during 1950; as a result, many good clonal cultivars were released in 1970. These newly developed cultivars had contributed much to the modern Japanese tea industry. There were 70 registered varieties, and few of them were Ooiwase, Yabukita, Surugawase, Sayamakaori, Yamakai, Kurasawa, Kanayamidori, Okuhikari, and Sawamizuka (Takeo 1992).

2.10 Bottlenecks of Tea Breeding

Although conventional tea breeding is well-established and contributed much for tea improvement over the past several decades, it is time-consuming and labor-intensive. The bottlenecks of conventional breeding are (1) perennial nature, (2) long gestation periods, (3) high inbreeding depression, (4) self-incompatibility, (5) unavailability of distinct mutant of different biotic and abiotic stress, (6) lack of distinct selection criteria, (7) low success rate of hand pollination, (8) short flowering time (2–3 months), (9) long duration for seed maturation (12–18 months), and (10) clonal difference of flowering time and fruit-bearing capability of some clones (Mukhopadhyay et al. 2016). Similarly, though vegetative propagation is an effective method of tea propagation, it is limited by several factors, such as (1) slower rates of propagation; (2) unavailability of suitable planting material due to winter dormancy, drought in some tea-growing areas, and so on; (3) poor survival rate at nursery due to poor root formation of some clones; and (4) seasonal dependent rooting ability of the cuttings. Therefore, to overcome the problems related to tea breeding, scientists across the world started finding some alternatives through biotechnological approaches which are discussed subsequently.

2.11 Conclusion

Conventional breeding in tea is well-established and contributed significantly. Several region-specific clones and bi-parental hybrids have been developed across the tea-growing regions of the world. However, several important aspects of tea breeding are very initial stages which need to be strengthened. Some of them are (1) association mapping has tremendous potential to identify the QTL particularly in tea, (2) development of large-scale molecular markers and their utilization in marker-assisted breeding, and (3) although pseudo-testcross has been utilized for developing the initial linkage map, it is necessary to develop the bi-parental mapping population for various applications where progress is very much limited compared to other similar woody perennials such as coffee and eucalyptus.

References

Ackerman WL (1971) Genetics and cytological studies with *Camellia* and related genera. Technical Bull. No 1427. USDA, US Gov Print Office, Washington, DC, p 115

Ahmed N, Singh ID (1993) A technique for rapid identification of ploidy level in tea. Two Bud 40:31–33

Amma S (1974) Characteristic of tetraploid tea induced from gamma irradiated Yabukita variety. Study Tea 46:1–6

Anon (1973) Annual report. Tea Research Institute of Sri Lanka, Talawakelle, pp 38–39

Anon (1979) Annual report. Tea Research Institute of Sri Lanka, Talawakelle, p 64

Anon (1999) Annual report. Tea Research Foundation of Kenya, Kericho, pp 45–50

Arfin T, Semangun P (1999) Tea industry in Indonesia. In: Jain NK (ed) Global advance in tea science. Aravali Book International Pvt Ltd, New Delhi, pp 65–72

Bakhtadze KE (1935) Methods of tea selection. Sov Sub-Trop 2:9–15

Banerjee B (1992a) Botanical classification of tea. In: Wilson KC, Clifford MN (eds) Tea cultivation to consumption. Chapman and Hall, London, pp 25–51

Banerjee B (1992b) Selection and breeding of tea. In: Willson KC, Clifford MN (eds) Tea cultivation to consumption. Chapman and Hall, London, pp 53–86

Barbora BC, Barua DN, Bera B (1996) Tea breeding at Tocklai. Two Bud 43:3–9

Barua DN (1989) Science and practice in tea culture. Tea Research Association, Jorhat, Assam, pp 56–58

Bezbarua HP (1968) Genetic improvement of tea in North East India - its problem and possibilities. Indian J Genet 28:126–134

Bezbaruah HP (1971) Cytological investigation in the family theaceae-I. Chromosome numbers in some *Camellia* species and allied genera. Caryologia 24:421–426

Bezbaruah HP (1975) Development of flower, pollination and seed set in tea in North-East India. Two Bud 22:25–30

Bezbaruah HP (1976) Aneuploidy in tea. Nucleus 19:167–169

Bezbaruah HP (1991) Tea breeding in North East India. In: Proceedings of the International Symposium on Tea Science, vol 34. The Organizing Committee of ISTS, Shizuoka

Bezbaruah HP, Gogoi SC (1972) An interspecific hybrid between tea (*Camellia sinensis* L.) and *C. Japonica* L. Proc Ind Acad Sci B76:219–220

Bhattacharjee H, Singh ID (1994) Storage of tea seed. Two Bud 41:32–34

Blakeslee AF, Avery AG (1937) Methods of inducing doubling of chromosome in plants treated with colchicines. J Hered 28:394–411

Chang HT (1991a) *Camellia atuberculata*. Acta Sci Nat Univ Sunyatseni 30:90

Chang HT (1991b) *Camellia fascicularis*. Acta Sci Nat Univ Sunyatseni 30:81

Chang HT (1998) Genus Camellia. In: Chang HT, Ren S-X (eds) Theaceae, flora republicae popularis sinicae, vol 49(3). Science Press, Beijing, 6–194

Chang HT, Bartholomew B (1984) Camellia. Timber Press, Portland, OR

Chang HT, Liang SY (1994) *Camellia achrysantha*. Guangxi Forest Sci 23:52

Chaudhuri TC (1979) Studies on the morphology and cytology of the progenies of triploid tea (*Camellia sinensis L.*). Dissertation, Assam Agricultural University, Jorhat, p 176

Chaudhuri TC, Bezbaruah HP (1985) Morphology and anatomy of the aneuploid and polyploidy tea {*Camellia sinensis (L.)* O. Kuntze}. J Plant Crop 13:22–30

Chen S, Ye D (1989) Cytological studies on polyploid tea. J Tea Sci 9:117–126

Chen L, Zhou ZX (2005) Variations of main quality components of tea genetic resources [*Camellia sinensis (L.) O. Kuntze*] preserved in the China National Germplasm Tea Repository. Plant Food Hum Nutr 60:31–35

Chen L, Yu FL, Tong QQ (2000) Discussion on phylogenetic classification and evolution of Sect Thea. J Tea Sci 20:89–94

Cohen Stuart CP (1929) Research on leaf yielding capacity of tea plants (Dutch). Arch Tree Cult Ned Ind 4:276–288

Das SK, Sabhapondit S, Ahmed G, Das S (2013) Biochemical evaluation of triploid progenies of diploid 3 tetraploid breeding populations of *Camellia* for genotypes rich in catechin and caffeine. Biochem Genet 51:358

Datta M, Agarwal B (1992) Intervarietal differences in karyotype of tea. Cytologia 57:437–441

Deka A, Deka PC, Mondal TK (2006) Tea. In: Parthasarathy VA, Chattopadhyay PK, Bose TK (eds) Plantation crops-I. Naya Udyog, Calcutta, pp 1–148

Do DN, Luong DV, Nguyen CD, Hoang ST, Le HT, Han JE, Park HS (2019) A new yellow Camellia (Theaceae) from central Vietnam. Korean J Plant Taxon 49:90–95

Du YY, Chen H, Zhong WL, Wu LY, Ye JH, Lin C et al (2008) Effect of temperature on accumulation of chlorophylls and leaf ultrastructure of low temperature induced albino tea plant. Afr J Biotechnol 7:1881–1885

Dung LV, Son HT, Ninh T, Ninh PH (2016) *Camellia quangcuongii* (*Theaceae*), a new species from Vietnam. J Jpn Bot 91:226–230

Duong DT (2011) Research on morphological, ecological and growth characteristics, and cutting propagation for *Camellia tamdaoensis* Ninh et Hakoda. Dissertation, Vietnam Forestry University

Fukusima E, Iwasa S, Endo N, Yoshinari T (1966) Cytogenetic studies in *Camellia*. I. Chromosome survey in some *Camellia* species. Jap J Hort 35:413–421

Furukawa K, Tanaka J (2004) 'Makura-Ck2': a tea strain with a high somatic embryogenesis. Breed Res 6:109–115

Ghosh Hajra N (2001) Tea cultivation: comprehensive treatise. International Book Distribution Co., Lucknow, pp 22–27

Gu Z, Xiao H (2003) Physical mapping of the 18S-26S rDNA by fluorescent in situ hybridization (FISH) in *Camellia reticulata* polyploid complex (Theaceae). Plant Sci 164:279–285

Gunasekara MTK (2000) Anatomical characteristics of polyploid tea cultivars. Annual report, vol 164. Tea Research Institute of Sri Lanka, Talawakelle

Gunasekara MTK, Ranatunga MAB (2003) Polyploidy in tea (*Camellia sinensis* L.) and its application in tea breeding: a review. Sri Lanka J Tea Sci 68:14–26

Hakoda N, Kirino S (2007) New species of genus Camellia in Vietnam. Int J Camellia 39:54–57

Hanson L, Mcmahon KA, Johnson MAT, Bennett MD (2001) First nuclear DNA C-values for another 25 angiosperm families. Ann Bot 88:851–858

Haridas P, Balasubramanian S, Netto LA, Ganesh Uma M, Mohan Kumar P (2006) Studies on improving planting material in tea (*Camellia sinensis L.*). J Plant Crop 34:243–249

Hu KM, Zhang YM, Wang JF, Xie TH, Hu KM, Zhang YM, Wang JF, Xie TH (2003) Comparison on the population dynamics and leafhopper resistance on different tea cultivars. J Tea Sci 23:57–60

Hu R, Wei S, Liufu Y, Nong Y, Fang W (2019) *Camellia debaoensis* (Theaceae), a new species of yellow *Camellia* from limestone karsts in South-western China. Phyto Keys 135:49–58

Huang H, Tong Y, Zhang QJ, Gao L-Z (2013) Genome size variation among and within *Camellia* species by using flow cytometric analysis. PLoS One 8:64981–64995

Hwang YJ, Okubo H, Fujieda K (1992) Pollen tube growth, fertilization and embryo development of *Camellia japonica L.* X *C. chrysantha* (Hu) Tyyama. J Jap Soc Hort Sci 60:955–961

Janaki Ammal EK (1952) Chromosome relationship in cultivated species of *Camellia*. The American Camellia Year Book. American Camellia Society, Fort Valley, GA

Jayasuriya P, Govindarajulu V (1975) Chromosome number of some tea clones. Planters Chron LXXX:185–186

Jeong B, Song Y, Moon Y, Han S, Bang J, Kim J, Kim J, Park Y (2005) Tea tree breeding plans for the tea Industry in Korea. In: International Tea Symposium-2005, Organised by Tea Res Inst, Chinese Acad Agric Sci, China Tea Science Society November 11–15, pp 322–332

Joshi R, Poonam, Gulati A (2011) Biochemical attributes of tea flowers (*Camellia sinensis*) at different developmental stages in the Kangra region of India. Sci Hort 130:266–274

Karasawa K (1932) On triploid tea. Bot Mag 46:458–460

Katsuo K (1966) Methods of inducing the polyploidy tea plant by colchicines treatment of the axillary bud. Study Tea 33:1–4

Kondo K (1975) Cytological studies in cultivated species of *Camellia*. Dissertation, Univ NC, Chapel Hill, p 260

Kondo K (1977) Chromosome number in the genus *Camellia*. Biotropica 9:86–94

Kondo K, Parks CR (1979) Giemsa C-banding and karyotype of *Camellia* C-banned karyotypes. Am Camellia Y Book 34:40–47

Kondo K, Parks CR (1980) Giemsa C-banding and karyotype of *Camellia*. In: Proc Intern Camellia Cong Kyoto, pp 55–57

Koskey JK, Wachira FN (2000) The use of plastid chloroplast count technique to determine ploidy levels in tea. Tea 21:15–18

Kulasegaram S (1980) Technical development in tea production. Tea Q 49:157–183

Le NNH, Uematsu C, Katayama H, Nguyen LT, Tran N, Luong DV, Hoang ST (2017) *Camellia tuyenquangensis* (Theaceae), a new species from Vietnam. Korean J Plant Taxon 47:95–99

Lee SL, Yang TUA (2019) *Camellia chinmeii*, a new species of *Camellia* sect. Para *Camellia* in Taiwan. Taiwania 64(3):321–325

Li X, Ye T, Huang Q, Fu D, Zhang C, Zeng L (2005) Study on distant hybridization for commercial tea production. In: 2005 International Symposium on Innovation in Tea Science and Sustainable Development in TEA INDUSTRY, TRA, CAAS, China Tea Science Society, November 1–11, Hangzhou, China, pp 389–395

Liang GL, Zhou CQ, Lin MJ, Chen JY, Liu JS (1994) Karyotype variation and evolution of sect. Thea in Guizhou. Acta Phytotaxon Sinica 32:308–315

Lin YS, Wu SS, Lin JK (2003) Determination of tea polyphenols and caffeine in tea flowers (*Camellia sinensis*) and their hydroxyl radical scavenging and nitric oxide suppressing effects. J Agric Food Chem 51:975–978

Linag KL, Fujianica FL (1988) Camellia lanceisepala. Int J Camellia 3:525

Liu S, Gao J, Chen Z, Qiao X, Huang H, Cui B, Zhu Q, Dai Z, Wu H, Pan Y, Yang C, Liu J (2017) Comparative proteomics reveals the physiological differences between winter tender shoots and spring tender shoots of a novel tea (Camellia sinensis L.) cultivar evergrowing in winter. BMC Plant Biol 17:206

Luu HT, Luong VD, Nguyen QD, Nguyen TQT (2015) *C. sonthaiensis* (Theaceae), a new species from Vietnam. Ann Bot Fenn 52:289–295

Luu HT, Gioi T, Nguyen QD, Cuong NH (2018) A new species of the family. Theaceae Central Vietnam 40(4):23–28

Manh TD, Thang NT, Son HT, Thuyet DV, Trung PD, Tuan NV, Duc DT, Linh MT, Lam VT, Thinh NH, Phuong NTT, Do TV (2019) Golden camellias: a review. Arch Curr Res Int 16:1–8

Ming TL, Bartholomew B (2007) Theaceae. In: Wu ZY, Raven PH, Hong DY (eds) Flora of China. 12: Hippocastanaceae through Theaceae. Science Press, Missouri Botanical Garden Press, Beijing; St Louis, MO, pp 366–478

Ming TL, Zhang WJ (1993) Acta Bot Yunnanica 15(1):12

Mondal TK (2008) Tea. In: Kole C, Hall TC (eds) A compendium of transgenic crop plants: plantation crops, ornamentals and turf grasses. Blackwell Publishing Ltd, London, pp 99–112. ISBN No. 978-1-405-16924-0

Mondal TK (2009) Tea. In: Prydarsini M, Jain SM (eds) Breeding plantation tree crops tropical species. Springer, New York, NY, pp 545–587

Mondal TK (2011) Camellia. In: Kole C (ed) Wild crop relatives: genomics and breeding resources plantation and ornamental crops. Springer, New York, NY, pp 15–40

Mondal TK, Bhattacharya A, Laxmikumaran M, Ahuja PS (2004) Recent advance in tea biotechnology. Plant Cell Tissue Organ Cult 75:795–856

Mondal TK, Rawal HC, Bera B, Kumar PM, Choubey M, Saha G, Das B, Bandyopadhyay T, Ilango V, Sharma TR, Barua A, Radhakrishnan B, Singh NK (2019) Draft genome sequence of a popular Indian tea genotype TV-1 [*Camellia assamica* L. (O). Kunze]. BioRxiv:762161

Morinago T, Fukushima E, Kano T, Maruyama Y, Yamasaki Y (1929) Chromosome number in cultivated plants. Bot Mag 43:569–594

Mukhopadhyay M, Sarkar B, Mondal TK (2013) Omics advances in Tea (*Camellia sinensis*). In: Bhar D (ed) Omics applications in crop science. CRC Press, Taylor and Franschis Group, Boca Raton, FL, pp 347–366. ISBN:978-1-4665-8582

Mukhopadhyay M, Mondal TK (2016) Biotechnology of tea. In: Bag N, Bag A, Palni LMS (eds) Tea: technological initiatives: some initiatives. NIPA, New Delhi, pp 301–328. ISBN: 978-93-85163-37

Mukhopadhyay M, Mondal TK, Chand PK (2016) Biotechnological advances in tea (*Camellia sinensis* [L.] O. Kuntze): a review. Plant Cell Rep 35(2):255–287

Nagata T, Sakai S (1984) Differences in caffeine, flavanols and amino acids contents in leaves of cultivated species of *Camellia*. Jap J Breed 34:459–467

Nagata T, Sakai S (1985) Purine base pattern of *Camellia irrawadiensis*. Phytochemistry 24:2271–2272

Nesumi A, Ogino A, Yoshida K, Taniguchi F, Maeda Yamam M (2012) 'Sunrouge', a new tea cultivar with high anthocyanin. J Agric Res Quart 46:321–328

Ng'etich WK, Wachira FN (1992) Use of a non-destructive method of leaf area estimation in triploid and diploid tea plants (*Camellia sinensis*). Tea 13:11–17

Nguyen LT, Tran N, Chiyomi U, Hironori K, Luong DV, Hoang ST, Nguyen KD, Nguyen HV, Thai TC (2018) Two new species of *Camellia* (Theaceae) from Vietnam. Korean J Plant Taxon 48:115–122

Ninh T (2007) *Camellia megasepala*. Int Camellia J 39:58
Ninh T, Hakoda N (1998) *Camellia crassiphylla*. Int Camellia J 30:76
Ninh T, Ninh LNH (2014) The yellow Camellias of the Tam Dao National Park. Int Camellia J 45:122–128
Ninh T, Rosmann JC (1998) *Camellia cucphuongensis*. Int Camellia J 30:71
Ogino A, Tanak J, Taniguchi F, Yamamoto MP, Yamada K (2009) Detection and characterization of caffeine less tea plant originated from inter-specific hybridization. Breed Sci 59:277–283
Orel G, Wilson PG (2010a) A new species of *C.* sect. *Stereocarpus* (Theaceae) from Vietnam. Novon 20:198–202
Orel G, Wilson PG (2010b) *C. luteocerata* sp. nov. and a new section of *C.* (*Dalatia*) from Vietnam. Nor J Bot 28:281–284
Orel G, Wilson PG (2012a) *Camellia cherryana* (theaceae), a new species from China. Ann Bot Fenn 49:248–254
Orel G, Wilson PG (2012b) *C. cattienensis*: a new species of *C.* (sect. *ArchaeC.*: Theaceae) from Vietnam. Kew Bull 66(4):565–569
Orel G, Marchant AD, Curry AS (1985) Molecular investigation and assessment of George Orel, A.D. Marchant 64 *C. azalea* C. F. Wei 1986 (syn. *C. changii* and A.S. CurryYe 1985) as potential breeding material. Int Camellia J 39:64–75
Orel G, Wilson PG, Curry AS, Luu HT (2012) *C. inusitata* (Theaceae), a new species forming a new section (*Bidoupia*) from Vietnam. Edinb J Bot 69:347–355
Orel G, Wilson PG, Curry AS, Truong LH (2013) *C. oconoriana* (THEACEAE), a new species from Vietnam. Edinb J Bot 70:439–447
Orel G, Wilson PG, Truong LH (2014a) *Camellia curryana* and *C. longii* spp. nov. (Theaceae) from Vietnam. Nord J Bot 32:42
Orel G, Wilson PG, Curry AS, Truong LH (2014b) Four new species and two new sections of Camellia (Theaceae) from Vietnam. Novon 23:307–318
Osone K (1958) Studies on the breeding of triploid plants by diplodising gamete cells. Jap J Breed 8:171–177
Parks CR, Kondo K (1974) Breeding studies in the genes *Camellia* (Theaceae). I. A Chemotaxonomic analysis of synthetic hybrid and backcross involving *Camellia japonica* and *C. saluensis*. Brittonia 26:321–332
Pham TV, Luong VD, Averyanov LV, Trinh NB, Nguyễn TTL (2019) *Camellia velutina* (Theaceae, Sect. Chrysantha), a new species from northern Vietnam. Pak J Bot 51:1441–1446
Prakash O, Sood A, Sharma M, Ahuja PS (1999) Grafting micropropagated tea (*Camellia sinensis* (L.) O. Kuntze) shoots on tea seedling - a new approach to tea propagation. Plant Cell Rep 18:137–142
Ranatunaga MAB, Gunasekare MTK (2002) Identification of polyploid marker in tea (*Camellia sinensis* L.). In: Proc Annual Sessions Sri Lanka Assoc Adv Sci, p 38
Rashid A, Chowdhary M, Badrul Alam AFM (1985) Studies on the progenies of a cross between diploid and tetraploid tea. Sri Lanka J Tea Sci 54:54–61
Sarmah PC, Bezbaruah HP (1984) Triploid breeding in tea. Two Bud 31:55–59
Satyanarayan N, Sharma VS (1982) Biometric basis for yield prediction in tea clonal selection. In: Proc. Placrosym IV, December 3–5, 1981, Mysore, India, pp 237–243
Satyanarayan N, Sharma VS (1986) Tea (*Camellia L.* spp) germplasm in south India. In: Srivastava HC, Vatsya B, Menon KKG (eds) Plantation crops: opportunity and constraints. Oxford IBH Publishing Co, New Delhi, pp 173–179
Sealy JR (1958) A revision of the genus *Camellia*. R. Hortic Soc, London, pp 58–60
Sebasthiampillai AR (1976) A simple technique for the polyploids in tea. Tea Q 46:12–15
Seurei P (1996) Tea improvement in Kenya: a review. Tea 17:76–81
Sharma VS, Ranganathan V (1986) Present status and future need of tea research. In: Srivastava HC (ed) Plantation crops, vol II. Oxford and IBH Publishing Co, New Delhi, pp 37–50
Sharma VS, Satyanarayana N (1987) UPASI clones. Planter Chronic 81:28–33

Sharma VS, Venkataramani KS (1974) The tea complex. I. Taxonomy of tea clones. Proc Ind Acad Sci 53:178–187

Simura T, Inabe T (1952) Studies on polyploidy of tea plants. Tokai-Kinki Natl Agric Exp Stn Res Prog Rep 1:1–14

Singh ID (1980) Non-conventional approaches in the breeding of tea in North East India. Two Bud 27:3–6

Singh ID (1984) Advances in tea breeding in North-East India. In: Proceedings of Placrosym. IV, pp 88–106

Singh ID (1999) Plant improvement. In: Jain NK (ed) Global advances in tea. Aravali Book International (P) Ltd, New Delhi, pp 427–448

Sivapalan P, Gnanapragasam NC, Kathiravetpillai A (1995) Field guide book. Tea Research Institute of Sri Lanka, Talawakelle, pp 5–12

Souladeth P, Tagane S, Yahara T (2019) Flora of Nam Kading national protected area V: two new species of C. (Theaceae), C. namkadingensis and C. rosacea. Thai For Bull Bot 47:82–90

Su SK, Chen SL, Lin XZ, Hu FL, Shao M (2000) The determination of ingredient of tea (*Camellia sinensis*) pollen. Apicult China 51:3–5. (in Chinese)

Su MH, Yang SZ, Hsieh CF (2004) The identity of *C. buisanensis* Sasaki (Theaceae). Taiwania 49:201–208

Takeda Y (1990) Cross compatibility of tea (*Camellia sinensis*) and its allied species in the genus *C.* JARQ 24:111–116

Takeda Y (2002) Genetic analysis of tea gray blight resistant in tea plants. JARQ 26:143–150

Takeo T (1992) Chemistry of tea. In: Willson KC, Cliford MN (eds) Tea: cultivation to consumption. Chapman and Hall, London, pp 413–457

Takyu T, Takeda Y, Nagatomi S (2003) Trichomeless mutant in tea. Tech News Int Rad Breed 67:2

Tanaka T, Mizutani T, Shibata M, Tanikawa N, Parks CR (2005) Cytogenetic studies on the origin of *C.* × *vernalis*. V. Estimation of the seed parent of *C.* × *vernalis* that evolved about 400 years ago by cpDNA analysis. J Jap Soc Hort Sci 74:464–468

Tavadgiridze SK (1979) Biology of growth and development in some polyploid forms of tea obtained by colchicines treatment and of irradiation. Subtropicheska Lenltry 3:137–139

Thirukkumaran G, Gunasekare MTK (2001) Use of pollen morphology and physiology to different ploidy level of tea (*Camellia sinensis*) clones. Proc Jaffna Sci Assoc 9:6–7

Tien DM (1993) Tea industry in Vietnam. In: Proc Intl. Symp. Tea Sci. Human Health. January 11–14, 1993, Tea Research Association, Calcutta, pp 103–106

Timoshenko MT (1936) The selection of tea for its chemical composition. Sov Sub-Trop 1:25–31

Toan NV, Tao NV (2005) Tea breeding selection by hybridization method in Vietnam. In: International Tea Symposium 2005. Organised by Tea Research Institute, Chinese Academy Agric Sci. China Tea Sci Soc, November 11–15, Hangzhou, China

Tran N, Luong VD (2013) *C. dilinhensis*: a new species from Vietnam. In: Li JY, Li ZH, Luo YJ, Fan ZQ (eds) Proceedings of the 3rd International Academic Forum on Yellow C.s, Nanning, Guangxi, China, 21–23 Feb: 79. Intl *Camellia*. Soc, Nanning, Guangxi, China

Truong LH, Gioi T, Dat NQ, Cuong NH (2018) A new species of the family theaceae from central Vietnam. Acad J Biol 40:23–28

Tubbs FR (1932) A note on vegetative propagation of tea by green shoot cuttings. Tea Q 5:154–156

Tunstall AC (1931a) A note on the propagation of tea by green shoot cuttings. Quart J Indian Tea Assoc 4:49–51

Tunstall AC (1931b) Experiment on vegetative propagation of tea by green shoot cuttings. In: Bulletin Tocklai Experimental Station, pp 113–114

Vijayan K, Zhang WJ, Tsou CH (2009) Molecular taxonomy of *C.* (Theaceae) inferred from nrits sequences. Am J Bot 96:1348–1360

Visser T (1969) Tea *Camellia sinensis* (L.) O. Kuntze. In: Ferwerdu EP, Wit F (eds) Outlines of perennial crop breeding in the tropics. Wageningen, Veenaran and Zonen, pp 459–493

Wachira FN (1994) Triploidy in tea (*Camellia sinensis*): effect on yield and yield attributes. J Hort Sci 69:53–60

Wachira FN, Kiplangat JK (1991) Newly identified Kenyan Polyploid tea strains. Tea 12:10–13
Wachira FN, Muoki RC (1997) Nucleolar and nucleolus organizer regions in tea as visualized by silver staining. Afr Crop Sci J 5:253–258
Wachira FN, Ng'etich WK (1999) Dry-matter production and partition in diploid, triploid and tetraploid tea. J Hortic Sci Biotechnol 74:507–512
Wang Y, Luo F, Li CH, Wang YC, Tang XB, Wang Y, Luo F, Li CH, Wang YC, Tang XB (2003) Selection of a tea accession Tianfu 28 with high quality and resistance. South-West China J Agric Sci 16:61–64
Wang X, Chen L, Yang Y (2011) Establishment of core collection for Chinese tea germplasm based on cultivated region grouping and phenotypic data. Front Agric China 5:344–350
Wang L, Yue C, Cao HL, Zhou YH, Zeng JM, Yang YJ et al (2014) Biochemical and transcriptome analyses of a novel chlorophyll-deficient chlorina tea plant cultivar. BMC Plant Biol 14:352
Wellensiek SJ (1933) Floral biology and technique of crossing with tea. Arch Thea Cult 12:27–40
Wellensiek SJ (1934) Research on quantitative tea selection. I. The Pajoeng reform see garden in Tjihirocan (Dutch). Arch Theecult Ned Ind 8:9–37
Wight W (1938) Recent advance in the classification and selection of tea plant. In: Proc 2nd Tocklai Annual Conference Tockali, Assam, India, p 38
Wight W (1939) Report. Indian Tea Association. Sci Dept Tocklai, Assam, pp 22–24
Wight W (1956) Genetic basis of yield. In: Proc 13th Tocklai Ann Conf., Assam
Wight W (1962) Tea classification revised. Curr Sci 31:298–299
Wight W, Barua PK (1957) What is tea? Nature 179:506–507
Wood DJ, Barua DN (1958) Species hybrids of tea. Nature 181:1674–1675
Wu CT (1964) Studies on hereditary, variation and morphology of pubescence on the young shoots of tea plants (China). Bull Pinchen Tea Exp Stn 20:1–23
Yang YJ, Yang SJ, Wang YS, Zeng JM, Yang YJ, Yang SJ, Wang YS, Zeng JM (2003) Selection of early budding and high quality green tea cultivar. J Tea Sci 23:9–15
Yongming Y (1999) Agrotechnology of tea in China. In: Jain NK (ed) Global advances in tea science. Aravali Books International (P) Ltd, New Delhi, pp 481–500
Yoshida K, Takeda Y (2006) Evaluation of anthracnose resistance among tea genetic resources by wound-inoculation assay. JARQ 40:379–386
Yu F, Xu N (1999) Tea germplasm resources of China. In: Jain NK (ed) Global advances in tea science. Aravali Books International (P) Ltd, New Delhi, pp 393–412
Zhan Z, Ke N, Chen B (1987) The cytology of tea clonal cultivars Fujian shuixian and their infertile mechanism. In: Proc Intl Tea Quality. Human Health Symp China, vol 46
Zhao OD (2019) New synonyms in *Camellia* (Theaceae): *Camellia cucphuongensis*, *C. cylindracea* and *C. vidalii*. Phytotaxa 419:100–104
Zhao D, Parnell JAN, Hodkinson TR (2017) Typification of names in the genus *Camellia* (Theaceae). Phytotaxa 292:171–179

Micropropagation

3

3.1 Introduction

The importances of in vitro multiplication are well-known. It is especially important for woody perennial plants such as tea and its wild species. Several aspects of micropropagation have been discussed below.

3.2 Need for Micropropagation

Although vegetative propagation is an effective method of propagation, yet it is limited by several factors in tea and other related species, such as (1) slower rates of growth and propagation due to perennial nature; (2) unavailability of suitable planting material due to winter dormancy and drought in some tea-growing areas; (3) poor survival rate at nursery due to poor root formation of some clones; and (4) seasonal dependent rooting ability of the cuttings.

Therefore, micropropagation technique appears to be an ideal choice for circumvention of the problems related to conventional propagation. Additionally, due to faster rate of multiplication, it is highly apt for a newly developed tea cultivar which owes high demand in the industry and hence need to be supplied in large quantity within a short span of time. Additionally, central to any successful transgenic technology is an efficient in vitro regeneration protocol. While an efficient regeneration protocol is essential for introduction of the foreign gene into plant tissues, micropropagation is important for the transfer of large number of genetically modified plants to the field within a short span of time (Bantawa et al. 2010; Mondal 2011).

T. K. Mondal, *Tea: Genome and Genetics*,
https://doi.org/10.1007/978-981-15-8868-6_3

3.3 Tea

Several reviews on micropropagation of tea and related species have been published in the past (Kato 1989; Vieitez et al. 1992; Dood 1994; Das 2001; Mondal 2003, 2004). It is evident from the literature that till the late 1980s, the thrust of tea micropropagation was on increasing the rate of multiplication. However, the emphasis on enhancing survivability of micropropagated tea at hardening stage was given only during the early 1990s. Studies on field performance tea and commercial exploitation of micropropagation started only at the beginning of the new millennium which is covered in the forthcoming discussion. Several factors from tissue type to media composition influence the success of micropropagation.

3.3.1 Explants

Fundamentals to the establishment of in vitro culture are the type, origin, and availability of explants throughout the year (Bantawa et al. 2009). Generally, shoot tips and nodal segments with dormant axillary buds of either juvenile or adult origin of current year growth are commonly used as explants for tea micropropagation (Vieitez et al. 1992). While Iddagoda et al. (1988) as well as Jha and Sen (1992) had used zygotic embryos, immature and mature cotyledons for the induction of adventitious buds, Kato (1985) employed epidermal layers of stem segments, stem segments without epidermal layer, and intact stem segments for shoot regeneration. Among these, callus from the epidermal layers produced buds more rapidly than the callus from other origin. Flower stalks, stem pieces, and leaf petioles had been attempted for the induction of adventitious buds without success barring callus formation (Sarwar 1985).

3.3.2 Initiation and Multiplication

Maximum efforts had been made to standardize the media formulation for increasing the shoot multiplication rate. The most common basal medium had been either full or half-strength MS salts (Murashige and Skoog 1962). However, there were also few reports on the use of other media such as woody plant medium (WPM, Llyod and McCown 1980) and Heller's (1953) medium (Table 3.1). Nakamura (1987a) compared various basal media, namely, MS, B_5 medium (Gamborg et al. 1968), as well as Nitsch and Nitsch (1969) and concluded that MS was the best for tea shoot multiplication. Tahardi and Shu (1992) achieved axillary shoot proliferation on WPM with various concentration of thidiazuron (TDZ) within 10–12 weeks. Though several researchers had reported that MS medium was most suitable for initiation of multiple shoots, yet half-strength MS salts were also reported for multiplication and shoot proliferation in tea (Phukan and Mitra 1984; Banerjee and Agarwal 1990; Agarwal et al. 1992). Manipulation of vitamin compositions along with organic and inorganic salts of MS was found to be useful for initiation and multiplication of

Table 3.1 Micropropagation of *Camellia* spp.

Species/cultivar	Explant	Medium				Response	Reference
		Bud break and shoot initiation	Multiplication	Regeneration via shoot organogenesis	Rooting		
C. sinensis , TV-1	Term-l bud and nodal stem segment of field grown plant	MS + IAA (0.25) + BA (4) + CW (10%) + YE (200)	½ MS + IAA (0.5) + BA (4.0) + CW (10%) + YE (200)	–	½ MS + AA (2) + IBA (7) + Sucrose (4%) IBA (7)	Shoot proliferation and rooting (60%)	Agarwal et al. (1992)
C. sinensis (L.) O. Kuntze cv. TRI-2025, CY9 and PK2	Shoot tip (4–5 cm) and nodal segment of mature plant	MS + modified vit + IBA (0.01) + GA_3 (0.1) + BA (1) for TRI-2025 and MS + modified vit + IAA (0.1) + BA (1) for others	MS + modified vit + IBA (0.01) + GA_3 (0.1) + BA (2) for TRI-2025 and MS + modified vit + IAA (0.1) + BA (2)	–	–	Axillary shoot proliferation (2–4/explant)	Arulpragasam and Latiff (1986)
C. saluenensis × *C. japonica*	Shoot tip	–	–	–	½ MS	Root induction and plantlet regeneration	Beretta et al. (1987)
C. sinensis (L.) O. Kuntze	Nodal segment	MS + BA (1.12) + NAA (0.2)	MS + BA (2.25) + IBA (0.2)	–	½ MS + NAA (3.6)	Axillary shoot multiplication and rooting	Bag et al. (1997)
Camellia sinensis (L.) O. Kuntze	Cotyledon	MS + BA (1.12) + NAA (0.2)	MS + BA (2.25) + IBA (0.2) + Phytogel (0.2%)	–	1/3 MS + NAA (3.6) + Sucrose (50 mM)	Prolonged subculturing around 12 weeks in larger container (500 mL flask) gives better result	Bag et al. (2019)
C. sinensis, TV-1	Shoot tip and nodal segment (0.7–1.0 cm) of mature	MS + YE (200) + CM (10%) + IAA (0.5) + BA (4)	½ MS + CM (10%) + IAA (0.25) + BA (4)	–	½ MS + modified vit + IBA (7)	Rooting (60%) of in vitro grown shoot	Banerjee and Agarwal (1990)

(continued)

Table 3.1 (continued)

Species/cultivar	Explant	Medium: Bud break and shoot initiation	Multiplication	Regeneration via shoot organogenesis	Rooting	Response	Reference
	field-grown plant						
Camellia sinensis (L.) O. Kuntze BT2, BT5, TV23	Nodes and shoot tips	MS + BAP (3) + IAA (0.5)	MS + BAP (3) + IAA (0.5) + GA_3 (0.5)	–	IBA (300) for 30 min	Maximum hardening of BT2 with 32% achieved	Begum et al. (2015)
Camellia sinensis (L.) O. Kuntze BT2	Shoot tips and nodal segments	MS + BAP (2) + IBA (0.1)	MS + GA_3 (0.5 mg) + BAP (2) + IBA (0.1)	–	IBA (500) for 30 min	Ex vitro rooting of microshoots is cost-effective; 80% survival was achieved	Boonerjee et al. (2013)
C. saluenensis × *C. chrysantha*	Seedling shoot tip	MS + Kin (1) + 2iP (1) + GA_3 (1) + IAA (1) + PVP (10 g/L)	MS + Kin (1) + 2iP (1) + GA_3 (1) + IAA (1) + PVP (10 g/L)	–	–	Axillary shoot development	Creze and Beauchesne (1980)
C. japonica, Purple Dawn	Shoot tip and nodal segments of 10-year-old tree	½ MS + BA (1)	½ MS + BA (1) + GA_3 (5)	–	–	Axillary shoot development	Carlisi and Torres (1986)
C. japonica	Shoot tip and nodal segment of 3–4-year-old seedling	MS + Kin (1) + 2iP (1) + GA_3 (1) + BA (1) + IAA (1) + A (20) + PVP (10 g/L)	MS + Kin (1) + 2iP (1) + GA_3 (1) + BA (1) + IAA (1) + A (20) + PVP (10 g/L)	–	–	Axillary shoot development	Creze and Beauchesne (1980)

C. sinensis L. cv. Kolkhida, kymyn, Tun-zui	Zygotic embryo (1–2 mm)	½ MS + modified vit + GA_3 (0.1)	½ MS + modified vit + BA (2) + IAA (0.2) + AA (1) + 2% sucrose	–	½ MS + IBA (0.5–2.0)	Axillary bud proliferation (6–8) and rooting of shoots	Iddagoda et al. (1988)
(*Camellia sinensis* (L.) O. Kuntze) (clone Iran 100)	Nodal segments	MS + BAP (1) + 3% sucrose (w/v) + agar (0.8%; w/v)	–	MS + BAP (3) + GA_3 (0.5).	IBA (300) for 30 min followed by transfer to ½ MS	Around 72.3% rooting was achieved and 65% survival at field level	Gonbad et al. (2015)
C. sinensis L.	In vitro micro shoot	MS + BA (6) + NAA (0.5) + GA_3 (2)	MS + BA (6) + NAA (0.5) + GA_3 (2)	–	(a) Shoots were treated with IBA (50) before putting in soil mixture (b) Liquid medium ½ MS + IBA (5) in dark (c) Filter paper bridge ½ MS + IBA (0.5) (d) Solidified agar with MS + . IBA (0.5)	Rooting: 97% for medium 1, 50% for medium 2, and 2.44% for medium 3	Jain et al. (1991), Jain et al. (1993)
C. sinensis (L.) O. Kuntze T-78	Shoot tip and cotyledonary node from germ-ted seedling; nodal segment of field-grown plant	½ MS + CW (10%) + CH (1 g/L) + BA (5) + IBA (0.1)	½ MS + CW (10%) + CH (1 g/L) + BA (1) + IBA (0.1)	–	MS + IBA (100) for 10 d and then transferred to MS	Axillary bud proliferation (8 and 35 per shoot tip and cotyledonary node, respectively), in vitro rooting (80–90%)	Jha and Sen (1992)

(continued)

Table 3.1 (continued)

Species/cultivar	Explant	Medium				Response	Reference
		Bud break and shoot initiation	Multiplication	Regeneration via shoot organogenesis	Rooting		
C. sinensis cv. Yabukita	Epidermal layer, intact stem segment, stem segment without epidermal layer	MS + IBA (4) + BA (2) from all three explants	–	MS + IBA (0.5) + BA (10)	½ MS + IBA (0.5)	Shoot organogenesis (20% from epidermal layer)	Kato (1985)
C. sinensis with 30 different clones.	Shoot apex	MS + BA (1) + IBA (1) + GA_3 (0.01)	–	–	–	Shoot proliferation	Kuranuki and Shibata (1992)
C. sinensis with 30 different clones	In vitro cutting	–	½ MS + BA (1) + IBA (0.01) + GA_3 (10)	–	–	Shoot proliferation	Kuranuki and Shibata (1993)
C. oleifera Abel	Hypocotyl, cotyledonary node, and radicle explants	½ MS + BAP (2) + IAA (0.1)	½ M + S + BAP (2) + IBA (0.05)	WPM + NAA + GA_3	½ MS + 3.5% perlite, IBA (1) + NAA (2).	Survival rate of 90.0%	Li et al. (2016)
C. japonica (cv. Reine des Beautes, cv. Lelie, cv. David Bocchi)	Young shoots	WPM + 25 g/L sucrose and solidified with 2.5 g/L Phytagel	2 mg/L TDZ + 0.5 mg/L Kn + 1 mg/L GA_3	Cv David Bocchi showed the highest percentage of callus formation (23.7) compared on WPM + 2 mg/L TDZ + 0.5 mg/L Kn + 1 mg/L GA	–	–	Malyarovskaya and Samarina (2017)
C. sinensis cv. Ch-hybrid	Nodal segment from in vitro grown seedling	MS + TDZ (1.1) + NAA (1.86)	MS	–	Treatment with IBA (500) for 30 min	Axillary shoot development	Mondal et al. (1998)

C. sinensis cv. CH 14 INTA and CH 318 INTA	Uninodal segments and axillary buds	½ MS + BAP (1)	½ MS + Kn (1) + GA_3 (1)	–	¼ MS + IBA (6)	Shoot multiplication, rooting	Molina et al. (2013)
C. sinensis cv. Yabukita	Axillary bud	MS + BA (1) + GA_3 (1)	MS + BA (1) + GA_3 (1)	–	½ MS + IBA (3) or MS + NAA (1 and 10)	Axillary shoot proliferation, rooting	Nakamura (1987a)
C. sinensis (L.) cv. Yabukita	Shoot tip of mature tea bush	MS, B_5, (Nitsch and Nitsch 1969)	MS + BA (1) + GA_3 (5)	–	½ MS + IBA (3)	Shoot multiplication, rooting	Nakamura (1987b)
C. sinensis with 23 cultivars.	Stem segment	MS + IAA (0.01–1.0)	MS + IAA (0.01–1.0) + GA_3 (1–5)	–	–	Adventitious bud formation (21.2%) from callus	Nakamura (1989)
C. sinensis	Axillary buds	–	½ MS + IBA (0.1) + BA (1) + GA_3 (5)	½ MS + IAA (1) + GA_3 (5)	–	Callus, differentiation of roots, differentiation of adventitious buds, and axillary shoot proliferation	Nakamura (1990)
C. sinensis cv. Yabukita	Stem segment of in vitro shoot	½ MS + IBA (0.1) or ½ MS + 2iP (30)	½ MS + IBA (0.1) + BA (1) + GA_3 (5)	–	–	Proliferation of shoot	Nakamura (1990)
C. sinensis UPASI-3	Nodal explant	½ MS	½ MS + BA (3)	–	¼ MS (liquid) + IBA (3)	Shoot multiplication and rooting (60%)	Pandidurai et al. (1996)
C. sinensis (L.) O. Kuntze	Shoot tip, nodal segment of field-grown plant	MS + YE (0.1%) + CM (10%) + NAA (2) + BA (6) for indirect regeneration (via callus) MS + IAA (2) + Kin (8) for direct regeneration	For indirect regeneration (via callus) ½ MS + IAA (1) + Kin (3)	–	–	4–5 shoot buds from NAA + BA-treated callus and 3–4 shoot buds from IAA + Kin-treated explants	Phukan and Mitra (1984)

(continued)

Table 3.1 (continued)

Species/cultivar	Explant	Medium				Response	Reference
		Bud break and shoot initiation	Multiplication	Regeneration via shoot organogenesis	Rooting		
C. sinensis	Nodal segment of in vitro shoots	½ MS + modified vit + IAA (1) + Kin (3) + YE (0.15%) + CW (10%)	½ MS + modified vit + IAA (1) + Kin (3) + YE (0.15%) + CW (10%)	–	½ MS + IBA (8)	Shoot multiplication and rooting (60–70%)	Phukan and Mitra (1990)
C. sinensis cv. Banuri-96	Nodal segment from field-grown plants	½ MS	½ MS + IAA (0.2) + BA (2.5)	–	Grafted on seedlings of tea plants	83.3% graft union	Prakash et al. (1999)
C. sinensis	Embryo of TRI 3013 (female parent) × DT 95 (male parent)	MS + BAP (3) + IBA (0.5)	MS + BAP (3) + IBA (0.5)	–	IBA (50) for 3 h	Ex vitro rooting of tea microshoots with simultaneous acclimatization reduced the cost of production of micropropagated plant by 71% compared to that of micropropagation with an in vitro rooting step	Ranaweera et al. (2013)
C. sinensis UPASI-3, UPASI-9	Nodal segment	MS + BA (5) + CM (10%)	MS + GA_3 (1) + BA (0.5)	–	MS + IAA (1)	Shoot proliferation and rooting (18%)	Rajkumar and Ayyappan (1992)
C. sinensis UPASI-9	Nodal explant (1.0–1.5 cm)	MS + BA (0.5)	MS + BA (5)	–	–	Axillary shoot multiplication	Rajasekaran and Mohankumar (1992)

C. japonica	Shoot apex and nodal segment	MS + modified vit + BA (1) + IAA (0.1)	MS + modified vit + BA (1)	–	½ MS + BA (1)	Shoot proliferation, rooting (91%)	Samartin et al. (1984)
C. japonica	Shoot tip from 2 to 3 month old seedling	MS + BA (1)	–	–	½ MS with modified vit + IBA (1 mg/L)	–	Samartin et al. (1986)
C. sasanqua, Onigoromo Thunb.	Shoot tip and nodal segment	–	–	–	Gamborg's (B_5) medium	Shoot proliferation and rooting (100%) for in vitro rooting. 95% for ex vitro rooting	Samartin (1991)
C. sinensis (TV-1, T-78, UPASI-9 and Kangra Jat)	Nodal segment from in vitro raised cultures	MS (liquid static cultures) + TDZ (1.1)	MS (liquid static cultures) + TDZ (1.1)	–	IBA treatment (30 min) before placing onto sand in closed jars under in vitro conditions	88% rooting	Sandal et al. (2001)
C. reticulata, "Captain Rawes"	Term-l shoot tip and node	Heller's (1953) macro + $(NH_4)_2SO_4$ (132) + MS micro + MS vit + BA (2) + Z (2) + IBA (0.01) + 2iP (2)	WPM + BA (2) + Z (2) + 2iP (2) + IBA (0.01)	–	½ WPM + dipping in IBA solution (1 g/L) for 30 min.	Axillary shoot proliferation and rooting	San-Jose and Vieitez (1990)
C. reticulata cv. Captain Rawes	Shoot tip and stem node of adult trees	Heller's (1953) macro + $(NH_4)_2SO_4$ (132) + MS micro + MS vit + BA (2) + Z (2) + IBA (0.01) + 2iP (2)	WPM + BA (2) + Z (2) + 2iP (2) + IBA (0.01)	–	½ macro WPM + full micro + vit + 6% sucrose	Axillary shoot development and plantlet regeneration (horizontal position of explants was better than vertical position for shoot multiplication)	San-Jose et al. (1991)

(continued)

Table 3.1 (continued)

Species/cultivar	Explant	Medium				Response	Reference
		Bud break and shoot initiation	Multiplication	Regeneration via shoot organogenesis	Rooting		
C. reticulata, Captain Rawes	In vitro leaf	Heller's (1953) macro + $(NH_4)_2SO_4$ (132) + vit + BA (2) + Z (2) + IBA (0.01) + 2iP (2)	WPM + BA (2) + Z (2) + 2iP (2) + IBA (0.01)	WPM + BA (2) + IBA (1)	½ WPM + 6% sucrose	Adventitious shoot regeneration and rooting (61.5%) from shoots of axillary origin	San-Jose and Vieitez 1992
C. sinensis L.	Flower stalk, leaf piece, petiole, immature cotyledon, stem piece, and shoot tip	MS + 2,4-D (4.42), NAA (3.72) + BA (2.25), Kin (2.15)	MS + 2,4-D (10) + YE (0.2%)	Buds with 2–3 leaves were initiated on MS + BA (2.25)	–	Callus and shoot-bud regeneration (50%) Explant browning were checked by lower concentration of MS salts (1/20)	Sarwar (1985)
C. sinensis cv. Banuri-96	Single node cutting of in vitro shoot	½ MS + 8% agar	½ MS + IAA (0.2) + BA (2.5)	–	–	71.6% root induction and 73% field survival	Sharma et al. (1999)
C. sinensis	In vitro cutting	MS + BA (1) + IBA (0.01) + GA_3 (5)	–	–	–	Shoot proliferation	Shibata and Kuranuki (1993)
C. oleifera	Immature zygotic embryo and cotyledon	–	–	MS + BA (4) + NAA (2)	–	Adventitious buds and plantlet regeneration	Tian-Ling (1982)
C. sinensis, TRI-2025	Axillary bud	MS + BAP (2.2)	WPM + TDZ (1.1)	WPM + GA_3 (0.5–2.5)	NAA (5) for 7 d in the dark, followed by transfer of the shoots to an auxin-free medium in the light	–	Tahardi (1994)

C. sinensis, TRI-2025	Axillary bud	Modified WPM + BA (2.25) + IBA (1.01) + TDZ (0.01)	Modified WPM + BA (2.25) + IBA (1.01) + TDZ (0.01)	–	–	Shoot proliferation	Tahardi and Shu (1992)
C. sasanqua, Day Dream	Shoot tip, stem segment	MS + BA (1) + NAA (0.1) for juvenile plant MS + BA (1) for adult material	½ MS + modified MS vit + NAA (0.1) + BA (2) + GA_3 (5)	–	½ MS + modified vit	Axillary shoot proliferation and plantlet regeneration	Torres and Carlisi (1986)
C. × *williamsii*, Debbie	Internode segment	WPM macro + MS micro + MS vit + BA (0.5) + IBA (0.01) + Glucose (30 g/L)	–	MS + IBA (0.1) + . TDZ (0.67)	–	–	Tosca et al. (1996)
C. japonica, Alba Plena	Shoot tip and nodal segment	–	–	–	WPM macro after dipping in IBA (1 g/L) for 15 min followed by 12 d darkness	Rooting of micropropagated shoot (87%)	Vieitez et al. (1989a)
C. japonica cv. Alba Plena	Shoot tip (2–4 cm), nodal segment, and whole shoot of field-grown plant	Heller's (1953) macronutrient increased by factor 1.25 + $(NH_4)_2SO_4$ (132) + MS micronutrient + BA (1) + IBA (0.01) + myoinositol (100) + Jacquiots vit (Gautheret 1959)	WPM + BA (2) + Z (2) + 2iP (2) + IBA (0.01)	–	WPM after giving IBA (1 g/L) treatment for 15 min	Shoot multiplication and rooting (76%)	Vieitez et al. (1989b)
C. sinensis var. *assamica* (Cinyuruan-143, Kiara-8, and TRI-2025) and *C. sinensis* var. *sinensis* (Tambi and Tambi Jingga)	Shoot tips (4–5 cm)	½ MS + 3% Sucrose for establishment. MS + BAP (3) and GA_3 (0.5)	MS + BAP (3) + GA_3 (5) + IBA (0.1)	MS + BAP (3) and GA_3 (0.5)	–	Shoot tip culture was better than cotyledon-derived explants	Widhianata and Taryono (2019)

(continued)

Table 3.1 (continued)

Species/cultivar	Explant	Medium				Response	Reference
		Bud break and shoot initiation	Multiplication	Regeneration via shoot organogenesis	Rooting		
C. oleifera	Lateral bud of adult tree	–	–	–	–	Axillary shoot development	Yan et al. (1984)

Figures in parenthesis without assigned units denote concentrations in mg/L
A, adenine; AA, ascorbic acid; B_5, Gamborg et al. (1968) medium; CH, casein acid hydrolysate; CW, coconut water; macro, macronutrients; micro, micronutrients; PVP, polyvinylpyrrolidone; vit, vitamins; WPM, woody plant medium (Llyod and McCown 1980); YE, yeast extract

axillary shoots of tea (Arulpragasam and Latiff 1986). Among the different plant growth regulators, addition of 6-benzyladenine (BAP) (1–6 mg/L) and indole-3-butyric acid (IBA) (0.01–2.0 mg/L) in the culture medium had been suited well for both shoot initiation and subsequent multiplication. Apart from tea, BAP (1–4 mg/L) was also found to be the best cytokinin in different related species such as *C. japonica* (Vieitez et al. 1989a), *C. oleifera* (Tian-Ling 1982), *C. reticulata* (San-Jose and Vieitez 1990; San-Jose et al. 1991), and *C. sasanqua* (Torres and Carlisi 1986). Kato (1985) succeeded in achieving indirect organogenesis via callus phase from three different explants, namely, epidermal layer of stem segment, intact stem segment, and stem segment without epidermal layer using BAP (10 mg/L) augmented medium. Although 2,4-dichlorophenoxyacetic acid (2,4-D) and α-naphthaleneacetic acid (NAA) were found to induce callus, these were ineffective for the growth and development of tea shoots (Nakamura 1988). Auxin such as NAA in combinations with BAP produced either callus or induced 4–5 shoot buds per explants within 8–12 weeks (Phukan and Mitra 1984; Bag et al. 1997). Picloram and 2,4,5-trichlorophenoxyacetic acid (2,4,5-T) had been used successfully for the elongation of tea shoots by several researchers (Arulpragasam and Latiff 1986; Nakamura 1987a, b, 1989; Iddagoda et al. 1988; Jain et al. 1991). Higher concentrations of these auxins (10 mg/L) were found to be critical for shoot elongation of 30 different cultivars of tea (Kuranuki and Shibata 1993). Importance of indole-3-acetic acid (IAA, 0.1–2.0 mg/L) and kinetin (0.21–8.0 mg/L) for both induction and multiplication of axillary shoots were emphasized by several researchers in tea (Phukan and Mitra 1984; Sarwar 1985; Das and Barman 1988) as well as in *C. sasanqua* (Torres and Carlisi 1986), *C. japonica* (Creze and Beauchesne 1980), and some of their hybrids (Creze and Beauchesne 1980; Lammerts 1958). It was clearly evident that MS was superior to WPM in all combinations.

The effect of TDZ on micropropagation of tea was studied details. Mondal et al. (1998) had shown that extremely low concentrations of TDZ (0.22×10^{-6}–0.22 mg/L) alone were effective in inducing shoot bud proliferation and maintaining high rates of shoot multiplication on hormone-free media up to 24 subcultures. On the other hand, higher concentrations of BAP (0.22–2.2 mg/L) and its continued presence were required to initiate and sustain shoot proliferation. Although all the explants produced callus on higher concentration of TDZ (1.1, 2.2 and 3.3 mg/L) in combination with either 2,4-D, NAA, or IBA at concentration ranging (1–3 mg/L) in MS, maximum response (98%) of shoot proliferation was observed with a combination of TDZ (1.1 mg/L) and NAA (2 mg/L). The number of shoots formed was higher in explants initiated on medium containing TDZ when compared to medium supplemented with BAP, but multiplication rates were more or less similar (i.e., 2–3 times) after each subculture. Since very low concentrations of TDZ are used only at the initial phase, the overall higher cost of TDZ was compensated in overall cost. TDZ thus appeared to be a potent cytokinin like growth factor for tea micropropagation with high proliferation rates. Despite of the fact that inclusion of liquid culture step for bulking microshoots may be cost-effective, it is generally limited by a low-oxygen environment coupled with the production of hyperhydric structures. Barring these limitations, establishment of cultures in liquid medium is an important step towards automation (Aitken-Christie et al. 1995). An efficient liquid culture system for tea shoot proliferation was standardized (Sandal et al. 2001). They found that MS along with TDZ (0.55–1.1 mg/L) was the best for shoot proliferation in un-agitated liquid medium. Of the different liquid volumes compared in 250 mL Erlenmeyer flasks, 20 mL was the most effective. However, with greater volume, hyperhydricity was induced. Therefore, use of 20 mL liquid medium with subculture periods at an interval of 6–8 weeks found to be cost- and labor-effective process in combination with the existing protocols of tea micropropagation involving solid media with subculture periods at 4 weeks interval. Similarly earlier, Carlisi and Torres (1986) found that liquid medium was better than solid medium for *C. japonica* when cultured on 1/2 MS fortified with BAP (1 mg/L) and GA_3 (5 mg/L). The significant role of growth adjuvants for micropropagation of tea was documented. These included coconut milk (Phukan and Mitra 1984; Sarathchandra et al. 1988; Banerjee and Agarwal 1990; Nakamura and Shibita 1990; Agarwal et al. 1992; Jha and Sen 1992; Rajkumar and Ayyappan 1992), yeast extract (Phukan and Mitra 1984, 1990; Sarwar 1985; Banerjee and Agarwal 1990), casein acid hydrolysate (Chen and Liao 1983; Jha and Sen 1992), serine and glutamine as nitrogen sources (Chen and Liao 1982), etc. Among the different carbon source, sucrose with the concentration between 15 and 60 g/L remains a unanimous choice for tea micropropagation. This was confirmed by Nakamura (1990) who found that 3–6% sucrose was the best for adventitious bud formation as compared to four other different sugars, viz., lactose, galactose, glucose, or maltose for adventitious bud formation. Therefore, it can be concluded that MS with low cytokinin specially BAP seems to be the best for micropropagation of tea.

Interestingly the effect of caffeine in in vitro growth of tea was studied in details. It was found that caffeine, which is abundantly present in tea leaves significantly, retards the growth/development of tea shoots, stem, and roots in vitro. This may be the reason why the replanting area growth of new plants are less and soil requires a rehabilitation as tea plants over the years may secrete the caffeine to the soils (Owuor et al. 2007).

3.3.3 Rhizogenesis

Establishment of microshoots in hardening phase depends upon the efficiency of rooting (Bantawa et al. 2011a). In tea, both in vitro and ex vitro root induction of microshoots were reported. While in vitro rooting of tea depends upon concentration and duration of auxin treatment, salt strength of basal medium, or physical condition of the cultures, ex vitro rooting was influenced by pH of the hardening media and relative humidity of the hardening chamber. Reduction of MS salt concentrations to half-strength favored not only root induction but also root elongation in tea (Kato 1985; Banerjee and Agarwal 1990). Generally, IBA (0.5–8 mg/L) was preferred over NAA in tea (Gunasekare and Evans 2000; Bidarigh and Azarpour 2011; Bidarigh et al. 2012) as well as in *C. japonica* (Samartin et al. 1984, 1986; Vieitez et al.

1989b; Bidarigh and Azarpour 2013), in *C. sasanqua* (Samartin 1991), and in *C. reticulata* (San-Jose and Vieitez 1990, 1992) as roots induced by NAA were shorter and thicker with accompanying calli making subsequent transplanting difficult. Although rooting occurred much later when treated with IBA, roots were long and fibrous (Nakamura 1987a; Banerjee and Agarwal 1990). Liquid medium with filter paper bridge was successfully used for in vitro rooting of tea (Tian-Ling 1982; Kato 1985; Nakamura 1987b). Jain et al. (1993) found that ex vitro rooting was better than all types of in vitro rooting methods such as liquid shake culture, agar-solidified medium, or filter paper bridges. They achieved 97% rooting in shoots whose cut ends were dipped in IBA (50 mg/L) for 2 h before transplanting into potting mixtures. Another important observation made by Banerjee and Agarwal (1990) was that low light with a low pH (4.5–4.6) was favorable for root induction in tea. This finding was in agreement with Nakamura (1987a), indicating that dark treatment after dipping the shoots in auxin promoted rooting in tea. In vitro rooting of tea also depends upon genotype. Murali et al. (1996) demonstrated that among the four different cultivars, the highest rooting was achieved with both UPASI-26 and UPASI-27 cultivars compared to two other cultivars, namely, UPASI-3 and BSB-1, with an IBA treatment (1 mg/L) for 30 min. They concluded that genotypic differences strongly influence rooting ability of tea, which require a fine tuning of the media to suit each individual cultivar.

In general, in vitro rooting of tea microshoots was achieved either by culturing on media containing low auxin for a longer time or with an "auxin-shock" treatment by initially culturing in high auxin concentration followed by the transfer to an auxin-free medium.

3.3.4 Hardening and Field Transfer

Perhaps the most critical phase of micropropagation is the establishment of in vitro plantlets to the greenhouse. Achievement of uniform plant growth and the high survival rate not only demand good greenhouse conditions but also require modification of the internal microclimate to match the local environmental need (Saha-Roy et al. 2010).

3.3.4.1 Conventional Hardening

Conventionally in vitro raised tea microshoots are hardened for 6 months in soil mixture containing various additives such as cow dung, soil rite etc. and then kept inside the indigenously developed poly-tunnel of various shapes and sizes. Arulpragasam et al. (1988) transferred 5–8 cm rooted plantlets to small plastic pots containing fumigated soil and kept in humid chamber for a period of 10 d. Plants were gradually acclimatized and planted in larger concrete pots. Das and Barman (1988) achieved better hardening on preconditioning plants at low temperature (22 °C) and low light intensity (250 $\mu mol/m^2/sec$) after transferring to soil sleeves. The plants were covered with poly-bags and kept in thatch house for few weeks and gradually exposed to sunshine before transferring to field. Generally, the standard procedure is to transfer the rooted plantlets of tea to potting mixtures containing peat and soil (1:1) under high humidity created by using misting or fogging units (Banerjee and Agarwal 1990; Agarwal et al. 1992; Jha and Sen 1992; Jain et al. 1993), but use of vermiculite and soil in equal ratio also worked very well (Tian-Ling 1982; Kato 1985). A detailed study on various parameters such as time of harvesting of microshoots, shoot size, soil pH, plant growth regulators, CO_2 enrichment, and light condition was carried out (Sharma et al. 1999). They found that soil pH coupled with CO_2 enrichment was the most critical factor for hardening to achieve high percentage of survival. Following this protocol, 300 plants were transferred to the field for planting (Sharma et al. 1999). In order to do large-scale multiplication, micropropagated tea shoots of 4–5 cm height were harvested and treated with IBA (500 mg/L) for 30 min before transferring to Hikkotray containing pre-sterilized sand and cow dung (1:1). These Hikkotrays were then kept in poly-tunnel with intermittent watering for 90 d inside an indigenously developed poly-house resulting in 90% survival. Thereafter, they were transferred to polythene sleeves for a year in the same poly-house, containing black virgin soil (Rajasekaran and Mohankumar 1992; Mondal et al. 1998; Boonerjee et al. 2013). Following this technique at Research and Development Department of Tata Tea Ltd, India, more than micropropagated 45,000 plants of 8 different tea cultivars were transferred to the field, and leaves had been harvested since then regularly to manufacture tea for the past several years (Mondal et al. 2004).

Physiological and biochemical responses were compared between field-grown plants of vegetatively propagated (VP) and tissue culture-raised (TC) plants. No significant variation was observed between them in terms of photosynthetic carbon assimilation rate. Carbon assimilation studies carried out with a radiotracer technique revealed that "Assam" cultivar UPASI-27 assimilated a higher amount of labeled carbon dioxide followed by UPASI-3. However, UPASI-27 was marginally better than UPASI-3 in terms of mobilization of assimilates to the growing sinks. Both UPASI-3 and UPASI-27 reassimilated higher quantities of photosynthates followed by BSB-1 and UPASI-26. Though there was a marginal variation in photosynthetic pigments of TC and VP plants, it was not statistically significant. Similarly, no significant variations were observed in certain substrates (polyphenols, catechins, and amino acids) and enzymes (polyphenol oxidase, peroxidase, and phenylalanine ammonia-lyase) except protease involved in the formation of quality constituents of made tea. However, clonal variation was evident with respect to photosynthetic pigments, substrates/enzymes. Under soil moisture stress, no significant variation was observed between VP and TC plants in terms of proline accumulation (Marimuthu and Rajkumar 2001). Effect of period of subculturing and size of container for hardening of tissue culture-raised tea was studied. It has been reported that around 12 weeks subculturing in larger container (500 mL), flask allows the tea plantlets to develop better root system and as well chlorophyll which helps the plant let to achieve 94% success in hardening in the field (Bag et al. 2019).

3.3.4.2 Biological Hardening

The biological hardening is an alternative concept for hardening the micropropagated plantlets. Like in any other plants, micropropagated tea plants often experience high mortality during or following laboratory-to-field transfer. Apart from various abiotic factors, one major cause of mortality of such "aseptically" raised plants is their sudden exposure to the soil microbial communities present in the rooting media. Tissue culture-raised plants, at least initially, are unable to resist sufficiently against the soil microbial attack. Therefore, microbial culture can be used to overcome the "transient transplant shock" and better development of the plants on the transfer to the soil. Although a number of organism antagonistic fungus such as *Trichoderma*, vesicular-arbuscular mycorrhiza, and *Piriformospora indica* were found to be suitable for micropropagated-raised woody plants (Singh et al. 2000), Pandey et al. (2000) found that two bacteria, namely, *Bacillus subtilis* and *Pseudomonas corrugata*, were suitable as microbial inoculants for hardening of micropropagated tea plants prior to transfer in open land. They found that bacterial inoculation enhanced survival rate of 100%, 96%, and 88% as against 50%, 52%, and 36% in corresponding control plants in rainy, winter, and summer seasons, respectively. Rhizoplane and rhizosphere soil analysis showed that the major biotic factor responsible for mortality of tea was the fungus *Fusarium oxysporum*. Thus biological hardening holds a greater promise for hardening of micropropagated tea shoots.

3.3.4.3 Micrografting as a Hardening Tool

Micrografting, a hardening tool of in vitro-raised shoots, has been used in a wide range of woody plants such as citrus, cherry, kiwifruit, pistachio, stone fruits, apple, grape, *Larix decidua*, and *Picea* spp. where in vitro-raised scions were directly grafted either onto in vitro-raised rootstocks under sterile conditions or in vivo-raised stocks (Banerjee et al. 2000). In tea, Prakash et al. (1999) reported the grafting of in vitro-raised tea shoots on seedlings at the greenhouse. Microshoots of tea cultivar Banuri-96 were either grafted on seedlings of the same cultivar (autograft) or on UPASI-9 seedlings (heterograft). Four important factors were investigated: (1) effect of PGRs, (2) assessment of compatibility, (3) effect of age of root-stock, and (4) season.

A higher amount of foliar development was observed when a pulse treatment of liquid formulation consisting of BAP (5 mg/L) and NAA (5 mg/L) was given at the graft union for 10 min. Also, a significantly higher percentage of survival was observed in autografts as compared to heterografts. Age of the rootstock was found to be the most critical factor. The success rate was much higher in 4-month-old seedlings than in either 2-month-old or 12-month-old seedlings. However, using the same technique, Mondal et al. (2005) reduced the hardening period up to 1 year of in vitro-raised tea shoots when grafted on 3-month-old tea seedlings. However, it was observed that the result of Mondal et al. (2005) differs in two aspects from the report of Prakash et al. (1999). Firstly, Mondal et al. (2005) achieved a higher rate of survival (98%) without any PGR treatment. Therefore, it was concluded that PGR play an insignificant role for the graft compatibility, which is an absolute necessity for commercial use of this technique to eliminate the cost of PGR. Secondly, Mondal et al. (2005) found that the growth of the grafted plants was much higher than the ungrafted in vitro-raised scion or seedling. One of the reasons for better growth of micropropagated plants than the vegetative counterpart was due to the higher root volume, which helped to absorb more water and nutrients from the soil. It was due to the pre-existing tap root system of the seedling. This was not only of great importance to reduce the hardening period of tissue culture plant but also for better root system to resist the plant during drought periods subsequently in the field. Micropropagated tea plants of the same cultivar as well as age required 12–-18 months time in the hardening phase through conventional hardening approach, whereas the same cultivar of micrografted one demanded only 6–8 months, which vitally reduced the hardening time by one planting season. This plays an important role in tea breeding where progress is slow due to the slow rate of propagation.

3.3.5 Field Performance of Micropropagated-Raised Plants

Several reports on micropropagation of tea stated that hardened plantlets were transplanted to potting mixture but only a few details of field performance are available. To date, no systematic study has been conducted to assess the growth performance of micropropagated tea plants grown in the field. Sharma et al. (1999) analyzed the nursery performance of 17-month-old micropropagated and vegetatively propagated tea plants of cultivar Banuari-96. No remarkable difference was noticed at nursery level except root induction time, which was earlier by 1 month in case of micropropagated shoots. They found that while average height and stem thickness at collar region were higher in vegetatively propagated plants, average number of leaves were twice (16 leaves/plant) in micropropagated plants in comparison to their in vivo counterparts (6 leaves/plant). However, reasons for such difference were not mentioned, but it seems that 17 months time was perhaps too early to detect the actual performance for this woody perennial.

In a comparative study of 8-year-old as well as 4-year-old, field-grown micropropagated and vegetatively propagated tea cultivars, namely, UPASI-9 and TTL-1, Mondal et al. (2004) reported that overall yields were comparable. While the yield of micropropagated UPASI-9 recorded 3656 kg made tea/ha/annum, vegetatively propagated plants of UPASI-9 yielded 3596 kg made tea/ha/annum after second cycle of pruning (8 years) under the environmental conditions of South Indian. On the other hand, micropropagated TTL-1 plants registered 2321 kg made tea/ha/annum, and their vegetative counterparts yielded 2383 kg made tea/ha/annum after first cycle of pruning (4 years) with 10,000 plants/ha as well as 22% out turn (a ratio which indicates that 100 kg green leaf produce 22 kg made tea in black tea manufacturing process) in both. Chemical analysis of made tea for thearubigin, theaflavin, total soluble solids, and total color evidenced no significant difference between the two types for both the cultivars, which is obvious as both are genetically similar and planted in the same environment. Although the different physiological parameters such as photosynthetic rate and chlorophyll content, etc.

remained the same, two different morphological variations were detected: (1) the number of lateral shoots produced after "centering" was significantly higher in micropropagated-raised plants as compared to vegetatively propagated plants. This is probably due to the effects of the various PGR treatments that the micropropagated-raised plants experienced under in vitro conditions, (2) secondly, the root volumes of micropropagated plants were also higher than vegetatively propagated plants. This may be perhaps due to the fact that micropropagated shoots were treated with IBA before transplanting to induce rooting, which may be responsible for better root development in the field.

3.3.6 Cold Storage and Cryopreservation

Low-temperature storage is an important technique for preserving the planting material. In tea long-term storage for various explants such as cut root, seed, and pollen was tested to reveal suitable storage techniques. It was found that in case of vegetative organ, cut roots were most suitable for long-term storage (Amma and Watanabe 1985).

Ballester et al. (1997) reported almost 100% survival in seven of the eight clonal shoot cultures of *C. japonica* and *C. reticulata* assayed, when stored at 4 °C for up to 12 months. Shoot tips of *C. japonica* encapsulated in alginate beads and stored at 4 °C survived for a shorter period of time than unencapsulated ones. Encapsulated material had survival rates of 75%, 50%, and 10% on 30 d, 60 d, and 75 d, respectively. Storage of artificial seeds made from nodal explants of tea at 4 °C reduced the time taken for bud sprouting in encapsulated buds than unencapsulated axillary buds by 15 d. Apart from retaining the viability of the micropropagules, low temperature also improved bud sprouting as well as proliferation efficiency of shoot buds of tea (Mondal et al. 2002).

Cryopreservation of tea shoot tips was standardized (Kuranuki and Sakai 1995) using both vitrification and alginate-encapsulation dehydration techniques. When the vitrification method was applied, shoot tips of 2 mm long from cold-hardened in vitro-grown plantlets were precultured for 2 d at 5 °C on 1/2 MS medium containing sucrose (68 g/L). Following preculturing, tips were dehydrated for 90 min at 0 °C in a highly concentrated vitrification solution consisting of 30% (w/v) glycerol, 15% (w/v) ethylene-glycol, and 15% (w/v) DMSO in 1/2 MS medium supplemented with sucrose (137 g/L). Then, these vitrified shoot tips were plunged directly in liquid nitrogen and kept there for 1 h. After rapid warming in a water bath at 30 °C, successfully vitrified shoot tips resumed growth in about 5 d and developed shoots without intermediary callus formation. The average rate of normal shoot formation was 60% about 1 month after plating on 1/2 MS culture medium. The assays performed by using the encapsulation dehydration technique (Kuranuki and Sakai 1995) indicated that shoot formation capacity of encapsulated dried shoot tips cooled to −196 °C was lower by as much as 20% than vitrified material (60% vs. 40%). Further, Kuranuki (2006) reported that 60% shoot formation rate was achieved after cryopreservation of tea multiple shoot tips using alginate-encapsulation dehydration techniques. She concluded that valuable cryogenic procedure of tea explants would be vitrification method for terminal buds and alginate-encapsulation dehydration technique for multiple shoot tips.

3.4 *Camellia* Species

Among the related species, *C. japonica* had been studied extensively due to its high demand of large number of cultivars. Vieitez et al. (1991) did a systematic study of micropropagation of *C. japonica*, which elaborately highlighted various factors affecting multiplication rate in in vitro and subsequently hardening processes. Depending on the species-specific requirements among the wild camellias, various factors that influence the micropropagation are briefly reviewed below.

3.4.1 *C. japonica*

In the late 1970s, the use of in vitro culture methods was suggested as a means to solve the constraints in propagation mainly due to shy rooting in vegetative cuttings (Bennett and Scheibert 1982). Since then, several protocols have been described for the micropropagation of *C. japonica*. The first elaborate report to regenerate plants from shoot tips and axillary buds was made by Creze and Beauchesne (1980), who took meristems with one or two leaf primordial of 0.5 mm long from 1-year-old rooted cuttings or from 3- to 4-year-old seedling and cultured them on a MS medium supplemented with adenine (20 mg/L), IAA (0.1 mg/L), and 1 mg/L each of Kn, BAP GA_3, as well as PVP (10 g/L). Though cultures were established and elongated to produce shoots more rapidly than shoot tips, no rooting and transfer to the soil were not described.

In *C. japonica*, buds of juvenile origin gave consistently better results in terms of both growth and vigor on MS as compared to other macronutrient formulae such as Lepoivre (Quoirin and Lepoivre 1977), Knop (Tabachnick and Kester 1977), Schenk and Hildebrandt (1972), and modified Heller (1953). However, regeneration from adult material of *C. japonica* cv. Alba Plena was poor on MS (Vieitez et al. 1989a). In a series of shoot multiplication experiments, Vieitez et al. (1989a) found that WPM was the best among the six macronutrient formula tested (modified Heller (1953), MS, half-strength MS, WPM, Gresshoff and Doy (1972), and Anderson (1984)). In contrast to Carlisi and Torres (1986), who found that full- and half-strength MS were the best for culturing of *C. japonica*, the observations recorded by Vieitez et al. (1989a) were poor in these media. The different responses observed by Carlisi and Torres (1986) were probably genotype-dependent. For *C. japonica*, the most widely used cytokinin was BAP. However, Creze and Beauchesne (1980) reported the importance of 2-iP (1 mg/L) as an essential component for shoot proliferation, and kn was also found to have no effect on shoot multiplication when used either alone or in combination with BAP (Samartin et al. 1984). The GA_3 (5–10 mg/L) was also used for proliferation of shoots in cv. Purple Dawn

(Carlisi and Torres 1986; Torres and Carlisi 1986). Among the auxins, IAA and IBA were used for shoot proliferation, but there were no reports on the use of NAA and 2,4-D for culturing of *C. japonica* (Creze and Beauchesne 1980; Vieitez et al. 1989b). Apart from PGR, another factor, which was found to be important, is the physical condition of the media. In general, liquid medium was more suitable than solid medium for shoot proliferation in *C. japonica* (Carlisi and Torres 1986; Vieitez et al. 1989a).

3.4.2 *C. oleifera*

Very little work has been done on this important oil-yielding species. Tian-Ling (1982) used MS medium supplemented with BAP (4 mg/L) and NAA (2 mg/L) for induction of adventitious buds leading to plantlet regeneration. In another study, lateral buds of adult trees were also used by Yan et al. (1984) for induction of axillary bud proliferation.

3.4.3 *C. reticulata*

Heller's (1953) macro elements with the addition of $(NH_4)_2SO_4$ (0.13 mg/L) in combination with MS vitamins were found to be the best for induction of axillary buds. WPM was also found to be superior than modified Heller (1953) as well as the recipes of Anderson (1984). A combination of BAP and zeatin had also been successfully used for promoting the growth and proliferation of axillary shoots (San-Jose and Vieitez 1990; San-Jose et al. 1991). Multiplication rates in terms of both number of axillary buds and the length of shoots could further be improved by horizontal placement of the explants (San-Jose and Vieitez 1990).

3.4.4 *C. sasanqua*

While Torres and Carlisi (1986) preferred MS medium, Samartin (1991) found B_5 (Gamborg et al. 1968) macronutrients supplemented with micronutrients of MS to be suitable for the growth and proliferation of axillary shoots. A combination of BAP and NAA was found to be the most suitable for shoot multiplication in both of these studies.

3.4.5 *Camellia* Hybrids

Despite the availability of limited information, the medium of Tukey (1934) was found to be the best for in vitro seedling growth of three different interspecific hybrids including *C. japonica* × *C. cuspidata*, *C. japonica* × *C. reticulata*, and *C. japonica* × *C. saluenensis* (Lammerts 1958). Creze and Beauchesne (1980) made the first attempt to regenerate plants from shoot tips and axillary buds from *C. saluenensis* and *C. chrysantha*, respectively.

3.4.6 Rooting and Hardening

Like other woody plants, rooting is a major limitation in micropropagation of *Camellia* species. Rooting of in vitro-raised shoots was achieved either upon continuous exposure to a low concentration of auxin or initially to a less exposure to a high auxin concentration followed by their transfer to an auxin-free medium. In vitro rooting of *Camellia* species had been reported by a number of workers (Table 3.1). Reduction of MS salt concentrations to half-strength favored both induction and elongation of rooting in *Camellia* species (Samartin et al. 1984, 1986; Kato 1985). However, Vieitez et al. (1989b) did not find any significant difference in rooting of in vitro-raised shoots of *C. japonica* cv. Alba Plena using half-strength MS medium. In *Camellia*, IBA (0.5–8 mg/L) had been shown to give better results than NAA for in vitro root initiation. Roots induced by NAA were shorter, thicker, and with accompanying calli, which were undesirable features for the subsequent transplanting. On the other hand, with IBA treatments, rooting occurred much later but were long and fibrous (Samartin et al. 1986). Liquid medium with filter paper bridge was beneficial for rooting in *C. oleifera* (Tian-Ling 1982; Kato 1985; Nakamura 1987b). Torres and Carlisi (1986) reported that a pulse treatment of shoots with IBA (500 mg/L) for 30 min before placing on a root induction medium gave best results in *C. sasanqua*. Beretta et al. (1987) obtained increased rooting in *C. saluenous* × *C. japonica* hybrids with IBA (1–2 g/L) treatment for 15 min. In comparison to other woody species, the *Camellia* species requires higher IBA concentration and longer immersion time, and such high treatments were not as deleterious for camellias as in other woody species like *Prunus avium* (Riffaud and Cornu 1981). Dark treatment after dipping the micropropagated shoots in auxin was reported to favor rooting in *C. japonica* (Samartin et al. 1986) and *C. reticulata* (San-Jose and Vieitez 1990). However, Samartin (1991), who also worked with *C. japonica*, did not find any significant difference between the effects of light and dark treatments on rooting. Rooting mixture alone influences the survival rate at nursery. While 75% survival of *C. japonica* was obtained in petrite/soil (1:1) (Samartin et al. 1984), a higher survival rate of 70–90% of the same species was achieved in soil/quartz (1:1) mixture (Samartin et al. 1986; Vieitez et al. 1989b). In *C. reticulate*, rooted shoots were transferred to pot containing 1:1 mixture of peat and quartz and placed in a plastic tunnel with mist/fog system in lamps to give a 16 h photoperiod, which recorded 80% survival (San-Jose et al. 1991). However, there is no report on either of any nursery performance or any field performance of micropropagated camellias presently.

3.5 Problems of Micropropagation

Like other woody perennials, major problems encountered in tea micropropagation are phenolic exudation from explants and microbial contamination in tissue culture medium, which are explained below.

3.5.1 Explant Browning

High phenols, for which tea is valued, are also exuded in the in vitro culture medium from the cut ends of explants, which undergoes enzymatic oxidation to form some toxic compounds (Bantawa et al. 2011b). These toxic compounds create problems by lowering down the pH of the tissue culture medium. Sarwar (1985) tried various chemicals such as ascorbic acid, catechol, L-cysteine, phloroglucinol, phenylthiourea, polyvinylpyrolidone-10, sodium diethyldithiocarbamate, sodium fluoride, and thiourea along with different strengths of MS inorganic salts. He found that reduced strengths of MS (1/20) prevented explants browning up to a greater extent. While ascorbic acid was used by several researchers (Iddagoda et al. 1988; Agarwal et al. 1992), polyvinylpyrrolidone was recommended by Creze and Beauchesne (1980) to prevent explant browning of in vitro cultures. Magnitude of polyphenols exudation varied between different cultivars of tea. Murali et al. (1996) categorized that polyphenol leaching under in vitro condition was highest in Cambod type followed by China type and was least with Assam type. Pandidurai et al. (1996) reported that successive transfer of the explants into a new culture medium at periodic intervals worked well to prevent explant browning of tea. Interestingly it had been observed that quality tea cultivars were difficult to establish in vitro as in the case of TTL-2 due to high phenolic content (Anon 1999). Being quality clone, TTL-2 must have higher amount of polyphenol, which made it difficult to establish under in vitro culture.

3.5.2 Microbial Contamination

In tea, explants are often taken from field grown plants that are heavily contaminated by various epiphytic and endophytic organisms, causing severe losses at every stage of micropropagation (Debergh and Vanderschaeghe 1988). Generally, cultures are pre-screened when explants from field-grown plants are used (Mondal et al. 2013). However, in tea, several strategies had also been employed to overcome these problems. Ogutuga and Northcote (1970) used 5–15 mm stem segments from the greenhouse grown tea plant and surface-sterilized with 70% ethanol followed by 7% sodium hypochlorite solution. Kato (1985) used stems with 3–4 leaves from the greenhouse-grown plants (cv. Yabukita) and sterilized with 7% calcium hypochlorite solution for 20 min. Arulpragasam and Latiff (1986) snapped off the leaf carefully with petioles as close as possible to the stem without any damage to the axillary buds. The shoots were surface sterilized in 10–15% Clorox solution for

15 min under constant agitation and inoculated in culture medium after rinsing with distilled water. Kuranuki and Shibata (1993) minimized the problems of endogenous bacteria by reducing the size of the explant to an apical meristem and a leaf primordium less than 0.5 mm. Rate of contamination varied upon the season of explant collection (Nakamura 1989), which was minimized by collecting the explants at the beginning of first seasonal flush when air temperature was relatively low and the plants were juvenile and vigorous. Haldeman et al. (1987) had shown significant reduction of bacterial as well as fungal contaminations with different concentrations of benomyl (1, 2, or 4 g/L) and rifampicin (10, 25, or 50 mg/L) without any phytotoxic effects. Das and Barman (1988) treated the explants with streptomycin sulfate (1%) for 1 min prior to inoculation in order to prevent contamination. However (0.05–1%) mercuric chloride solution (0.05–1%) was also used for surface sterilization of the explants (Rajkumar and Ayyappan 1992; Rajasekaran and Mohankumar 1992; Jha and Sen 1992; Agarwal et al. 1992). Recently Ali et al. (2018) using 16s rDNA identified seven gram-positive and two gram-negative bacteria which are predominant in in vitro culture of tea and suggested the proper antibiotics to control their growth.

3.6 Conclusion

With available information, it can be concluded that while extensive works have been done on micropropagation of tea, challenge remains on commercial application of micropropagation of tea. A wide range of media composition, explants, PGR, and varietal difference along with physiological stage of the explants has been studied. While MS is the most preferred basal medium, BAP is the most widely used PGR. Important aspects which remain uninvestigated are (1) field behavior of micropropagated tea specially after second round of pruning which is indicative for stabilized yield, (2) low-temperature long-term storage of tea shoot, and (3) lowering down the cost of production of micropropagation techniques including hardening so that it becomes comparable with vegetative propagation.

It is noteworthy to mention here that micropropagation of tea has not been commercially exploited. Very few micropropagated plants are in the field. Akula and Akula (1999) have rightly pointed out excellent reasons. The reasons are lose of juvenility of in vitro culture that hinder long-term production and lack of tap root system that makes micropropagated tea more vulnerable to drought. But the major point is the higher cost of production of the micropropagated tea plantlets which is seven times higher compared to vegetatively propagated plant, although it has been reduced much in recent years with the advancement of hardening protocols such as carbon enrichment, biological hardening, micrografting, specially designed hardening chamber, etc. that leads to higher survival rate. However, when a cultivar is developed that generates high demands of large planting material of superior clone including transgenic tea to supply within a short period of time to the growers, micropropagation is a better alternative practically for the commercial exploitation.

References

Agarwal B, Singh U, Banerjee M (1992) In vitro clonal propagation of tea (*Camellia sinensis* (L.) O. Kuntze). Plant Cell Tissue Organ Cult 30:1–5

Aitken-Christie J, Kozai T, Takayama S (1995) Automation in plant tissue culture. General introduction and overview. In: Aitken-Christie J, Kozai T, Smith MAL (eds) Automation and environmental control in plant tissue culture. Kluwer Academic Publishers, Dordrecht, pp 1–15

Akula A, Akula C (1999) Somatic embryogenesis in tea (*Camellia sinensis* (L) O Kuntze). In: Jain SM, Gupta PK, Newton RJ (eds) Somatic embryogenesis in woody plants, vol 5. Kluwer Academic Publishers, Dordrecht, pp 239–259

Ali M, Boonerjee S, Islam MN, Saha ML, Hoque MI, Sarker RH (2018) Endogenous bacterial contamination of plant tissue culture materials: identification and control strategy. Plant Tissue Cult Biotechnol 28:99–108

Amma S, Watanabe A (1985) Long term storage of germplasm of tea (*Camellia sinensis (L)* O Kuntz). JARQ 19:196–201

Anderson WC (1984) A revised tissue culture medium for shoot multiplication of rhododendron. J Am Soc Hort Sci 109:343–347

Anon (1999) Research and development department. Tata Tea Ltd, Munnar

Arulpragasam PV, Latiff R (1986) Studies on the tissue culture on tea (*Camellia sinensis* (L.) O. Kuntze). 1. Development of a culture method for the multiplication of shoots. Sri Lank J Tea Sci 55:44–47

Arulpragasam PV, Latiff R, Seneviratne P (1988) Studies on tissue culture of tea (*Camellia sinensis* (L.) O. Kuntze). 3. Regeneration of plants from cotyledon callus. Sri Lank J Tea Sci 57:20–23

Bag N, Palni LMS, Nandi SK (1997) Mass propagation of tea using tissue culture methods. Physiol Mol Biol Plants 3:99–103

Bag N, Palni LMS, Nandi SK (2019) An efficient method for acclimatization: *in vitro* hardening of tissue culture-raised tea plants (*Camellia sinensis* (L.) O. Kuntze). Curr Sci 117:288–292

Ballester A, Janeiro LV, Vieitez AM (1997) Cold storage of shoot cultures and alginate encapsulation of shoot tips of *Camellia japonica* and *Camellia reticulate* Lindley. Sci Hortic 71:67–78

Banerjee M, Agarwal B (1990) *In vitro* rooting of tea, *Camellia sinensis* (L.) O. Kuntze. Indian J Exp Biol 28:936–939

Banerjee AK, Agrawal DC, Nalawade SM, Krishnamurty KV (2000) Recovery of *in vitro* cotton shoots through micrografting. Curr Sci 78:623–626

Bantawa P, Saha Roy O, Ghosh P, Mondal TK (2009) Effect of bavistin and adenine sulphate on *in vitro* shoot multiplication of *Picrorhiza scrophulariiflora* pennell.: an endangered medicinal plant of Indo-China Himalayan Regions. Plant Tissue Cult Biotechnol 19(2):237–245

Bantawa P, Ghosh SK, Bhandari P, Singh B, Ghosh PD, Ahuja PS, Mondal TK (2010) Micropropagation of an elite line of *Picrorhiza scrophulariiflora*, Pennell, an endangered high valued medicinal plant of the Indo-China Himalayan region. Med Aroma Plant Sci Biotechnol 4:1–7

Bantawa P, Ghosh SK, Mondal TK (2011a) *In vitro* regeneration of a medicinal plant Nepalease Kutki (*Picrorhiza scrophulariiflora*) Pennell. Biol Plant 55(1):169–172

Bantawa P, DA Silva AJ, Ghosh SK, Mondal TK (2011b) Determination of essential oil contents and micropropagation of *Gaultheria fragrantissima*, an endangered woody aromatic plant of India. J Hortic Sci Biotechnol 86:479–485

Begum A, Ahmad I, Prodhan SH, Azad AK, Sikdar MBH, Ara R (2015) Study on *in vitro* propagation of tea [*Camellia sinensis* (L.) O *Kuntze*] through different explants. J Glob Sci 4:2878–2887

Bennett WY, Scheibert P (1982) *In vitro* generation of callus and plantlets from cotyledons of *Camellia japonica*. Int Camellia J 37:12–15

Beretta D, Vanoli M, Eccher T (1987) The influence of glucose, vitamins and IBA on rooting of *Camellia* shoots *in vitro*. In: Abstracts of symposium on vegetative propagation of woody species, Italy, p 105

Bidarigh S, Azarpour E (2011) The study effect of cytokinin hormone types on length shoot in vitro culture of tea (*Camellia sinensis* L.). World Appl Sci J 13:1726–1729
Bidarigh S, Azarpour E (2013) Study effect of IBA hormone levels on rooting in micro cuttings of tea (*Camellia sinensis* L.). ARPN J Agric Biol Sci 8:1–5
Bidarigh S, Hatamzadeh A, Azarpour E (2012) The study effect of IBA hormone levels on rooting in micro cuttings of tea (*Camellia sinensis* L.). World Appl Sci J 20:1051–1054
Boonerjee S, Hoque ML, Sarker RH (2013) Development of *in vitro* micropropagation system in tea plant [*Camellia sinensis* (l.) O. *kuntze*] using shoot tip and nodal segment explants. Tea J Bangladesh 42:21–30
Carlisi JC, Torres KC (1986) *In vitro* shoot proliferation of *Camellia* 'Purple Dawn'. Hortic Sci 21:314
Chen Z, Liao H (1982) Obtaining plantlet through anther culture of tea plants. Zhongguo Chaye 4:6–7
Chen Z, Liao H (1983) A success in bringing out tea plants from the anthers. China Tea 5:6–7
Creze J, Beauchesne MG (1980) *Camellia* cultivation *in vitro*. Int Camellia J 12:31–34
Das SC (2001) Tea. In: Parthasarathy VA, Bose TK, Deka PC, Das P, Mitra SK, Mohandas S (eds) Biotechnology of horticultural crops, vol 1. Naya Prokash, Calcutta, pp 526–546
Das SC, Barman TS (1988) Current state and future potential of tissue culture in tea. In: Proc. 30th Tocklai Conf. TRA Jorhat, pp 90–94
Debergh PC, Vanderschaeghe AM (1988) Some symptoms indicating the presence of bacterial contaminants in plant tissue culture. Acta Hortic 255:77–81
Dood AW (1994) Tissue culture of tea (*Camellia sinensis* (L.) O. Kuntze)-a review. Int J Trop Argic 12:212–247
Gamborg O, Miller R, Ojima K (1968) Nutrient requirements of suspension cultures of soybean root cells. Exp Cell Res 50:157–158
Gautheret RJ (1959) La culture des tissus végétaux: techniques et réalisations. Masson, Paris
Gonbad RA, Sinniah UR, Aziz MA, Mohamad R (2015) Influence of cytokinins in combination with GA_3 on shoot multiplication and elongation of tea clone Iran 100 (*Camellia sinensis* (L.) O. Kuntze). Sci World J 2015:943054. pp 1–9
Gresshoff PM, Doy CH (1972) Development and differentiation of haploid *Lycopersicon esculentum*. Planta 107:161–170
Gunasekare MTK, Evans PK (2000) *In vitro* rooting of microshoots of tea (*Camellia sinensis* L.). Sri Lanka J Tea Sci 66:5–15
Haldeman JH, Thomas RL, Mckamy DL (1987) Use of benomyl and rifampicin for *in vitro* shoot tip culture of *Camellia sinensis* and *Camellia japonica*. Hortic Sci 22:306–307
Heller R (1953) Recherches sur la nutrition minerale des tissus vegetaux cultives *in vitro*. Annales des Sciences Naturelles (Botanique) Biologie Vegetale 14:1–223
Iddagoda N, Kataeva NN, Butenko RG (1988) *In vitro* clonal micropropagation of tea (*Camellia sinensis* L.) 1. Defining the optimum condition for culturing by means of a mathematical design technique. Indian J Plant Physiol 31:1–10
Jain SM, Das SC, Barman TS (1991) Induction of roots from regenerated shoots of tea (*Camellia sinensis* L.). Acta Hortic 289:339–340
Jain SM, Das SC, Barman TS (1993) Enhancement of root induction from *in vitro* regenerated shoots of tea (*Camellia sinensis* L.). Proc Indian Natl Sci Acad 59:623–628
Jha TB, Sen SK (1992) Micropropagation of an elite Darjeeling tea clone. Plant Cell Rep 11:101–104
Kato M (1985) Regeneration of plantlets from tea stem callus. Jap J Breed 35:317–322
Kato M (1989) *Camellia sinensis* L. (Tea): *in vitro* regeneration. In: Bajaj YSP (ed) Biotechnology in agriculture and forestry, Medicinal and aromatic plants II, vol 7. Springer, Berlin, pp 83–98
Kuranuki K (2006) Developments of cryopreservation techniques for genetic resources of tea {*Camellia sinensis, (L.)* O. Kuntze}. Tech Bull Shizuoka Tea Exp Stat 3:1–63
Kuranuki K, Sakai A (1995) Cryopreservation of *in vitro* grown shoot tips of tea (*Camellia sinensis*) by vitrification. CryoLetters 16:345–352

Kuranuki Y, Shibata M (1992) Effect of concentration of plant growth regulators on the shoot apex culture of tea plant. Bull Shizuoka Tea Exp Stn 16:1–6

Kuranuki Y, Shibata M (1993) Improvement of medium components for *in vitro* cuttings of tea plant. 2. Optimum concentration of plant growth regulators. J Tea Sci 77:39–45

Lammerts WE (1958) Embryo culture in *Camellia* seed germination. In: Tourje EC (ed) *Camellia* culture. Southern California Camellia Society, Pasadena, CA, pp 171–174

Li Z, Tan X, Liu Z, Lin Q, Zhang L, Yuan J, Zeng Y, Wu L (2016) *In vitro* propagation of *Camellia oleifera* Abel. using hypocotyl, cotyledonary node, and radicle explants. Hortic Sci 51:416–421

Llyod G, McCown B (1980) Commercially feasible micropropagation of mountain laurel, *Kalmia latifolia* by use of shoot tip culture. Comb Proc Int Plant Prop Soc 30:421–427

Malyarovskaya V, Samarina L (2017) *In vitro* morphogenesis of ornamental shrubs *Camellia japonica* and *Hydrangea macrophylla*. Plant Tissue Cult Biotechnol 27:181–187

Marimuthu S, Rajkumar R (2001) Physiological and biochemical responses of micropropagated tea plants. In Vitro Cell Dev Biol-Plant 37:618–621

Molina SP, Rey HY, Pérez ML, Mroginski LA (2013) Plant regeneration of tea (*Camellia sinensis*) by *in vitro* culture of meristems, axillary buds and uninodal segments. Rev FCA UNCUYO 45:127–134

Mondal TK (2003) Micropropagation of tea (*Camellia sinensis*). In: Jain SM, Ishii K (eds) Micropropagation of woody trees and fruits. Kluwer Academic Publishers, Dordrecht, pp 671–720

Mondal TK (2004) Biotechnological improvements of tea. ISB News Lett:24

Mondal TK (2011) Camellia. In: Kole C (ed) Wild crop relatives: genomics and breeding resources planation and ornamental crops. Springer, New York, NY, pp 15–40. ISBN No. 978-3-642-21200-0

Mondal TK, Bhattacharya A, Sood A, Ahuja PS (1998) Micropropagation of tea using thidiazuron. Plant Growth Regul 26:57–61

Mondal TK, Bhattacharya A, Sood A, Ahuja PS (2002) Propagation of tea (*Camellia sinensis* (L.) O Kuntze) by shoot proliferation of alginate-encapsulated axillary buds stored at 4°C. Curr Sci 83 (8):941–944

Mondal TK, Bhattacharya A, Laxmikumaran M, Ahuja PS (2004) Recent advance in tea biotechnology. Plant Cell Tissue Organ Cult 75:795–856

Mondal TK, Parathiraj S, Mohankumar P (2005) Micrografting-A technique to shorten the hardening time of micropropagated shoots of tea {*Camellia sinensis (L)* O. Kuntze}. Sri Lank J Tea Sci 70:5–9

Mondal TK, Bantawa P, Sarkar B, Ghosh PD, Chand PK (2013) Cellular differentiation, regeneration, and secondary metabolite production in medicinal *Picrorhiza* spp. Plant Cell Tissue Org Cult 112:143–158

Murali KS, Pandidurai V, Manivel L, Rajkumar R (1996) Clonal variation in multiplication of tea through tissue culture. J Plant Crop 24:517–522

Murashige T, Skoog F (1962) A revised medium for rapid growth and bioassay with tobacco tissue cultures. Physiol Plant 15:473–497

Nakamura Y (1987a) Shoot tip culture of tea cultivar Yabukita. Tea Res J 65:1–7

Nakamura Y (1987b) *In vitro* rapid plantlet culture from axillary buds of tea plant (*C. sinensis* (L.) O. Kuntze). Bull Shizuoka Tea Exp Stn 13:23–27

Nakamura Y (1988) Effects of the kinds of auxins on callus induction and root differentiation from stem segment culture of *Camellia sinensis* (L.) O. Kuntze. Tea Res J 68:1–7

Nakamura Y (1989) Differentiation of adventitious buds and its varietal difference in stem segment culture of *Camellia sinensis* (L.) O. Kuntze. Tea Res J 70:41–49

Nakamura Y (1990) Effect of sugar on formation of adventitious buds and growth of axillary buds in tissue culture of tea. Bull Shizuoka Tea Exp Stn 15:1–5

Nakamura Y, Shibita M (1990) Micropropagation of tea plant (*Camellia sinensis L. O Kuntze*) through *in vitro* cuttings - effects of various hormones on growth of shoots from axillary buds. Jap J Agric Q 72:9–17

Nitsch JP, Nitsch C (1969) Haploid plants from pollen grains. Science 163:185
Ogutuga DBA, Northcote DH (1970) Caffeine formation in tea callus tissue. J Exp Bot 21:258–273
Owuor PO, Wachira FN, Obanda M, Waller GR (2007) Effects on caffeine on *in vitro* tea growth. J Plant Crop 35:36–38
Pandey A, Palni LMS, Bag N (2000) Biological hardening of tissue culture raised tea plants through rhizosphere bacteria. Biotechnol Lett 22:1087–1091
Pandidurai V, Murali KS, Manivel L, Rajkumar R (1996) Factors affecting *in vitro* shoot multiplication and root regeneration in tea. J Plant Crop 24:603–609
Phukan MK, Mitra GC (1984) Regeneration of tea shoots from nodal explants in tissue culture. Curr Sci 53:874–876
Phukan MK, Mitra GC (1990) Nutrient requirements for growth and multiplication of tea plants *in vitro*. Bangladesh J Bot 19:65–71
Prakash O, Sood A, Sharma M, Ahuja PS (1999) Grafting micropropagated tea (*Camellia sinensis* (L.) O. Kuntze) shoots on tea seedling-a new approach to tea propagation. Plant Cell Rep 18:137–142
Quoirin M, Lepoivre P (1977) Improved media for *in vitro* culture of *Prunus* sp. Acta Hortic 78:437–442
Rajasekaran P, Mohankumar P (1992) Rapid micropropagation of tea (*Camellia* spp). J Plant Crop 20:248–251
Rajkumar R, Ayyappan P (1992) Micropropagation of *Camellia sinensis* (L.) O kuntze. J Plant Crop 20:252–256
Ranaweera KK, Gunasekara MTK, Eeswara JP (2013) *Ex vitro* rooting: a low cost micropropagation technique for tea (*Camellia sinensis* (L.) O. Kuntz) hybrids. Sci Hortic 155:8–14
Riffaud JL, Cornu D (1981) Utilization de la culture *in vitro* pour la multiplication de merisiers adultes (*Prunus avium* L.) selectionnes en foret. Agronomie 1:633–640
Saha-Roy O, Bantawa P, Ghosh PD, Ghosh SK, Silva JATD, Mondal TK (2010) Micropropagation and field performance of 'Malbhog'(*Musa paradisiaca*, AAB group): a popular banana cultivar with high keeping quality of North East India. Tree For Sci Biotechnol 4:52–58
Samartin A (1991) Potential for large scale *in vitro* propagation of *Camellia sasanqua* Thunb. J Hort Sci 67:211–217
Samartin A, Vieitez AM, Vieitez E (1984) *In vitro* propagation of *Camellia japonica* seedlings. Hortic Sci 19:225–226
Samartin A, Vieitez AM, Vieitez E (1986) Rooting of tissue cultured camellias. J Hort Sci 61:113–120
Sandal I, Bhattacharya A, Ahuja PS (2001) An efficient liquid culture system for tea shoot proliferation. Plant Cell Tissue Organ Cult 65:75–80
San-Jose MC, Vieitez AM (1990) *In vitro* regeneration of *Camellia reticulata* cultivar 'Captain Rawes' from adult material. Sci Hortic 43:155–162
San-Jose MC, Vieitez AM (1992) Adventitious shoot regeneration from *in vitro* leaves of adult *Camellia reticulata*. J Hort Sci 67:677–683
San-Jose MC, Vidal N, Vieitez AM (1991) Improved efficiency of *in vitro* propagation of *Camellia reticulata* cv. captain leaves. J Hort Sci 66:755–762
Sarathchandra TM, Upali PD, Wijeweardena RGA (1988) Studies on the tissue culture of tea {*Camellia sinensis* (L.) O. Kuntze}. Somatic embryogenesis in stem and leaf callus cultures. Sri Lanka J Tea Sci 52:50–54
Sarwar M (1985) Callus formation from explanted organs of tea (*Camellia sinensis* L.). J Tea Sci 54:18–22
Schenk RU, Hildebrandt A (1972) Medium and techniques for induction and growth of monocotyledonous and dicotyledonous plant cell cultures. Can J Bot 50:199–204
Sharma M, Sood A, Nagar PK, Prakash O, Ahuja PS (1999) Direct rooting and hardening of tea microshoots in the field. Plant Cell Tissue Organ Cult 58:111–118

Shibata M, Kuranuki Y (1993) Improvement of medium components for *in vitro* cuttings of tea plant. 1. Effects of concentration of some components of MS medium and comparison between MS medium and woody plant medium. J Tea Sci 77:39–45

Singh A, Sharma J, Rexer KH, Varma A (2000) Plant productivity determinants beyond minerals, water and light: *Piriformospora indica* - a revolutionary plant growth promoting fungus. Curr Sci 79:1548–1554

Tabachnick L, Kester DE (1977) Shoot culture for almond and almond peach hybrid clones *in vitro*. Hortic Sci 12:545–547

Tahardi JS (1994) Micropropagation of tea through shoot proliferation from excised axillary buds. Menara Perkebunan 62:20–24

Tahardi JS, Shu W (1992) Commercialization of clonal micropropagation of superior tea genotypes using tissue culture technology. In: USAID/CDR Network meeting on tea crop biotech Costa Rica

Tian-Ling L (1982) Regeneration of plantlets in cultures of immature cotyledons and young embryos of *Camellia oleifera* Abel. Acta Biol Exp Sin 15:393–403

Torres KC, Carlisi JA (1986) Shoot and root organogenesis of *Camellia sasanqua*. Plant Cell Rep 5:381–384

Tosca A, Pondofi R, Vasconi S (1996) Organogenesis in *Camellia x williamsii*: cytokinin requirement and susceptibility to antibiotics. Plant Cell Rep 15:541–544

Tukey HB (1934) Artificial culture methods for isolated embryos of deciduous fruits. Am Soc Hortic Sci Proc 32:303–322

Vieitez AM, Barciela J, Ballester A (1989a) Propagation of *Camellia japonica* cv. Alba Plena by tissue culture. J Hort Sci 64:177–182

Vieitez AM, San-Jose MC, Ballester A (1989b) Progress towards clonal propagation of *Camellia japonica* cv. Alba Plena by tissue culture techniques. J Hort Sci 64:605–610

Vieitez AM, San-Jose MC, Vieitez J, Ballester A (1991) Somatic embryogenesis from Roots of *Camellia japonica* plantlets cultured *in vitro*. J Am Soc Hortic Sci 116:753–757

Vieitez AM, Vieitez ML, Ballester A, Vieitez E (1992) Micropropagation of *Camellia* spp. In: Bajaj YPS (ed) Biotechnology in agriculture and forestry, High tech and micropropagation III, vol 19. Springer, Berlin, pp 361–387

Widhianata H, Taryono (2019) Organogenesis responses of tea (*Camellia sinensis* (L.) O. Kuntze) var. *assamica* and *sinensis*. 1st International Conference on Bioinformatics, Biotechnology, and Biomedical Engineering, Bangkok. AIP Conf Proc 2099:020026-1–020026-9. https://doi.org/10.1063/1.5098431

Yan MQ, Ping C, Wei M, Wang YH (1984) Tissue culture and transplanting of *Camellia oleifera*. Sci Silvae Sin 20:341–350

4 Somatic Embryogenesis and Alternative In Vitro Techniques

4.1 Introduction

Genetic engineering through in vitro manipulation of cell seems to be an additional tool for overcoming some of the problems of tea breeding. The prerequisite step for such a technology is an efficient in vitro plant regeneration system. Somatic embryogenesis is considered to be the most efficient regeneration system of tea (Jain and Newton 1990). However, the efficacy of such a system for plant regeneration depends on the efficiency of multiplication and conversion rate of somatic embryos. The advantage of somatic embryogenesis is the development of adventitious embryos from explant tissue without an intervening callus phase which helps in maintaining genetic fidelity (Bano et al. 1991). Therefore, it has a tremendous potential in clonal propagation (Mondal et al. 2001a) and most importantly through genetic transformation of tea (Mondal et al. 1999, 2001b). It has been successfully used for artificial seed production (Mondal et al. 2000; Mondal 2002) and some interspecific hybrid crosses of *Camellia* (Nadamitsu et al. 1986), where immature somatic embryos were rescued and cultured before abortion. It can also be used for disease-free plant production and androgenic or haploid plant production of tea (Chen and Liao 1982). In tea, somatic embryogenesis is the only alternative pathway to conventional micropropagation due to little or no success on neoformation of adventitious buds or protoplast/suspension cultures (Deka et al. 2006). The various factors, which govern the somatic embryogenesis of tea and related species, have been summarized concisely in Table 4.1 and described below.

4.2 Induction

Induction of somatic embryogenesis depends upon several factors such as explant type, physiological stage of the explants, and media formulation including plant growth regulators. A judicious selection of all the factors may lead to a successful embryogenesis of tea.

T. K. Mondal, *Tea: Genome and Genetics*,
https://doi.org/10.1007/978-981-15-8868-6_4

Table 4.1 Somatic embryogenesis in *Camellia* species

		Medium				
Species/cultivar	Explant	Induction	Maturation	Germination	Multiplication	Reference
C. sinensis	Immature cotyledon	MS + BA (0.5) + IBA (0.5)	–	MS + BA (0.5) + NAA (5)	–	Abraham and Raman (1986)
C. sinensis TRI-2025	Nodal segment	MS + BA (0.5) + GA_3 (3)	–	–	MS0 + 1/2 macro salts	Akula and Dodd (1998)
C. sinensis	Mature seed	1/2 MS macro+ full micro MS+ AHS (100) + Gln (100)	1/2 MS macro+ full micro MS+ AHS (100) + Gln (100)	1/2 MS macro+ full micro MS+ AHS (100) + Gln (100)	–	Akula et al. (2000)
C. sinensis (L.) O. Kuntze	Cotyledon	Modified MS + BA (10) + YE (2)	–	Modified MS + BA (10) with 2% sucrose	–	Arulpragasam et al. (1988)
C. sinensis (L.) O. Kuntze	Callus derived from shoot apex	MS + BA (1) + mannitol (56 g/L) MS + BA (1) + hygromycin (10)	–	–	–	Aoshima (2005)
C. sinensis	Cotyledon	MS + BA (1.13) + NAA (22.7)	MS + BA (1.13) + NAA (22.7)	MS + BA (2.26) + NAA (22.7) + GA_3 (0.12)	–	Bag et al. (1997)
C. sinensis	Mature cotyledon, leaf	MS + BA (2) + 2,4-D (5)	–	MS + BA (2) + NAA (3)	–	Balasubramanian et al. (2000a)
Thea sinensis (L.), B-61	Cotyledon	MS + Kn (0.05) + 2,4-D (0.5)	1/2 MS + Kn (0.05) + AC (0.2%) + glucose (1.5%)	1/2 MS + Kn (0.05) + AC (0.2%) + glucose (1.5%)	–	Bano et al. (1991)
C. japonica	Cotyledon	MS + BA (1)	–	–	–	Barciela and Vieitez (1993)

C. japonica	Cotyledon from immature seed	MS with modified vit + BA (1–2) + IBA (0–2)	–	–	MS with modified vit + BA (1)	Barciela and Vieitez (1993)
C. japonica	Mature cotyledon	Knop + BA (0.5–5)	–	–	–	Bennett and Scheibert (1982)
C. sinensis	Immature cotyledon	Modified MS + BA (3) + NAA (2)	–	MS + BA (5)	–	Haridas et al. (2000)
C. sinensis, T-78	Cotyledon	MS + BA (10) + IBA (0.5) + A (80)	MS + BA (10) + IBA (0.5) + A (80)	B_5+ BA (3) + IAA (2)	–	Jha et al. (1992)
C. sinensis, Yabukita	Cotyledon	MS + BA (4) + IBA (2)	–	MS + BA (10) + IBA (0.5)	–	Kato (1982)
C. japonica	Mature cotyledon	MS + BA (0–5) + IBA (0–2)	–	MS + GA_3 (1)	–	Kato (1986a)
C. sasanqua	Mature cotyledon	MS + BA (0–10) + IBA (0–2)	–	–	–	Kato (1986b)
C. japonica	Primary somatic embryo	MS + GA_3 (1) + colchicine (0.1%)	–	MS + GA_3(1)	–	Kato (1989b)
C. sinensis, Yabukita, Sayamamidori, Benikaori, Akane	Immature leaves of in vitro-grown shoots	Liquid MS + 2,4-D (0.5)	MS + BA (10) + IBA (0.5) or MS + BA (4) + IBA (2)	MS + BA (10) + IBA (0.5) or MS + BA (4) + IBA (2) + GA_3 (1)	–	Kato (1996)
C. nitidissima	Mature cotyledons	WPM+ 2,4-D (1)	WPM + BAP (0.2) + NAA (0.2)	WPM+ BAP (0.2) + NAA (0.2)	WPM supplemented with IBA (5) + NAA (0.05)	Lü et al. (2013)
C. sinensis	Mature de-embryonated cotyledon	MS + BA (2) + IBA (0.2)	Modified MS + BA (2) + IBA (0.2) + Gln (1) + K_2SO_4	Modified MS + BA (2) + IBA (0.2) + Gln (1)		Mondal et al. (2001a)

(continued)

Table 4.1 (continued)

Species/cultivar	Explant	Medium				Reference
		Induction	Maturation	Germination	Multiplication	
C. vietnamensis X C. chrysantha	Mature cotyledon	MS + BA (3) + NAA (1)	–	MS + GA_3(1) + CM (10%)	–	Nadamitsu et al. (1986)
C. sinensis	Cotyledonary segment	NN	NN + N-Z-amino type-A (0.1%)	NN + N-Z amino type-A (2.6%)	–	Nakamura (1985)
C. sinensis with 13 cultivars, *C. japonica* with 3 cultivars, *C. sasanqua*, *C. brevistela*, *C. nokoensis*, *C. japonica* (cv. Kosyougatu) *x C. granthamiana*	Half sliced cotyledon	MS + BA (1–5)	–	–	–	Nakamura (1988a)
C. sinensis cv. Shan Chat Tien	Somatic embryo	MS + BAP (8)	–	MS + BAP (3)	BAP (3) and IBA (0.3)	Nguyen (2012)
C. sinensis and *C. assamica*	De-embryonated cotyledon segments	B_5 + NAA (0.1) + BA (2)	B_5 + NAA (0.1) + BA (2)	B_5 + NAA (0.1) + BA (2)	–	Paratasilpin (1990)
C. sinensis (L.) O. Kuntze	Mature cotyledon	MS + BAP (0.1–4.5)	MS + BAP (0.1–4.5)	MS + BAP (0.1–4.5)	MS + BAP (0.1–4.5)	Park et al. (1977)
C. japonica cv. Elegans	In vitro leaf segment	1/2 MS + DTT (2.5) + ferric citrate (2.5) in lieu of ferrous sulfate	1/2 MS + DTT (2.5) ferric citrate (2.5) in lieu of ferrous sulfate	Full-strength induction media + D-glucose (25 g/L) + BA (1) + IBA (0.1) + IAA (0.1)	–	Pedroso and Pais (1993)

C. reticulata	Mature and immature cotyledonary segment	MS + IBA (0.5–1) MS + BA (1) + IBA (0.5)	–	MS + GA_3 (3–5) + IAA (1–2)	–	Plata and Vieitez (1990)
C. reticulata cv. Mouchang	Immature zygotic embryo	–	–	–	MS + modified vit + BA (0.5) + IBA (0.1)	Plata et al. (1991)
C. sinensis, UPASI-10	De-embryonated immature cotyledonary segment	MS + PBOA (0.12) + BA (0.1)	MS + brassin (0.5)	MS + brassin (0.5)	–	Ponsamuel et al. (1996)
C. sinensis, UPASI-10	Cotyledon with/ without embryogenic axis	Modified MS + BA (0.25)	MS + BA (3–5) + CW (10%) + GA_3 (0.25–1)	MS + BA (3–5) + CW (10%) + GA_3(0.25–1)	–	Rajkumar and Ayyappan (1992)
C. sinensis, TRI-2024	Nodal segment and leaf	VW + CW (15%) MS + IBA (0.1) + BA (1) for nodal segment; MS + 2,4-D (1) + Kn (1) for leaf	–	–	Callus, embryo-like structure	Sarathchandra et al. (1988)
C. japonica 'Alba Plena' *C. reticulata* 'Mouchang'	In vitro leaf segment	MS + BA (8) + IBA (0.5)	MS+ BA (8) + IBA (0.5)	MS + GA_3 (2) + IAA (1)	Secondary somatic embryogenesis in MS+ BA (1) + IBA (0.1)	San-Jose and Vieitez (1993)
C. sinensis, Kangra Jat	Cotyledon slice	1/2 MS + BA (2) + IAA (0.2)	–	1/2 MS + BA (2) + IAA (0.2) + GA_3 (0.2)	–	Sood et al. (1993)

(continued)

Table 4.1 (continued)

Species/cultivar	Explant	Medium				Reference
		Induction	Maturation	Germination	Multiplication	
C. sinensis, UPASI-9, UPASI-10, and ATK-1 and "Cambod" variety UPASI-17	Excised cotyledons	MS + glutamine (200)	MS + ABA (5) + PEG 3%	MS + GA_3 (5) and BAP (2)	MS + GA_3 (3)	Suganthi et al. (2012)
C. sinensis, cv. Yabukita	De-embryonated cotyledon	1/2 MS media + BAP (2)	1/2 MS Kn (0.1) + ABA (0.1)	1/2 MS liquid	Pre-sterilized soil–sand (3:1) mix in the greenhouse with a low mortality rate	Tahardi et al. (2000)
C. sinensis, cv. Yabukita	Primary globular embryo	WPM liquid + BAP (1) + IAA (5)	WPM liquid + BAP (0.1) + ABA (0.5)	WPM liquid	–	Tahardi et al. (2003)
C. japonica	Immature and mature zygotic embryo	MS with modified vit + BA (1–2) + IBA (0–2)	MS with modified vit + BA (1–2) + IBA (0–2)	MS with modified vit + BA (1) + (0.1) IBA + GA_3 (5) + IAA (2)	Secondary embryogenesis in MS + GA_3 (1–2)	Vieitez and Barciela (1990)
C. japonica cv. Alba Plena	In vitro root segment	MS with thiamin (1) + nicotinic acid (0.1) + pyridoxine–HCl (0.1)	MS with thiamin (1) + nicotinic acid (0.1) + pyridoxine–HCl (0.1)	MS + GA_3(5) + IAA (1)	–	Vieitez et al. (1991)
C. sinensis	Cotyledon slice	Modified MS	Modified MS	Modified MS + Kn (1.8) or BA (1)	–	Wachira and Ogado (1995)
C. sinensis, Chyi-Men and Pyng-Shoei	Mature cotyledon	MS+ Kn (10) + IAA (1)	–	–	–	Wu et al. (1981)
C. sinensis	Mature cotyledon	–	–	–	–	Yan and Ping (1983)
C. oleifera	Cotyledon	–	–	–	–	Yan et al. (1984)

C. japonica x *C. chrysantha*	Immature zygotic embryos	MS+ Kn (0.1–0.5) + NAA (0.5–1.0) + YE (1)	–	AM +2ip (0.2–0.5) + GA_3 (5) + PVP (5 g/L)	–	Yamaguchi et al. (1987)
C. chrysantha	Cotyledon	MS + BA (1) + NAA (0.2–0.5)	–	–	–	Zhuang and Liang (1985b)
C. reticulata	Mature whole cotyledon	MS + BA (1) + NAA (0.2)	–	B_5 or liquid MS + BA (0.1–0.2) + IAA (0.1–0.5)	–	Zhuang and Liang (1985a)
C. sasanqua	Whole cotyledon	MS + BA (1) + NAA (0.2–0.5)	–	MS + BA (2) + IAA (0.5) + ABA (0.2) + Gln (500) or MS + GA_3 (1)	–	Zhuang et al. (1988)

Figures in parenthesis denote concentrations in mg/L (unless otherwise stated)

A adenine, *ABA* abscisic acid, *AC* activated charcoal, *AHS* adenine hemisulfate, *AM* Anderson (1984) basal medium, *CW* coconut water, *DTT* dithiothreitol, *Gln* glutamine, *MS0* MS (Murashige and Skoog 1962) basal medium without added growth regulators, *NN* Nitsch and Nitsch (1969) medium, *PBOA* phenylboronic acid, *PVP* polyvinylpyrrolidone, *vit* vitamins, *VM* Vacin and Went (1949) medium, *YE* yeast extract

4.2.1 Explants

Although induction of androgenesis in tea began with pollen as explants (Chen and Liao 1983), later mature cotyledons or zygotic embryos became more popular as choice of explants (Table 4.1). Use of immature cotyledons (Abraham and Raman 1986; Nakamura 1988a; Bano et al. 1991), de-cotylenated embryos (Nakamura 1985; Paratasilpin 1990; Mondal et al. 2000; Mondal 2007), de-embryonated cotyledons (Rajkumar and Ayyappan 1992; Ponsamuel et al. 1996), nodal cuttings (Akula and Akula 1999), juvenile leaves (Sarathchandra et al. 1988), and leaf stalk (Hua et al. 1999) for the production of somatic embryos was also documented. Although somatic embryogenesis has been reported from various explants of ornamental camellias, most of the workers have, however, used slices of mature cotyledon or zygotic embryos (Table 4.1). In *C. japonica* and *C. reticulata*, immature cotyledons and zygotic embryos, roots, stems, and leaves had been used for the induction of somatic embryogenesis (Plata and Vieitez 1990; Vieitez and Barciela 1990; Plata 1993; Pedroso and Pais 1993; Zhuang and Liang 1985a). Dark period of at least 14 weeks appeared to be necessary for somatic embryogenesis from in vitro leaf for *C. reticulata* (San-Jose and Vieitez 1993). Vieitez et al. (1991) reported somatic embryogenesis from the in vitro roots of *C. japonica* clones cv. "Alba Plena."

4.2.2 Physiological Stages and Genotypic Variations

Seeds of different tea varieties mature at different times. Therefore, the physiological stage and genotypic variation are inter-related; hence, both play important roles for the successful induction of somatic embryos from cotyledons (Vieitez 1994; Mondal 2008). However, systematic study on the effect of physiological maturity of different cultivars of tea is scanty. Nakamura (1988a) found that the optimum timing for cotyledon culture of *C. sinensis* was late September to mid-October. This was when seeds were physiologically matured in Shizuoka, Japan. In that report, a high percentage of embryogenesis (45–50%) was observed during this period in contrast to 15% from immature seeds collected during August. Paratasilpin (1990) reported that while China type produced 28% variation of induction of somatic embryogenesis, Assam type could produce 2% only. Mondal et al. (2000) found that seed maturity of four different Indian tea cultivars was attained at different periods. Primary somatic embryos were induced from de-embryonated cotyledons of tea on MS supplemented with NAA (2.5 mg/L) and BA (0.2 mg/L). Among them, the best time for induction of somatic embryogenesis for cultivar TG 270/2/B and UPASI-9 was during the monsoon (July–August) with embryogenic responses of 11.1% and 50.0%, respectively. For the cultivar Tuckdah-78, the best time was during autumn (September–October) with the response of 44.1%, whereas the highest number of somatic embryo inductions in cultivar Kangra Jat was registered at 66.7% during winter (November–December). Nakamura (1988a) screened 13 Japanese tea cultivars and observed a variation between 0 and 50% somatic embryogenesis.

Among the cultivars, the best response was reported in 'Yabukita'. She also reported the formation of 28% somatic embryos in *C. sinensis*, against 2% in *C. assamica*. Genotypic difference also plays a major role in the induction of embryogenesis. In the same study, she also screened four *Camellia* species including Japanese tea cultivars, among which the embryogenic response varied between 0 and 50%. Of all the cultivars screened, the best response was reported in 'Yabukita' and 'Kurasawa' tea cultivars. Among the other *Camellia* species, high differentiation rates of somatic embryos were obtained with *C. japonica* (48–58%), *C. sasanqua* (9–81%), and *C. brevistela* (93%). Kato (1996) also used various cultivars of tea and reconfirmed the highest frequency of embryogenesis in cultivar 'Yabukita'. All these works collectively indicated that somatic embryogenesis was controlled by genetic factors in *Camellia.*

Successful induction of somatic embryos from cotyledon explants depends upon the physiological maturity of the cotyledons. In *C. japonica*, Vieitez and Barciela (1990) collected seeds in July, September, and October to determine the right stage for induction of somatic embryogenesis. They achieved 94% embryogenesis in seeds collected in September in contrast to 20% of those collected in October under the climatic condition prevailing at Spain. The seeds collected in September were fully grown but were still immature suggesting a transitory dormancy stage.

A protocol for the induction of direct somatic embryogenesis was developed for *C. japonica* to achieve a controlled induction of somatic embryogenesis on the specific region of leaf. Leaves isolated from 12-week-old shoots were immersed in IBA (1.0 g/L) for 20 min, followed by incubation on diluted MS modified basal medium for 11 d in dark (Pedroso and Pais 1993). The cells from a cultured leaf responded differently to the same culture conditions forming embryos, roots, and non-morphogenic calli. Shoot and leaf ages also affected in vitro leaf response. So, not all leaves from the same shoot were induced embryos after the induction treatment. Those that did not were referred to as non-embryogenic induced leaves. Direct embryo formation occurred only in defined regions of the leaf blade. Direct root formation only occurred in a well-defined region of the midrib, whereas calli were preferentially formed on the leaf base. These results suggested the existence of differences in morphogenic competence according to leaf regions. A leaf region-specific response was also observed when leaves from 1-year-old shoots were sectioned transversely in slices (2–4 mm) or sections and cultured. Explant region-specific embryogenic competence also occurred in stem segments and cotyledons of this species (Pedroso and Pais 1995a). Plantlet regeneration was successfully achieved from all the culture systems developed (Pedroso and Pais 1993, 1995a, b). The *C. japonica* leaf culture system enabled a comparison between embryogenic and non-embryogenic leaves isolated from the same shoot and the study of embryogenic and non-embryogenic leaf regions within the same leaf. For these reasons, this culture system was used as an experimental model system with the subsequent studies searching for markers of direct somatic embryogenesis in dicotyledonous species.

4.2.3 Basal Media and Growth Regulators

The type, concentration, and time of application of different growth regulators in culture media have been extensively worked out. Though MS was the most commonly used medium for the induction of tea somatic embryos (Table 4.1), WPM (Llyod and McCown 1980) and Nitsch and Nitsch (1969) media had also been used by researchers. The type, concentration, and time of application of different PGRs have also been extensively worked out. In general, a high cytokinin-to-low auxin ratio or low cytokinin alone was found to be necessary for the induction of somatic embryos in *Camellia* including tea; even reduction or omission of cytokinin in subsequent subculturing is also known to be successful. BAP (0 to 10 mg/L) had been widely used for tea and related species such as *C. japonica* (Bennett and Scheibert 1982; Beretta et al. 1987; Barciela and Vieitez 1993), *C. sasanqua* (Kato 1986b; Zhuang et al. 1988), *C. chrysantha* (Zhuang and Liang 1985b), and *C. reticulata* (Zhuang and Liang 1985a), but in certain reports, Kn (0.05 to 10 mg/L) had also been used in the induction medium (Wu et al. 1981; Bano et al. 1991; Wachira and Ogado 1995). Among the different auxins, IBA (0 to 2 mg/L) was used mostly for somatic embryo induction in tea. However, different concentrations of NAA (Paratasilpin 1990; Bag et al. 1997; Balasubramanian et al. 2000a), 2,4-D (Das and Barman 1988; Bano et al. 1991), and IAA (Wu et al. 1981; Sood et al. 1993) were also used to elicit a better response for the induction of somatic embryos in *C. sinensis*. Ponsamuel et al. (1996) used some novel auxins, namely, tetraphenyl boron (TPB), phenylboronic acid (PBOA), etc., for the induction of somatic embryos from immature cotyledons of *C. sinensis* cultivar UPASI-10. They found that TPB (0.35 g/L) and PBOA (0.12 g/L) were best for inducing somatic embryos. Mondal et al. (2000) also found that half-strength MS salts containing NAA (2.5 mg/L) and BA (0.2 mg/L) were the best for the induction of somatic embryo. Among the different genotypes, maximum of 66.7% somatic embryos were formed by cultivar Kangra Jat. Addition of ABA (7.5 mg/L) to the induction medium significantly enhanced the rapid formation of somatic embryos without callusing from mature cotyledon of tea cultivar 'TRI-2025' within 2 weeks of culture initiation (Akula et al. 2000).

In general, direct somatic embryos in *Camellia* can be obtained on a wide range of culture conditions such as full- to quarter-strength modified MS, sucrose (15–30 g/L), D-glucose, or combinations of both auxin (0–10 mg/L) and cytokinin (0–10 mg/L) in liquid, semi-solid, or solid medium (Mondal 2009). Light was an important parameter for somatic embryo formation, especially from stem and leaf explants (direct and indirect). Somatic embryos either did not form in the dark or their number was significantly lower than that for cultures under photoperiod. Successful conversion of cotyledon-derived embryos into plants ranged from 35 to 79%, depending on the culture medium used (Pedroso-Ubach 1994). Identical results were obtained for leaves, cultured on modified MS medium supplemented with sucrose (20 g/L) or D-glucose (25 g/L), 2,4-D (1.0 mg/L), and kinetin (0.1 mg/L) (Pedroso-Ubach 1991). Leaves cultured on MS liquid medium supplemented with BAP (1.0 mg/L) and 2,4-D (0.5 mg/L) produced clusters of 3–23 somatic embryos/leaf. Globular embryos in a less frequently (2–3%), detached from the remaining differentiated leaf tissues and developed singularly in the liquid culture. Only the embryos arising in clusters developed into plantlets. Among the PGRs, cytokinin such as BAP (0–10 mg/L) had been widely used for *Camellia*, though Vieitez et al. (1991) claimed that zeatin (1 mg/L) in combination with BAP and IBA was essential for the induction of somatic embryos in *C. japonica* cv. Alba Plena. Auxin such as IBA (0–2 mg/L) was widely used in the induction medium for *Camellia*, though NAA had also been used in different concentrations in different species such as 0–2 mg/L in *C. reticulata* (Zhuang and Liang 1985a), 0.2–0.5 mg/L in *C. sasanqua* (Yamaguchi et al. 1987), 0.5–1.0 mg/L in *C. japonica C. chrysantha*, and 1 mg/L in *C. vietnamensis*, as well as in *C. chrysantha* (Nadamitsu et al. 1986). The other species in which embryogenesis was found are *C. brevistela* (Nakamura 1988a), *C. chrysantha* (Zhuang and Liang 1985b), *C. oleifera* (Yan et al. 1984), and *C. sasanqua* (Nakamura 1988b; Zhuang et al. 1988), and with new hybrids whose development was hindered by poor fertility or embryo viability (Nadamitsu et al. 1986; Yamaguchi et al. 1987). In all these reports, somatic embryogenesis was achieved using MS medium containing a cytokinin (usually BAP) with or without auxin (usually NAA). However, none of these studies involved comprehensive experiments to determine optimum conditions for embryogenesis.

4.2.4 Growth Adjuvants

Growth adjuvants play an important role; nevertheless, it seems that in tea their requirement is less, which may be due to fact that cotyledons of tea have a high degree of inherent embryogenic capacity (Vieitez 1994) and hence do not demand any growth adjuvant. However, this had not restrained researchers from using adjuvant like yeast extract (Arulpragasam et al. 1988), coconut milk (Sarathchandra et al. 1988; Rajkumar and Ayyappan 1992), adenine sulfate (Jha et al. 1992), betaine (Akula et al. 2000), etc. The positive result of yeast extract had also been found in some *Camellia* hybrids (Yamaguchi et al. 1987).

4.3 Secondary Embryogenesis

Exploitation of embryogenesis through a biotechnological approach demands a system of repetitive/recurrent or secondary somatic embryogenesis, which is specially important for genetic transformation as foreign genes can be introduced into primary embryos and made to multiply subsequently (Mondal et al. 2001a). Two different growth patterns for secondary embryogenesis have been reported in *Camellia*, (1) somatic embryo-to-somatic embryo, commonly known as repetitive embryogenesis, more frequent in tea as well as other *Camellia* species and (2) callus-to-somatic embryo, in which the multiplication of somatic embryos depends upon subculturing of callus (Vieitez 1994; Mondal 2011). However, the various works of synchronized secondary somatic embryo formation and their long-term maintenance of embryogenic capacity are discussed below.

Abraham and Raman (1986) used a single medium containing different concentrations of BAP and NAA for the induction, germination, and maintenance of somatic embryos. Kato (1986a) used GA_3 (1 mg/L) for the induction of secondary embryos from isolated primary embryos. However, Jha et al. (1992) obtained secondary embryos on cotyledons of primary embryo within 6–8 weeks on MS or B_5 medium supplemented with BAP (0.1–3 mg/L), IBA (0.1 mg/L), IAA (0.1–2 mg/L), and GA_3 (5.0 mg/L). Histological observations indicated that an embryogenic callus was induced from a special zone around vascular bundle of primary embryo (Balasubramanian et al. 2000a). Apart from growth regulators, nutritional effect of nitrate salts of potassium and ammonium, together with different concentrations of sulfate salts of aluminum, potassium, magnesium, and ammonium on secondary somatic embryogenesis, was investigated (Mondal et al. 2001a). The primary embryos, which were induced on de-embryonated cotyledon, were used for secondary embryogenesis. Each responsive explant (primary embryo) evinced about 10–15 secondary embryos. Among the different concentrations of nitrates, it had been found that MS but reduced potassium nitrate (800 mg/L) as well as ammonium nitrate (825 mg/L) along with potassium sulfate (260 mg/L) produced maximum number of secondary embryo within 2 weeks (i.e., 20–25 secondary embryos per primary embryo in 91.6% responsive explants). Each of these yellow, heart-shaped secondary embryos weighed about 4–5 mg and developed synchronously to the shiny, green torpedo-shaped structure within 3 weeks of culture. In a further experiment, different factorial combinations of L-glutamine, BAP, and IBA were investigated, and the maximum number of synchronous secondary embryos developed on MS medium supplemented with BAP (2 mg/L), IBA (0.2 mg/L), and L-glutamine (1 g/L). Following this formulation, synchronous globular, heart-, and torpedo-shaped somatic embryos were developed which later germinated to an extent of 52%. Furthermore, high embryogenic capacity of this culture could be maintained on the above medium for more than 4 years, highest so far among the published reports.

In *C. japonica*, secondary embryogenesis had been reported primarily from embryos derived from a wide range of initial explants, i.e., from cotyledons or from excised embryos (Kato 1986a; Vieitez and Barciela 1990), from roots (Vieitez et al. 1991), or from in vitro leaves (San-Jose and Vieitez 1993). Primary embryo upon transfer to MS medium with or without growth regulators gave rise to secondary embryogenesis within 3–4 weeks. In general, growth regulators used for *C. japonica* have higher concentration of BAP along with lower concentration of IBA.

In *C. reticulata*, high-frequency (65%) secondary embryogenesis was found on cotyledonary and hypocotyl region of isolated primary embryos (Plata and Vieitez 1990). This response was evinced on MS medium containing BAP (0.5 mg/L) and IAA (0.5 mg/L). Plata et al. (1991) studied the anatomical sequence of events, which led to the differentiation of secondary embryogenesis in *C. reticulata* cv. Mouchang. They found that embryogenesis occurred mainly on the hypocotyl region of primary embryos. Histological monitoring revealed that secondary embryos apparently had a multicellular origin from embryogenic areas originating in both epidermal and subepidermal layers of hypocotyl region. This morphogenic competence was related to the presence of relatively undifferentiated cells in superficial layers of the hypocotyl of the primary embryo.

4.4 Bioreactor Technology for Secondary Embryogenesis

Applications of bioreactor technology in micropropagation facilitate large-scale continuous production of propagules. Although there are reports on various aspects of embryogenesis, bioreactor technology was not properly explored initially. Later, Akula et al. (2000) were pioneers in this field with the only approach until now for developing a bioreactor system for repetitive embryogenesis in tea. Uniform globular embryos were induced on nodal explants of cultivar 'TRI-2025' for developing a modified temporary immersion system (TIS). The relative efficiency of different conventional methods, namely, roller drum and liquid medium shaker, for multiplying embryos was compared with the temporary immersion method. The highest rate of multiplication of secondary embryos was 24-fold which was achieved using the TIS. Further, by controlling the immersion cycles, they achieved more consistent, synchronized multiplication and embryo development with a high level of plant recovery. A one-step computer-programmed immersion protocol based on a single, simple medium with no growth regulators was developed, enabling multiplication, maturation, germination, and plant recovery within 17 weeks. Plantlets recovered through this method were hardy, with 2–5-cm-long shoots with a minimum of 2–4 lush green leaves with well-formed taproot. Callus formation, hyperhydricity, and other developmental abnormalities were not observed at any stage of the process. Harvesting of secondary embryos was continued up to 2 years from the original set of experiment by regular subculturing after every 6 weeks. Plantlets produced using this method was successfully acclimatized to glasshouse conditions. This protocol avoided culture transfers and thus minimized the risk of contamination and reduced labor costs. Therefore, this technique was the first significant step for the commercial implication of bioreactor technology for large-scale production of tea somatic embryos.

4.5 Maturation and Germination

The success of embryogenesis also depends on high germination rate and subsequent plant regeneration, which occurs through series of events, such as synchronized embryo maturation followed by cotyledon expansion, hypocotyl–root axis elongation, and emergence of the shoot (Webster et al. 1990). However, it had been noticed that as compared to other woody plants, induction of somatic embryos on tea cotyledon was easier due to inherent capacity but they had low germination rates, specially on hormone-free culture media (Vieitez 1994). Perhaps, this is due to the lack of physiological understanding of the somatic embryogenic pathway. Two common problems that are often encountered with tea somatic embryos are precocious and abnormal germination due to lack of storage reserve and improper balance of endogenous hormone profile coupled with vulnerability of somatic embryo to desiccation. Therefore, various factors such as supply of external nutrient and PGRs, which will improve the conversion rate of tea somatic embryos, are discussed below.

4.5.1 Sugars

Source and concentration of sugars play an important role for maturation and germination of somatic embryos. Sugars served dual purpose of acting as a carbon source as well as an inducer of desiccation tolerance during maturation and germination of somatic embryos (Lecouteux et al. 1993) by acting as an osmotic agent (Tremblay and Trembly 1995). One way to resist desiccation by preventing osmotic changes could be achieved by addition of an osmotic stabilizer to the culture medium. This was generally accomplished by inclusion of an osmoprotectant such as sucrose. The sucrose concentration in the culture medium and duration of culture treatment affected the germination rate of tea somatic embryos (Mondal et al. 2002). They registered significantly higher germination percentage (35.2%) at 3% sucrose after 5 weeks of treatment, whereas no response was observed at any of the higher sucrose concentrations (6 or 9%) with various periods of treatment (3, 4, and 5 weeks). This may due to the fact that food reserves in the form of readily available carbohydrates during maturation are important for subsequent successful germination of zygotic embryos within seeds. Thus, it is not surprising that the embryos cultured on 3% sucrose for periods of 5 weeks germinated normally to the extent of 35.2% but declined thereafter.

4.5.2 Desiccation

Desiccation treatment promotes germination capacity for somatic embryos (Roberts et al. 1990). In tea somatic embryos were still found to be desiccation sensitive and were adversely affected when desiccated. Probably the recalcitrant behavior of tea seed was reflected in the somatic embryos also. In tea, drying and death coupled with poor germination (1.2–1.6%) of somatic embryos were observed within a week when the globular embryos were desiccated for 3 weeks under high relative humidity created by different chemicals, i.e., 60% relative humidity by $MgNO_3$, $6H_2O$ or 90% relative humidity by $ZnNO_3$, $7H_2O$. Therefore, it was clearly evident that to achieve the successful maturation, one should use the osmotic protectant in the culture media to prevent the tea somatic embryo from desiccation (Mondal et al. 2002).

4.5.3 Plant Growth Regulators and Additives

Among the different PGRs, ABA had been widely used for the maturation of somatic embryos of woody plants particularly because it inhibited precocious germination by conferring desiccation tolerance (Roberts et al. 1990) and promoting accumulation of storage lipids (Avgioglu and Knox 1989) and proteins (Roberts et al. 1990). The report of Mondal et al. (2002), however, indicated that none of the tea somatic embryos germinated after ABA (0.5–62.5 mg/L) treatment applied for 3–5 weeks. Similarly GA_3 (0.5–5 mg/L), which promoted germination in several woody species and hybrids of *Camellia*, was not significantly effective for tea. Although Kato (1986a) found that BAP (10 mg/L) along with IBA (0.5 mg/L) was useful for the induction of the concomitant development of shoots and roots in tea embryos, a much lower concentration of BAP (1 mg/L) and IBA (0.1 mg/L) could still induce germination of *C. japonica* somatic embryos (Vieitez and Barciela 1990; Vieitez et al. 1991; Pedroso and Pais 1993), and BAP (1 mg/L) along with IBA (0.5 mg/L) could induce germination of *C. reticulata* somatic embryos (Plata and Vieitez 1990; Plata et al. 1991). However, Jha et al. (1992) obtained 20% germination of tea somatic embryos on B_5 medium (Gamborg et al. 1968) containing BAP (3 mg/L) and IAA (2 mg/L). Thus it seems that a high cytokinin stringently BAP to low auxin such as IBA or IAA is the best for somatic embryo germination of tea as well as related species. Influence of other growth regulators such as brassin, a synthetic analog of a naturally occurring brassinolide (Maugh 1981), was tested by Ponsamuel et al. (1996). They achieved about 50% germination of *C. sinensis* somatic embryos on 0.8% agar solidified MS medium containing brassin (0.48 g/L). The positive influence of brassin on germination rate may be due to the inherent sensitivity of the tissue to this compound as brassinolide had been reported to be present in tea leaves (Ikekawa 1991). The effect of different concentrations of maltose and trans-cinnamic acid (t-CA) alone or in combination was evaluated to improve the germination efficiency of tea somatic embryo (Mondal et al. 2002). While none of the untreated somatic embryos germinated, significantly a low percentage (3–6%) of germination was observed at all concentration of either maltose or t-CA alone. On the other hand, normal and significantly high germination (70.6%) was observed when the embryos were cultured on a medium supplemented with both maltose (4%) and t-CA (3 mg/L) for 4 weeks followed by their transfer to MS medium containing GA_3 (1.5 mg/L). The success achieved with maltose and its superiority over sucrose are probably due to difference in its breakdown products. Since maltose is broken down more slowly than sucrose, it can provide a readily metabolizable carbon source (glucose) to the embryos, which probably lack any reserved carbohydrate. This further reflects on the inefficiency of the developing embryos in accumulating reserves during the embryo maturation phase. Cinnamic acid is an important compound on a wide variety of metabolic pathways such as flavonoid biosynthesis, the phenolic synthesis, and most importantly the synthesis of malonyl-CoA, the key precursor of fatty acid synthesis pathway. However, the inability of either maltose or t-CA alone for germination indicated that the requirement of both sugars and fatty acid was essential during embryo maturation and subsequent germination of tea somatic embryo.

4.6 In Vivo Embryogenesis

Hitherto, emphasis has been given to manipulate the nutrient composition, growth regulators in culture medium, physical conditions of incubation, and other stress treatments to induce somatic embryos. However, induction of in vivo embryogenesis of tea could be achieved without using the conventional tissue culture media. In an attempt to germinate tea seeds under sterile conditions inside steel boxes containing moist sand for the use of protoplast isolation, Mondal et al. (2001c) observed the induction of embryogenesis on cotyledon surface of mature tea seeds under in vivo conditions at 28 °C. The induction was, however, influenced by genotype, time of seed collection, and temperature of incubation. Selected regime of temperature (24 °C, 28 °C, and 32 °C) for incubating the seeds was examined so as to determine the effect on embryogenesis in vivo among the three cultivars of tea, namely, Kangra Jat, UPASI-9, and T-78. At 32 °C, no response of embryogenesis was evident in any of the three genotypes irrespective of the time of seed collection, except a very low germination frequency (1.6%), which occurred in Kangra Jat seeds. At 24 °C also, small percentage (4.2%) of Kangra Jat seeds germinated, but there was no development of somatic embryos. However, when the seeds were incubated at 28 °C, both germination and adventive embryos were observed. The percentage of germination was lower in comparison to that of embryo formation. Incidentally, a considerable amount of seeds remained unresponsive, irrespective of the treatments. Further, the formation of somatic embryos per seed also depends upon genotype. While the maximum number of somatic embryos (15 to 16 per seed) was observed after 90 d of incubation at 28 °C in UPASI-9, that of Kangra Jat and T-78 showed 10 to 12 and 6 to 8 such embryos per seed, respectively (data not shown). The maximum number of seeds showing formation of embryogenic response was observed in UPASI-9 compared to two other cultivars. The response in UPASI-9 seeds collected in September was 23.3%, while that of T-78 collected in the month of October was 16.5% and Kangra Jat collected in the month of November was 11.7%. After another 42 d, the embryos underwent to a typical embryogenic pathway as evidenced by cup-shaped structures. One important observation was that the seeds, which follow normal germination, did not show embryogenesis. Histological evidence confirmed somatic embryogenesis. However, a considerable amount of seeds remained unresponsive irrespective of the treatments. Therefore, we concluded that the induction of embryogenic response was not only dependent upon the external supply of nutrients and growth regulators but also on temperature. At a particular temperature, mature tea seeds are capable of expressing their somatic embryogenic potential even under in vivo conditions. Bhatia et al. (1986) observed in vivo response of callusing and adventitious shoot formation from de-embryonated *Arachis hypogea* cotyledons inoculated on moist sand and cotton wool in enamel trays. The greater influence of temperature on embryogenesis forced us to propose a hypothesis that, in *Camellia*, the induction of in vivo embryogenesis must be regulated by temperature-dependent gene. Though the reason for this observation is not clear presently, seeds of *Camellia* appear to have a considerable inherent capacity for embryogenesis. Thus, at the right physiological stage, with appropriate levels of internal hormone and appropriate moisture profile and temperature of the substrate under sterile conditions, the tea seeds are able to produce embryos without any exogenous nutrient.

4.7 Hardening and Field Transfer

Information on hardening of tea somatic embryo-derived plants is scant in the literature. It may be due to the fact that the establishment of somatic embryo-derived plantlets is easy compared to micropropagated plants due to the presence of taproot system in the former. Generally, sand or soil with some additives such as peat or vermiculite in the presence of high humid condition is sufficient for initial establishment. Although Wu et al. (1981) were the first to transfer the plantlets to soil, the composition of soil mixture and other conditions used by them was not mentioned clearly. Kato (1989a) grew healthy tea somatic seedlings under natural conditions by transferring them into a mixture of vermiculite and soil (1:1), and they were covered under the plastic. On the other hand, Jha et al. (1992) hardened tea plantlets for 8 weeks in quarter-strength MS salts before transplanting to pots containing a mixture of peat and soil (1:1) and achieved an acclimatization rate of 70%. Wachira and Ogado (1995) reported that multiple shoots differentiated from the germinated embryos were successfully rooted in autoclaved mixture of sand/peat (3:1) in small pots. Shoots were watered and kept under poly-tunnel to prevent excessive moisture loss as well as to maintain a high relative humidity. After 8 weeks, the small plants had developed root system and were transferred into 10-cm-long lay flat polythene sleeves containing a mixture of soil and sand in a ratio of 3:1. Subsequently, plants were shifted to a shaded nursery bed. In order to improve rooting, Ponsamuel et al. (1996) treated the plantlets with indole-3-acetonitrile (0.34 g/L), brassin (0.48 g/L), and phloroglucinol (1.2 g/L) in liquid MS medium for 15 d, and after profuse root proliferation, the plantlets were acclimatized in pots containing vermiculite. Eventually the plants were transplanted in the greenhouse. Akula and Akula (1999) transferred the small plantlets with a strong taproot and 4–6 leaves into small pots filled with pre-sterilized potting mixture (sand/peat/vermiculite: 1:2:1) and kept them into a greenhouse with misting facility, at 80–95% humidity under low light. The new leaves were observed within 5–6 weeks when they were taken into bigger pots. Following this procedure, they achieved 90–95% survival rate, and more than 200 plantlets were transferred to the field at Indonesia. Tea plantlets with young leaves and stout roots, with a height of 4–5 cm, were taken from culture room to Hikkotrays containing pre-sterilized sand and cow dung (1:1). These Hikkotrays were then kept in poly-tunnel with intermittent watering for 60 d inside indigenously developed poly-house (90% survival rate). Later, they were transferred to polythene sleeves filled with black virgin soil and kept for further 1 year in the same poly-house (Mondal et al. 2002). Following these techniques, 3000 somatic seedlings of tea were produced at the Research and Development Department of Tata Tea Ltd., Kerala, India, which had been transferred to the field (Mondal et al. 2004).

4.8 Somaclonal and Gametoclonal Variation

Despite the fact that stable somaclones may be desirable for crop improvement specially in perennials, no effort has been made to generate the useful somaclones in tea. It may be due to the fact that tea is cross-pollinated, variations already exist among natural populations, and worldwide most of the breeders were engaged in identifying the superior variants. Secondly, in tea, the number of tissue culture-derived plants, which had been transferred to the field, might not represent a sizable amount to study the somaclonal or gametoclonal variants. Rajkumar et al. (2001) undertook a detailed study about somaclonal variation of tea. The plants were derived from somatic embryos, which were induced on cotyledonary tissues of clone UPASI-10. They identified five somaclonal variants in the field on the basis of morphological, physiological, and biochemical parameters. Interestingly based on the "membrane stability index" study, they concluded that the somaclonal variants would be tolerant to low temperature stress and thus suitable to grow at high altitude. There were few preliminary attempts earlier the work of Rajkumar et al. (2001) to identify the somaclonal variants among the micropropagated tea. A wide range of chromosomal variations was observed in callus derived from in vitro leaf and cotyledon explants of tea (Das 1992). The study revealed that more than 60% cells were diploids in 8-week-old calli with the remaining cells being triploids, tetraploids, or aneuploids with more than 60 chromosomes. However, Jha et al. (1992) reported that embryo-derived plants had chromosome number of $2n = 30$, suggesting an apparent genetic uniformity. Variation in plants regenerated from gametophytic tissue had been reported in some cases due to uncovering "residual heterozygosity" (Evans and Sharp 1986). Variation in the chromosome number of gametes or gametophytic tissue plays an important role in gametoclonal variation. This is evident from a range of aneuploids and mixoploids recovered from anther cultures of wheat, maize, and sexual hybrid of wheat and triticale (Hu 1983). In tea, Chen and Liao (1983) examined the ploidy level of anther-derived plantlets. They found that 9 out of 12 such plantlets were aneuploids and most of them had chromosome number of $2n = 18$–22. Further, morphological variations were not mentioned.

A study has been undertaken to characterize 15 field-grown somaclonal variants derived from cotyledonary tissues of UPASI-10 using morphological, physiological, and biochemical characters. Although variants were derived from UPASI-10, a very few variants possessed unique "Chinary" characters, while others exhibited "Assam" characters. However, no variant showed identical morphological characters aligning with the parent. Somaclonal variants showed distinct variation in terms of photosynthetic carbon assimilation, stomatal conductance, and diffusion resistance. Proline accumulation and water use efficiency showed marginal variations among the variants. Two variants, namely, SE 8 and SE 10, recorded higher values of "membrane stability index" denoting their tolerant nature against stress. Variants SE 2 and SE 13 were segregated distinctly representing their black tea characters (Thomas et al. 2006).

4.9 Origin and Morphology of Somatic Embryos

The somatic embryo originates from either single cell or group of cells, which depends upon the plant species. The histological and anatomical aspects of somatic embryo origin in *Camellia* had been studied (Plata and Vieitez 1990; Vieitez et al. 1991). Barciela and Vieitz (1993) made a detailed study on the origin and anatomical development of *C. japonica* somatic embryos differentiated on cotyledon sections. They studied through computer-aided image analysis for cytological quantification the measurements of cell starch and protein contents as the stained cell areas by staining with periodic acid–Schiff (PAS) as well as by mercuric bromophenol blue, respectively. It had been observed that small protuberances or nodules began to appear on the abaxial epidermis of the cotyledons from 7-day-old in vitro tissue. The nodules continued to develop, and by the 30th day, they were 4–6 mm in diameter and became moderately prominent. After 2 months of culture, embryos were 6–8 mm long and became ready to isolate for either germination or secondary embryogenesis. It was found that only the abaxial surface of the cotyledon explants was morphologically competent and had multicellular origin. To determine whether the embryogenic nodules could be maintained indefinitely in culture, they were isolated from the initial cotyledons, removed from in vitro produced somatic embryos, and then cultured for 6 months with monthly transfer to a fresh medium. The parenchymatic tissue of the nodule failed to proliferate or grow and turned progressively necrotic. The above histological analysis suggested that the nodules associated with the occurrence of somatic embryogenesis can be considered as small localized callus tissue, which was necessary for the redetermination of embryogenic cells. Histological observations of embryogenesis in *C. reticulata* (Plata and Vieitez 1990) suggested that somatic embryos developed directly from cotyledon without any apparent callus phase. However, the differentiation of the embryos was nevertheless related to the developing swollen parts, swollen whitish areas, or compact bulging tissue of cotyledon explant (Plata and Vieitez 1990). Such swellings might be equivalent to the nodules observed in *C. japonica* (Vieitez and Barciela 1990; Barciela and Vieitz 1993).

The morphology of the *Camellia* somatic embryos was influenced by the concentration of cytokinin in the medium. In the case of *C. japonica* (Vieitez and Barciela 1990; Vieitez et al. 1991), most embryos could be classified into the following two clearly distinct types, (1) seed-like embryos, which were yellowish-white with large cotyledons alike to mature zygotic *Camellia* embryos, and (2) bud-like embryos, which were green with cotyledons resembling true leaves. They generally developed in media with relatively high BAP concentration. Ammirato (1985) stated that reasonably high levels of cytokinins partially or totally inhibited the development of somatic embryo cotyledons and the shoot apex grows out to form the first mature leaves so that the somatic embryo looked more like a shoot. The observed bud-like embryos may be an example of such cytokinin-induced premature shoot emergence. Anomalies such as polycotyledon, hypertrophy, or fascination were also observed to various extents among both seed-like and bud-like embryos, but both kinds were genuinely bipolar having both shoot and root meristems.

4.10 Biochemical Changes of Somatic Embryogenesis

Low germination and poor conversion into normal plantlets were the major limitations to somatic embryogenesis in tea (Mondal et al. 2002). The main cause for poor germination of embryos was their failure to complete normal stages of embryogeny that was generally common to zygotic embryos (Sharma et al. 2004). Embryos in angiosperm develop through several changes, namely, histo-differentiation and endosperm formation, followed by embryo maturation and finally enter into the desiccation stage. The maturation on the other hand starts with cessations of cell division and increasing of cell size due to reserve accumulation. This is followed by gradual depletion of metabolism and finally terminates by drying as the embryos enter into the desiccation stage. They then become metabolically inactive and tolerant to desiccation. However, the ability to tolerate desiccation is gradually acquired during the maturation phase itself (Vieitez 1994). It had been found several biochemical changes took place during the development of somatic embryos which are discussed below.

Changes of ABA, starch, total soluble sugars (TSS), proteins, and phenols were studied in the somatic embryos at different stages of development (globular, heart, torpedo, and germinating embryos) in order to investigate the accumulation of storage reserves to prevent the precocious germination. It had been found that after ABA treatment with 5.0 mg/L for 14 d, the starch and protein contents of the embryo were increased by several-folds with a simultaneous increase in TSS. ABA treatment at the heart stage improved the germination. ABA treatment prior to or after heart stage did not improve somatic embryo germination (Sharma et al. 2004). In the normal seed formation of tea, ABA content was found to be maximum at maturation than in any other stages (Nagar and Sood 2006; Preeti et al. 2004). Therefore, it was concluded that external application of ABA at a specific dose was capable of circumventing the developmental blocks of improper reserve accumulation to a certain extent, provided that ABA was supplemented only at the early maturation stage. Further, Bhattacharya et al. (2002) showed that ABA accumulation was maximum at heart stage of tea somatic embryo. This analysis suggested that ABA application earlier or later than heart stage was ineffective. Free ABA was converted into the inactive or bound form during late maturation in zygotic embryos of tea (Bhattacharya et al. 2002).

Mondal et al. (2002) observed adverse effects of ABA application at the both globular and torpedo stages of somatic embryo development. Thus, while no, or negligible, germination was observed in the globular and torpedo stages of somatic embryos, more than 50% of the heart stage somatic embryos developed normally, germinated, and finally converted into healthy plantlets after ABA treatment. Therefore they concluded that while tea somatic embryos failed to accumulate storage reserves during early stage of embryo development, ABA at later stages might help in triggering the process of reserve accumulation, thereby leading to high germination. This finding was also in accordance with studies of other researchers examining different plant species (Tian and Brown 2000).

The role of endogenous IAA, soluble proteins, and RNA in the development of tea seeds was also investigated. Unlike other orthodox seeds, the level of free IAA in tea embryos also remained high even at full maturity. The total RNA content remained high in the stages with high moisture content but declined with progressive decline of moisture content (Bhattacharya et al. 2004). Various polyamines (PA), namely, putrescine (Put), spermidine (Spd), and spermine (Spm), had played an important role in embryogenic processes. Studies on PA pattern and biosynthesis had shown that somatic embryo formation, development, and conversion were affected by free PA levels in *C. japonica* (Pedroso and Pais 1993). Further, it had been suggested that the levels of endogenous PAs present in the embryogenic and non-embryogenic regions were not responsible to determine the response of embryogenesis in *C. japonica* (Pedroso and Pais 1993). It had been found that PA patterns were identical for both embryogenic and non-embryogenic leaf regions until the formation of proembryo (day 20) of the in vitro leaf, differing from each other in soluble and insoluble conjugated Put and soluble conjugated Spd contents upon formation of globular embryos (day 45). The absence of significant differences in the ratios of free Put/free PA between both leaf regions was due to the fact that the non-embryogenic leaf regions were also morphogenically competent (rhizogenic). Endogenous PA–protein binding was highest when mitotic activity was greatest, suggesting a positive relation between PA binding proteins and mitotic activity. In fact, conjugated PA in the embryogenic leaf regions kept on increasing during the entire period of culture, while in the non-embryogenic leaf regions, conjugated PA did not change significantly after 20d of culture (Pedroso and Pais 1994). The slightly higher free Put/free PA ratio recorded for embryogenic leaf regions in *Camellia* at 20th and 45th d, compared with the non-embryogenic ones, might be associated with the embryonic development occurring in these leaf regions. Apart from this, starch also played a critical role in embryo development in *Camellia* (Pedroso and Pais 1994). Primary embryo cortical cells underlying globular secondary embryos contained abundant starch, but these deposits declined as the secondary embryo developed. The starch deposits of the secondary embryo itself were also declined as it developed and were found only around the apical meristems in the cotyledonary stage.

A strongly starch-positive zone was always detected during the formation of the nodules (7–9 d), and the PAS-staining proportion of these areas was about four times that of the equivalent region in the initial cotyledon. Starch deposits declined in the nodule parenchyma adjacent to preembryogenic cells, reaching a minimum upon the development of somatic embryos, when the parenchyma had few starch grains and thinner cell walls than the initial cotyledon tissue. On the other hand, in nodules producing no embryos, very few starch grains were observed (Pedroso and Pais 1994).

4.11 Histological and Ultrastructural Changes During Embryogenesis

4.11.1 Direct Somatic Embryogenesis

Histological study of direct somatic embryogenesis from leaf tissue enabled the establishment of a time sequence of physiological and ultrastructural changes occurring during the induction of embryogenesis in the *C. japonica* (Pedroso-Ubach 1994). It had been noticed that Ca^{++} concentration played a major role during embryogenesis as its concentration varied during embryogenesis (Pedroso and Pais 1994). Cell wall became thick in embryogenic cell followed by accumulation of starch and calcium oxalate crystals in the surrounding parenchyma cells, up to 15 d after induction (until initial divisions for proembryo formation). However, both starch and calcium oxalate crystals disappeared after in the onset of somatic embryogenesis. The changes in Ca^{++} level recorded during induction of somatic embryogenesis in *C. japonica* could be related to changes in the levels of linked and precipitated Ca^{++}. An increase in free calcium and calmodulin after induction of direct embryogenesis was observed which became maximum at globular stages but decreased thereafter (Roberts et al. 1992). Histochemical tests, including enzymatic digestion with PGS-lipase (a cutinase), showed the presence of cutin and callus on the embryogenic parenchyma cells of cell walls after 3 to 15 d (Pedroso-Ubach 1994).

4.11.2 Secondary Embryogenesis

Barring the increased sizes, primary embryos did not undergo any changes till the 12th day, when the newly formed secondary embryos became visible. Secondary embryos occurred mainly in cluster on the hypocotyl of the mother (primary) embryos, without any intermediate callus formation. After that, secondary embryogenesis passed through typical globular, heart, and torpedo stages before reaching the cotyledonary stage. After 6 to 7 weeks of culture, 70% of the primary embryos produced secondary embryos at an average rate of 10 per productive embryo. Secondary embryos over 7 mm long were used for a further embryogenic cycle or transferred to germination medium to develop into plantlets. Smaller secondary embryos had much less embryogenic capacity and tend to suffer necrosis in both embryogenic and germination media. Histological monitoring revealed that secondary embryos apparently had a multicellular origin from embryogenic areas originating in both epidermal and subepidermal layers of the hypocotyl region. Microcomputer image analysis was applied for quantifying cytological events associated with somatic embryogenesis. This method showed an increasing gradient in the nucleus-to-cell area ratio from differentiated cells passing through preembryogenic cells to embryogenic cells. The formation of embryogenic areas was preceded by accumulation of starch in the surrounding cortical cells. The cells underlying globular secondary embryos still contained abundant starch, but it declined as the secondary embryos developed (Plata et al. 1991).

4.12 Electron Probe X-Ray Microanalysis: A Tool for Early Diagnosis of Embryogenesis

A protocol to study the early embryogenic competence of *C. japonica* leaf was developed by electron probe X-ray microanalysis. In vitro-derived leaves isolated from 12-week-old shoots were immersed in IBA (1 g/L) for 20 min followed by incubation of 11 d in darkness on diluted MS basal medium (Pedroso and Pais 1993). These leaves upon culturing produced different regions with varying response such as embryogenesis, rooting, callogenesis, and non-morphogenic response. The results suggested the existence of difference in morphogenic competence in different regions of leaf. Plantlet regeneration was successfully achieved from all the culture systems developed (Pedroso and Pais 1993, 1995a, b). Later Pedroso and Pais (1994) demonstrated that significant fluctuations of Ca, C, O, K, Na, P, Fe, S, and Mg levels which were detected in the cells from embryogenic and non-embryogenic region of the same leaf were analyzed with X-ray spectra. Thus a positive correlation was established between the content of element in the tissue region and in vitro response such as embryogenesis, rhizogenesis, and callogenesis. Therefore, in *C. japonica*, electron probe X-ray microanalysis can be used for early detection of embryogenic competence or its absence.

4.13 Alternative In Vitro Techniques

Apart from micropropagation and somatic embryogenesis, several other in vitro approaches were standardized which have been summarized in Table 4.2 and discussed here.

4.13.1 Storage of In Vitro Culture

The potential of using in vitro systems for germplasm collection and conservation as well as for multiplication had been broadly discussed in several reviews (Kartha 1985; Engelmann 1997). A summary of different in vitro works on this aspect is tabulated in Table 4.3. The application of in vitro techniques to germplasm storage is of particular interest for the conservation of plants such as *Camellia* species that are normally propagated vegetatively due to the presence of recalcitrant seeds. The storage of *Camellia* seeds in genebanks was problematic because *Camellia* seeds were classified as recalcitrant (Pence 1995). They were sensitive to low temperatures as well as desiccation and were unable to retain their viabilities through long-term storage (Kato 1989a). Even when they were maintained under moist conditions at 3–5 °C, their viabilities were relatively short-lived (Salinero and Silva-Pando 1986).

Table 4.2 A brief description of some cell culture techniques used in tea

Name of technique	Objectives	Remarks	Reference
Somaclonal and gametoclonal variation	Development of mutant resistant to different stresses	No commercial success	Rajkumar et al. (2001)
Artificial seed	Storage of propagule	Maximum 60 d storage was possible without loss of germination	Janeiro et al. (1995) and Mondal et al. (2002)
Protoplast culture	Hybrid plant production	Regeneration was not possible	Balasubramanian et al. (2000b)
Anther culture	Haploid plant production	Regeneration was not achieved	Saha and Bhattacharya (1992) and Raina and Iyer (1992)
Suspension culture	Secondary metabolite production	Commercially exploited	Orihara and Furuya (1990) and Matsuura et al. (1991)
Cryopreservation	Long-term storage of propagule	Not exploited further either academic or commercial purposes	Kuranuki and Yoshida (1991)

Table 4.3 Summary of studies on the cold storage and cryopreservation of embryogenic and embryonic axis cultures of *Camellia*

Species	Explant	Storage method	Response/remarks	Reference
C. japonica	Somatic embryo clusters	Short- to medium-term storage at 2–4 °C	Reduced embryogenic competence after 6 months; improved germination capacity after 2 months	Janeiro et al. (1995)
	Encapsulated somatic embryos	Short- to medium-term storage at 2–4 °C	Reduced embryogenic competence after 2 months. 30–40% plant recovery after 2 months	Janeiro et al. (1997)
	Somatic embryos and encapsulated somatic embryos	Cryopreservation	No survival of frozen material	Janeiro et al. (1996)
	Embryonic axes from mature seeds	Cryopreservation after 2–3 h desiccation	100% survival, 40% plant recovery, and 18% somatic embryogenesis rate	Janeiro et al. (1996)
C. reticulata	Somatic embryo clusters	Cold storage at 2–4 °C	76% germination after 2 months	Chaudhury et al. (1991) and Chandel et al. (1995)

The most common method for preserving the genetic resources of species with recalcitrant seeds or those vegetatively propagated is as plants under field genebanks which is popularly known as ex situ conservation. Limited works have been carried out on cold storage and cryopreservation of *Camellia* with both material obtained ex vitro (seeds) and material cultured in vitro (somatic embryos, embryonic axes, and shoot apices).

4.13.2 Low-Temperature and Short-Term Storage

The potential use of low-temperature storage is numerous for in vitro conservation of germplasm and their exchange, hand-pollinated hybrids with reduced seed fertility, and genetically modified plants with sterile or unstable genotypes. However, efficacy of storage depends upon efficiency with which synthetic seeds can be stored at low temperature without reducing multiplication rate or germination efficiency. In tea, the first successful plant regeneration from syn-seed was established by Mondal et al. (2000). They investigated the effect of low-temperature/cold storage (4 °C) on germination efficiency of encapsulated somatic embryos up to 75 d derived from cotyledons of cultivar, Kangra Jat. Pale yellow- to white-colored somatic embryos (0.2–0.5 mm diameter) that developed directly on the surface of cotyledons were taken for syn-seed production. Among the different alginate concentrations, 3% and 4% alginate were found to be equally good for formation of clear, transparent, round-shaped beads, whereas increase or decrease in the alginate concentration resulted in abnormal beads. A maximum germination efficiency of 59% was achieved with naked somatic embryos kept in germination media in culture room at 25 ± 2 °C. However, germination rate of unencapsulated somatic embryos reduced more rapidly from 59% to 9.7% with increasing periods of storage time. Upon transfer to MS medium supplemented with GA_3 (1.5 mg/L), the beads cracked and bud sprouted within 8 weeks. Similar observation was made earlier in *C. japonica* somatic embryos (Janeiro et al. 1995). It was postulated that this is perhaps due to the naked embryo just excised from the embryo clusters which was unable to recover from "excision shock" while detaching from mother tissue coupled with less capability to resist drying than the embryo clusters. On the other hand, germination efficiency of unstored encapsulated tea somatic embryos was 34.3%, which was reduced to 7.7% up to 60 d of storage. No germination was registered in either unencapsulated or encapsulated somatic embryo stored for 75 d. The lower germination rate of encapsulated somatic embryos may be related to both oxygen deficiency and rapid drying in the gel bead (Redenbaugh et al. 1991). The reduction of germination efficiency recovered in alfalfa (Redenbaugh et al. 1987), *Asparagus cooperi*, (Ghosh and Sen 1994), and *Eucalyptus citriodora* (Muralidharan and Mascarenhas 1995) from cold storage was documented earlier. However, considerable amount of secondary somatic embryogenic response was observed irrespective of the duration of cold treatment. Therefore, it was concluded that tea somatic embryos were susceptible to cold. The germinated somatic embryos were grown into plants, successfully hardened, and transferred to the field.

It is well-documented that cold-treated somatic embryos of certain woody perennials were known to promote both normal maturation and germination capacity (Tulecke 1987). While 2 months of cold-treated tea somatic embryos at 4 °C improved germination, the plantlet conversion ability of secondary embryos increased significantly (up to 100%) as compared to control embryogenic lines, which accounted for 32–65% germination rate. The incidence of secondary somatic embryogenesis during germination also decreased after the cold treatment. While cold storage reduced the competence for secondary somatic embryogenesis in *Camellia*, it increased their capacity for conversion to plantlets (Mondal et al. 2000) which looks initially yellow sig green and later turned to green (Muslihatin et al. 2018). In order to evaluate the effect of cold storage on bud sprouting response, encapsulated nodal explants were stored under both at 25 ± 2 °C (tissue culture room condition) and at 4 °C for varying sets of 15 d interval up to 60 d. Storage of artificial seeds at 4 °C did not show any adverse effects on shoot proliferation efficiency; rather the time taken for initial sprouting was reduced by 15 d as compared to the ones that were stored at 25 °C. The bud sprouting efficiency (90.2–92%) of both non-encapsulated and encapsulated nodal explants was maintained up to 45 d but declined to 45.9% and 50%, respectively, when stored up to 60 d (Mondal et al. 2000). Therefore, it was concluded that while somatic embryos were unsuitable for low-temperature storage, nodal explants not only were suitable but also sprouted buds with cold treatment earlier than the untreated one in tea.

In an another study, it had been found that syn-seeds of tea stored up to 30 d under room temperature germinated but their viability declined significantly thereafter. Half-strength MS medium was found to be better over full-strength MS medium on germination, while during growth and development of the germination, the trend was reversed. Among the PGRs, BAP influenced germination rate. Physical state of the nutrient medium showed a significant variation between liquid and solid state. Germination rate of syn-seeds was significantly higher in acidic pH, while near-neutral pH enhanced the growth rate of germinated synthetic seeds (Mariya John and RajKumar 2006).

The effects of short- to medium-term cold storage on the maintenance of embryogenic capacity and germination of somatic embryos of *Camellia* were investigated (Janeiro et al. 1995). Four embryogenic lines were used: three belonging to *C. japonica* (1, 2, and SY-89) and one to *C. reticulata* cv. Mouchang. Lines 1 and 2 of *C. japonica* which were used to study survival and the preservation of embryogenic capacity of somatic embryos induced directly on the roots of in vitro-grown plantlets (Vieitez et al. 1991). The effect of cold storage on the germination of *Camellia* somatic embryos into whole plants was studied in greater detail (Janeiro et al. 1995). Cold treatment for 8 weeks significantly improved the secondary embryogenesis, but improvement varied on genotypes. Depending on the genotypes, the shoot and root length of the germinated plantlets were also significantly increased by 2 months of cold treatment. However, the incidence of secondary embryogenesis during germination also decreased after cold treatment.

Janeiro et al. (1995) found that 2 months of storage of somatic embryo clusters at 4 °C significantly enhanced germination and plant production, whereas germination of individually cold-stored naked embryos was reduced (Janeiro et al. 1996). This apparent discrepancy might be due to individual embryos responded relatively poorly because having been excised from embryogenic clusters just before cold storage, they had not yet recovered from the stress of excision. Alternatively, individual embryos may be more susceptible to damage by drying than embryo clusters.

The potential uses for artificial seeds are numerous including storage, handling, and delivery of elite germplasm. The possibility of using cold storage to preserve synthetic *Camellia* seeds was investigated (Janeiro et al. 1996). In that study, the effects of cold storage of *C. japonica* somatic embryos on the maintenance of embryogenic competence and germination of encapsulated embryos were determined. Somatic embryos were encased in sodium alginate (3%) beads made in MS basal medium with 3% sucrose. The beads were then stored for 1–2 months in darkness at 2–4 °C. After 1 month, the encapsulated embryos exhibited a significant reduction in both survival rate and competence for secondary embryogenesis; however, further, 60 d of cold storage had little reduction effect. The survival and secondary embryogenesis rates were 68% and 69%, respectively, when placed in the maintenance medium following 60 d of storage at 4 °C which were still acceptable. However, the productivity (number of secondary embryos per responsive encapsulated embryo) was dramatically reduced from 62.6% for unstored encapsulated embryos (control) to 5.4% secondary embryos indicating the negative influence of cold on secondary embryo formation. The reduced competence for secondary embryogenesis of cold-stored encapsulated embryos of *Camellia* appears to reflect increased maturity, since their capacity for germination is better preserved than their embryogenic competence. In this respect, short- or medium-term cold storage of syn-seeds destined for germination appears to be feasible as long as a 30–50% fall in plant recovery rate could be tolerable. In contrast, cold storage alone could not be used to maintain embryogenic competence, since the productivity of cold-stored encapsulated embryos was seriously reduced.

Ballester et al. (1997) reported almost 100% survival frequencies in seven of the eight clonal shoot cultures of *C. japonica* tested, when stored at 2–4 °C for up to 12 months. Shoot tips of *C. japonica* encapsulated in alginate beads and stored at 2–4 °C survived for a shorter period of time than unencapsulated ones. Encapsulated material had survival rates of 75, 50, and 10% after 30, 60, and 75 d of storage at 4 °C, respectively.

4.13.3 Cryopreservation

It has been seen that tissues may be stored in liquid nitrogen at −196 °C for long duration. This would be of great value in the conservation of woody plants for which vegetative propagation is of prime importance such as tea. The preservation of embryonic axes in liquid nitrogen was attempted in *C. sinensis* (Chaudhaury et al.

1990; Chandel et al. 2005). While the intact seeds could not be cryopreserved due to their large size as well as high moisture content, the excised embryonic axes could be preserved successfully by desiccating them up to 13% moisture (Chaudhury et al. 1991). The cryopreserved embryonic axes could be revived and germinated by placing them simply on a moisten filter paper for 20–25 d followed by culturing on nutrient medium leading to healthy seedlings of 5–6 cm height. Kuranuki and Yoshida (1991) reported that the excised embryonic axes were able to tolerate not only the cold but also desiccation.

The feasibility of using cryopreservation to maintain the embryogenic competence was investigated in *Camellia* species (Janeiro 1996; Janeiro et al. 1996). In these studies, somatic embryos of *C. japonica* (2–5 mm) were subjected to several protective pretreatments to prevent the formation of ice crystals inside the cells. Following pretreatment, half of the somatic embryos in each experiment (controls) were placed directly in maintenance medium (MS medium supplemented with 1 g/L BAP and 0.1 g/L IBA), and the other half was placed in sterile 2 mL polypropylene cryovials and immersed in liquid nitrogen for 24 h. After 10 weeks, they found that no frozen somatic embryos survived regardless of the desiccation period; however, survival rate of unfrozen embryos was 100% after 15 min of desiccation treatment, which was further reduced to 53% after 2 h of desiccation indicating an acceptable tolerance of these somatic embryos to dehydration (Janeiro 1996). Janeiro et al. (1996) also investigated the feasibility of cryopreservation of *C. japonica* embryonic axes. The explants were isolated from mature seeds. After sterilization of seeds, the embryonic axes were excised from the cotyledons with 1–2 mm of petiole to protect the plumule and were either used as such or dehydrated for 1.5 or 3 h in sterile laminar air flow. Half the material was placed directly in MS maintenance medium (controls), and the other half was placed in cryovials and frozen in liquid nitrogen for 24 h before being transferred to the same MS medium. They found that the capacity of *Camellia* embryogenic axes to produce somatic embryos, especially on the hypocotyl region, was maintained and even enhanced after the stress produced by cryo-exposure. For preserving the elite cell lines with higher content of metabolite, the cryopreservation of suspension culture cells by vitrification in tea has been reported. It was found that 4 d preculture on 60% PVS2 (DMSO, ethylene glycol, and glycerol) in freezing bath for 20 min was best. Interestingly, cryopreservation of suspension culture cells warmed at 40 °C in water bath had the additional benefits (Liu et al. 2009).

4.14 Organogenesis

Adventitious shoot regeneration via callus phase from in vitro leaf explants of tea has been standardized. Callus and rhizogenesis were obtained on MS medium supplemented with varied concentrations of 2,4-D (2.5, 5.0, 7.5, and 10.0 mg/L). Adventitious shoot buds developed indirectly on leaf explants after prolonged culture for 16 weeks on MS medium supplemented with 2,4-D (10 mg/L). GC analysis of the medium as well as the tissues at different stages of development concluded that critical level of 2,4-D in the tissue was responsible for morphogenesis. Shoot buds developed on rhizogenic calli, only when 2,4-D declined to undetectable or negligible concentrations in the tissue probably due to detoxification and metabolism. Alternatively, shoot buds could also be evoked when rhizogenic calli were transferred to medium supplemented with low concentration of 2,4-D (1.5 mg/L). The adventitious nature of the shoots was confirmed through histological studies (Sandal et al. 2005).

4.15 Caulogenesis

In order to produce callus, in vitro-derived leaves of Assam tea were cultured on modified MS medium containing 2,4-D (0.1–1.0 mg/L) and BAP (0.1–1.0 mg/L) or Kn (0.1–1.0 mg/L). Cultures were incubated for 6 weeks at 30 ± 2 °C and 16 h photoperiod at light intensity of 2000 Lux. It was found that the cultured leaves produced compact callus. The amount of callus produced varied from 45% to 65%, depending on the concentrations of 2,4-D, BAP, and Kn in the medium. The highest percentage (65%) of explants producing callus was obtained on modified MS medium supplemented with 2,4-D (1 mg/L) in combination with Kn (0.1 mg/L) within 6 weeks of treatment (Tripetch et al. 2009).

The effect of ABA and excess BAP on the formation of callus was investigated. Callus was formed and grew well when explants were cultured on MS basal medium supplemented with thiamine–HCl (1.25 mg/L); pyridoxine–HCl (0.625 mg/L); nicotinic acid (0.625 mg/L); IAA (30 mg/L); NAA (30 mg/L); Kn (0.1 mg/L); myoinositol (100 mg/L); as well as 3% (w/v) sucrose. After 2 months, the calli were transferred to a modified B_5 medium in order to regenerate plant. As a result, rhizogenesis occurred in the transferred calli to B_5 medium after 1 month. Subsequently, the calli were transferred to the aforesaid B_5 medium supplemented with ABA (2 mg/L) and excess BAP (400 mg/L) to form shoot. The calli turned green and showed differentiation of globular and heart-shaped embryos when transferred to the modified B_5 medium, without formation of shoot. These findings showed that the applied concentration of ABA may cause inhibition of conversion of globular and heart-shaped embryos to shoot. The increased level of BAP, however, was not able to ameliorate the effect of ABA (Ghanati and Ishka 2009).

Rana et al. (2016) did an elaborate study for callus induction. They tried five explants (namely, leaf, stem, cotyledon, hypocotyl, and root) of tea (cv. Nong Kangzao), a Chinese genotype. Initially mature seed was germinated on 2,4-D (2 mg/L) in combination with BA (1 mg/L), and then different explants were excised from the plantlets. Among these, cotyledon-derived callus was best to produce callus which was found to be suitable for long-term proliferation on the B_5 medium supplemented with 2,4-D (0.25 mg/L) and kinetin (0.1 mg/L).

4.16 Protoplast Culture

Trans-generic gene transfer through protoplast and production of somatic hybrids or cybrids is of prime importance to tea. However, the success of achieving a good yield of protoplasts depends upon the source material, which is mostly a "cell suspension culture." Although Bagratishvili et al. (1979) were first to establish tea suspension cultures, Nakamura (1983) successfully isolated the protoplasts with a yield of 3.1×10^5/g of fresh weight from the suspension culture of tea leaves and flower petals. This was further improved by Kuboi et al. (1991) who attained a yield of 87% viable protoplasts with a yield of 3×10^7/g of fresh weight from young leaves of cultivar 'Yabukita'. The problems of endogenous bacterial contamination and polyphenol oxidation during protoplast preparation from in vitro leaves were solved with the addition of PVP in the isolating media (Purakayastha and Das 1994). Additionally, Purakayastha and Das (1994) reported that macerozyme including cellulase along with other factors such as enzyme concentration and osmotica and their concentrations was important for increasing yield. However, further development of micro-calli could not be achieved. Effects of different synthetic auxins were tried to induce the callus from tea stem tissue, and it was found that picloram (241 mg/L) alone induced most callus formation (Frisch and Camper 1987). Balasubramanian et al. (2000b) used different explants such as leaves, cotyledons, and embryogenic and morphogenic calli for culture and isolation of protoplasts and reported in vitro leaves to be the best explants for a good yield of viable protoplasts provided fructose was used as an osmoticum. Despite the fact that success has always eluded protoplast culture and somatic hybridization, these techniques have tremendous potential in the improvement of tea. Many of the wild relatives of tea have agronomically important biotic and abiotic stress-tolerant trait, which can be incorporated into the cultivated variety of tea through somatic hybridization or cybridization. Somatic hybridization also has potential for the production of caffeine-free tea through the fusion of protoplasts of caffeine-free but aroma-rich *C. luetacense* or *C. irrawadiensis* with that of cultivated tea. Thus protoplast culture had tremendous potential for varietal improvement of tea (Sriyadi 1987).

Cell suspension culture of tea was also used as a system to study the insecticide metabolism. Initially callus was developed from the leaf, and then friable callus was made on MS along with 2,4-D (1 mg/L) and kinetin (0.1 mg/L) which was kept in liquid media with constant shaking. Later six different insecticides (5 μg/mL each thiamethoxam, imidacloprid, acetamiprid, imidaclothiz, dimethoate, and omethoate) were added and incubated with this media for certain period of times. After that HPLC and GC–MS were used to quantify the insecticide (Jiao et al. 2019). Simultaneously using the similar approach, metabolism of thiamethoxam was also studied in tea suspension culture. Further the metabolomics change due to thiamethoxam was also studied using HPLC and GCMS. This study helped to understand the interaction mechanisms of pesticides with plant cells (Ge et al. 2019).

4.17 Anther Culture

Microspore culture presents a number of potential advantages mainly in relation to in vitro selection strategies and to genetic studies for developing doubled-haploid mapping population, etc. Tea being highly heterozygous and heterogeneous, production of homozygous diploid plants is of great importance in tea improvement (Mukhopadhyay and Mondal 2016). The pioneer attempts of anther culture of tea were made by Katsuo (1969) and by Okano and Fuchinone (1970) who could produce roots from anther-derived callus. However, the first successful plants from tea anthers of cultivar Fuyun No-7 were produced by Chen and Liao (1982). Though they worked on nine different tea cultivars, plants were produced only from 'Fuyun-7' on N6 medium supplemented with Kn (2 mg/L), 2,4-D (0.5 mg/L), L-glutamine (800 mg/L), and serine (100 mg/L), followed by subculturing on N6 medium supplemented with zeatin (2 mg/L), adenine (20 mg/L), and lactoalbumin hydrolysate (10 mg/L). On this medium, the calli continued to proliferate into shiny masses or shoots. These shoots were subsequently rooted on medium containing IAA (0.1 mg/L). While three out of the four plants were haploids, the rest were aneuploids with a chromosome number 2n = 18. Later, Saha and Bhattacharya (1992) used NAA (0.1 mg/L), 2,4-D (0.1 mg/L), Kn (0.1 mg/L), sucrose (7%), and L-glutamine (400 mg/L) which produced up to globular structures from protoplast-derived callus but failed to differentiate further. However, the differentiation of true pollen embryos and regeneration of haploid plants were described by Raina and Iyer (1992) and Shimokado et al. (1986). A systematic study was made at the Tea Experimental Station, Assam, India, on haploid production of tea plants using anthers of TV-1 and TV-13 cultivars. Uninucleate stage of the pollen was found to be the best for induction of haploid calli. They established a correlation between developmental stage of pollens and color of the anthers for selecting appropriate stage of pollen for optimum response, though regeneration from the calli was not reported but cytologically confirmed to be haploid ($n = 15$) (Laskar et al. 1993). Pedroso and Pais (1994) tested 17 different media combination based on MS and N6 with various concentrations of carbon source, growth regulators, and amino acids such as serine as well as glutamine for *C. japonica*. The embryogenic calli were achieved when microspores were cultured on 2,4-D (1 mg/L) and Kn (0.1 mg/L) and subsequently on MS supplemented with BAP (0.5 mg/L). However, further growth was ceased at maturation stage. In an attempt to regenerate haploid plants from anthers of five different Sri Lankan tea clones, Seran et al. (1999) reported that the highest response (98%) in terms of micro-calli formation was achieved in 1/2 MS medium supplemented with 2,4-D and BAP in the presence of light with the cultivar TRI-2043. Other cultivars which produced callus were TRI-2023, TRI-2024, TRI-2025, and TRI-777 in a descending order. Determination of ploidy levels in the callus cells showed that the frequency of haploid cells was greater (68%) than that of diploid cells (6%). However, plantlets could not be regenerated. Thus it seems that although several attempts have been made to regenerate haploid plant in tea, success remains up to mere development of micro-calli.

In *C. japonica*, embryogenesis was induced from microspore. Among the various media composition and PGR formulation, MS along with 2,4-D (10 mg/L) and Kn (0.1 mg/L) was reported to be the best. The development of microspore-derived proembryos was obtained in MS medium supplemented with BAP (0.5 mg/L) and reached the highest level when the microspores were cultured in this medium. However, the development of microspore-derived embryos ceased at maturation stage (Pedroso and Pais 1994), and no further work on this area had been reported further. Recently, Hazarika and Chaturvedi (2013) induced haploid calli with significant antioxidant activity from unpollinated ovary of tea. A high cytokinin/auxin ratio, provided by BAP and 2,4-D, and high-temperature treatment (33 °C) for 10 d in the dark promoted maximum callus induction. Callus was maintained on MS medium containing BAP (5 mg/L) and IBA (2 mg/L) in the light at 25 °C. Well-developed tracheids were formed within 4 weeks in callus subcultured on MS medium containing TDZ (0.4 mg/L) and 2,3,5-triiodobenzoic acid (2.5 mg/L). Flow cytometric analysis revealed that most of the cells were haploid.

4.18 Secondary Metabolite Production

Tea is valued for pleasant taste and aroma, which is due to the presence of alkaloids, caffeine, and other methyl xanthenes such as theobromine and theophylline. These alkaloids have also been used as therapeutic and drugs which have been discussed in earlier section. The exploration of production of secondary metabolites in tea is very important so much so that the work started in way back 1821, when caffeine (1,3,7-trimethylxanthine) was first prepared in pure form from tea leaves (Spedding and Wilson 1964). Much later, Ogutuga and Northcote (1970) also produced caffeine from callus tissue of tea. Catechins too were produced in the cell cultures of tea as high as 30% (Hao et al. 1994). The formation of polyphenols in intact tea plants as well as in callus tissue was first reported by Forrest (1969) who found that the synthesis of simplest catechins and leucoanthocyanin was highly dependent on the original explants and inversely co-related with the growth rate of cultured cells. Koretskaya and Zaprometov (1975) also reported that polyphenol content depends on the addition of two kinds of precursor, namely, shikimic and quinic acids. However, a higher level of gallic acid inhibited the formation of phenolic compounds. Plant growth regulators such as NAA are found to be better than 2,4-D to increase total soluble phenolics, flavans, as well as the phenolic polymer lignins in callus culture (Zaprometov and Zagoskina 1979). Application of kinetin to the auxin-containing medium enhanced the synthesis of total soluble phenolics and flavans (Zaprometov and Zagoskina 1987). Bagratishvili et al. (1979) reported that NAA stimulated both cell growth and polyphenol production. ABA inhibited cell growth as well as the formation of all forms of soluble phenolic compounds (Zaprometov and Zagoskina 1987). The best carbon source for catechins and pro-anthocyanin production was reported to be 5% glucose. Orihara and Furuya (1990) reported that tea cells cultured on B_2K medium (MS basal with 2 mg/L IBA and 0.1 mg/L Kn) produced L-glutamine as main free amino acid and that addition of ethylamine to the medium increased theanine production. Matsuura et al. (1991) reported that a theanine accumulation level was remarkably high in the stem-derived callus when ethylamine–HCl was added to solidified MS along with BAP (4 mg/L). The effect of main inorganic constituents in medium on theanine accumulation was also investigated. Biotransformations of primary amines by *C. sinensis* cell culture were performed, and their γ-glutamyl derivatives (products C_{1-8}) were obtained in the yield. Their molecular weight showed that primary amines were combined with glutamic acid, and the characteristic fragment peaks suggested that primary amines were connected with γ-carboxylic acid in glutamic acid not to carboxylic acid (Furuya et al. 1990). Takemoto and Tanaka (2001) developed a method for producing (*S*)-(+)-α-phenyl-2-pyridylmethanol in 83% chemical yield with 86% optical yield by the repetitive use of immobilized tea cell culture. The tea cell culture showed similar capability for asymmetric bioreduction to that of *Catharanthus roseus* cell culture. In vitro cell culture of tea with high peroxidase activity for producing various chemically active compounds was also developed in tea (Takemoto et al. 2002a, b).

Callus and root suspensions from tea had been established to produce and accumulate caffeine and theobromine as secondary metabolites. Leaf fragments from a mature greenhouse tea plant, grown on MS medium supplemented with IAA (1 mg/L), produced roots, while leaf explants of the same plant, cultured on MS medium supplemented with 2,4-D (1 mg/L) and BAP (0.1 mg/L), gave rise to the formation of friable callus. Both callus and roots, when transferred to MS liquid medium supplemented with 2,4-D (1 mg/L) and BAP (0.1 mg/L), produced caffeine and theobromine, which were detected by TLC, UV, and GC (Shervington et al. 1998). The effect of UV-B radiation on the accumulation and tissue localization of phenolic compounds in two strains of callus cultures of tea plant was investigated. The strains differed in their morphological and physiological characteristics and biosynthetic capacity. UV-B radiation hampered culture growth, decreased the size of callus-forming cells, and promoted the accumulation of soluble and, to a lesser extent, polymeric forms of phenolic compounds, such as lignin. This accumulation was accompanied by an increase in the phenolic compound deposition in cell walls and intercellular space and by deposition of a lignin-like material on the surface of callus cultures. The strain characterized by an increased formation of phenolic compounds was more resistant to UV-B radiation as compared to that with lower phenolic productivity (Zagoskina et al. 2003).

The effects of cadmium on the growth of tea callus cultures derived from leaves, stems, and roots and on the formation, in these cultures, of phenolic compounds, including flavans and lignin, which are characteristic of the tea plant, were investigated. In the calli derived from leaves and stems, cadmium treatment decreased the biomass increment, while in the calli derived from roots, growth characteristics remained at the control level. Under the effect of cadmium, the content of phenolic compounds, including flavans, in the leaf calli decreased, while in the stem and root calli, it either increased or was close to a control one. The lignin content in the root and stem calli increased, but it did not change in the leaf calli. All this data demonstrated that the cadmium-induced changes in phenolic metabolism of the tea plant callus culture depended both on the cadmium concentration in the medium and on the origin of calli (Zagoskina et al. 2007).

To determine whether caffeine biosynthesis is controlled by the availability of purine precursors and/or methyl donors, Deng et al. (2008) examined the effect of some purine compounds on purine alkaloid accumulation, using tea callus cultures. No stimulation of caffeine biosynthesis was observed when the calli were cultured with adenosine, guanosine or hypoxanthine for 3 weeks. However, paraxanthine (0.1 mg/L) doubled the caffeine level relative to controls. Adenosine stimulated the growth of callus and reduced the caffeine concentration 3 months after treatment. These results indicated that methylation of xanthosine by 7-methylxanthosine synthase was the most plausible rate-limiting step of caffeine biosynthesis.

Shikimic acid can act as an inducer to produce secondary metabolites under in vitro condition. It had been found that shikimic acid (3.5 g/L) increased the synthesis of polyphenols, catechins, caffeine, and other secondary components in the suspension culture of tea (Muthaiya et al. 2013).

4.19 Embryo Rescue

Natural hybridization among the different species of *Camellia* is not very common due to cross incompatibility. Therefore embryo rescue is very important for *Camellia* for developing new hybrids. Breeders from all over the world have desired to develop yellow-flowered *Camellia* (Mukhopadhyay et al. 2013). The discovery of yellow-colored *C. chrysantha* generated great excitement among the *Camellia* growers and breeders as a potential source for a new range of *Camellia* floral colors. Although numerous interspecific hybridizations had been attempted, crossing of *C. chrysantha* with some other species was very difficult. In this regard, several cultivars of *C. japonica* that contributed to about 70% of the current horticultural needs and *C. chrysantha* with its potential for new color were thought to be especially important. However, probably due to the high phylogenetic distance between the two species, the interspecific hybridization was extremely difficult (Yoshikawa and Yoshikawa 1990). Hwang et al. (1992), therefore, did a systematic investigation to understand the nature of reproductive barrier between and *C. japonica* and *C. chrysantha* with intra- and interspecific crosses using two different lines of each species. They found that zygote formation and early embryo development were similar in intra- and interspecific crosses. Full size but empty ovules in mature capsules resulted from embryo abortion. Liang et al. (1986) reported that interspecific hybrid embryos of *C. pitardii* var. yunnanensis *x* *C. chrysantha* developed normally, reached torpedo stage, and differentiated normally. However, a complete successful protocol of embryo rescue will be immensely helpful to develop the long-awaited yellow-colored *Camellia* using *C. chrysantha* as a trait donor source.

4.20 Conclusion

Somatic embryogenesis is an inherent character of tea cotyledon, so much so that, it is being produced in vivo, without any culture media. There are excellent reports on induction, maturation, germination, and long-term multiplication including bioreactor approach of tea and other *Camellia* species. The advantage of somatic embryogenesis over conventional micropropagation is the presence of taproots in somatic embryo-derived seedlings, which help them better to combat drought. Significant amount of works have also been done related to the development of primary embryogenesis and secondary embryogenesis and their origin, morphology, as well as histochemical and biochemical mechanisms so much so that embryogenic region within a leaf has been identified. Cold storage and their recovery have also been studied in detail for both tea and *Camellia*. It is particularly important as tea seeds are recalcitrant in nature. However, efforts should be directed towards the commercialization of somatic embryogenesis, bioreactor technology, and field performance of somatic embryo-derived plants. Additionally regeneration from protoplast, anther-derived tissue for haploid plant production though important in tea breeding, yet little success have been achieved.

References

Abraham GC, Raman K (1986) Somatic embryogenesis in tissue culture of immature cotyledons of tea (*Camellia sinensis*). In: Somers DA, Gengenbach BG, Biesboor DD, Hackett WP, Green CE (eds) International congress of plant tissue and cell culture. University of Minnesota, Minneapolis, MN, p 294

Akula A, Akula C (1999) Somatic embryogenesis in tea (*Camellia sinensis* (L) O Kuntze). In: Jain SM, Gupta PK, Newton RJ (eds) Somatic embryogenesis in woody plants, vol 5, pp 239–259

Akula A, Dodd WA (1998) Direct somatic embryogenesis in a selected tea clone, TRI-2025 (*Camellia sinensis* (L).O. Kuntze) from nodal segment. Plant Cell Rep 17:804–809

Akula A, Akula C, Bateson M (2000) Betaine a novel candidate for rapid induction of somatic embryogenesis in tea (*Camellia sinensis* (L.) O. Kuntze). Plant Growth Regul 30:241–246

Ammirato PV (1985) Patterns of development in culture. In: Henke RR, Hughes KW, Constantin MP, Hollaender A (eds) Tissue culture in forestry and agriculture. Plenum, New York, pp 9–29

Aoshima Y (2005) Efficient embryogenesis in the callus of tea (*Camellia sinensis*) enhanced by the osmotic stress or antibiotics treatment. Plant Biotech 22:277–280

Arulpragasam PV, Latiff R, Seneviratne P (1988) Studies on the tissue culture of tea (*Camellia sinensis* (L.) O. Kuntze). 3. Regeneration of plants from cotyledon callus. Sri Lanka J Tea Sci 57:20–23

Avgioglu A, Knox RB (1989) Storage lipid accumulation by zygotic and somatic embryo in culture. Ann Bot 63:409–412

Bag N, Palni LMS, Nandi SK (1997) Mass propagation of tea using tissue culture methods. Physiol Mol Biol Plants 3:99–103

Bagratishvili DG, Zaprometov MN, Butenko RG (1979) Obtaining a cell suspension culture from the tea plant. Fiziol Rast 26:449–451

Balasubramanian S, Marimuthu S, Rajkumar R, Haridas V (2000a) Somatic embryogenesis and multiple shoot induction in *Camellia sinensis* (L.) O. Kuntze. J Plant Crop 28:44–49

Balasubramanian S, Marimuthu S, Rajkumar R, Balasaravanan T (2000b) Isolation, culture and fusion of protoplast in tea. In: Muraleedharan N, Rajkumar R (eds) Recent advance in plants crops research. Allied Publishers, New Delhi, pp 3–9

Ballester A, Janeiro LV, Vieitez AM (1997) Cold storage of shoot cultures and alginate encapsulation of shoot tips of *Camellia japonica* and *Camellia reticulata* Lindley. Sci Hortic 71:67–78

Bano Z, Rajaratnam S, Mohanty BD (1991) Somatic embryogenesis in cotyledon culture of tea (*Thea sinensis* L.). J Hort Sci 66:465–470

Barciela J, Vieitez AM (1993) Anatomical sequence and morphometric analysis during somatic embryogenesis on cultured cotyledon explants of *Camellia japonica* L. Ann Bot 71:395–404

Barciela J, Vieitz AM (1993) Anatomical sequence and morphometric analysis during somatic embryogenesis on cultured cotyledon explants of *Camellia japonica* L. Ann Bot 71:395–404

Bennett WY, Scheibert P (1982) *In vitro* generation of callus and plantlets from cotyledons of *Camellia japonica*. Camellia J 37:12–15

Beretta D, Vanoli M, Eccher T (1987) The influence of glucose, vitamins and IBA on rooting of *Camellia* shoots *in vitro*. In: Abstract from the symposium on vegetative propagation of woody species, Pisa, Italy, p 105

Bhatia CR, Murty GSS, Mouli C, Kale DM (1986) Nuclear techniques and *in vitro* culture for plant improvement. In: Proceedings on the International Symposium on nuclear techniques and *in vitro* culture for plant improvement. IAEE, FAO and UN, Vienna, 19–23 August, pp 419–427

Bhattacharya A, Nagar PK, Ahuja PS (2002) Seed development in *Camellia sinensis* (L.) O. Kuntze. Seed Sci Res 12:39–46

Bhattacharya A, Nagar PK, Ahuja PS (2004) Changes in endogenous indole-3-acetic acid and some biochemical parameters during seed development in *Camellia sinensis* (L.) O. Kuntze. Acta Physiol Plant 26:399–404

Chandel KPS, Chaudhury R, Radhamani J, Malik SK (1995) Desiccation and freezing sensitivity in recalcitrant seeds of tea, cocoa and jackfruit. Ann Bot 76:443–450

Chandel KPS, Chaudhury R, Radhamani J, Malik SK (2005) Desiccation and freezing sensitivity in recalcitrant seeds of tea, cocoa and jackfruit. Ann Bot 76:443–450

Chaudhaury R, Lakhanpal S, Chandel KPS (1990) Germination and desiccation tolerance of tea (*Camellia sinensis* (L.) O. Kuntze) seeds and feasibility of cryopreservation. Sri Lanka J Tea Sci 59:89–94

Chaudhury R, Radhamani J, Chandel KPS (1991) Preliminary observation in the cryopreservation of desiccated embryonic axes of tea (*Camellia sinensis* L.O.Kuntze) seeds for genetic conservation. Cryo Lett 12:31–36

Chen Z, Liao H (1982) Obtaining plantlet through anther culture of tea plants. Zhongguo Chaye 4:6–7

Chen Z, Liao H (1983) A success in bringing out tea plants from the anthers. China Tea 5:6–7

Das SC (1992) Non-conventional techniques of regenerating polyploids in tea. In: Proc. 31st Tocklai Conf., TRA, Jorhat, India, pp 26–30

Das SC, Barman TS (1988) Current state and future potential of tissue culture in tea. In: Proc. 30th Tocklai Conf. TRA Jorhat, India, pp 90–94

Deka A, Deka PC, Mondal TK (2006) Tea. In: Parthasarathy VA, Chattopadhyay PK, Bose TK (eds) Plantation crops-I. Naya Udyog, Calcutta, (ISBN no. 81-85971-97-8), pp 1–148

Deng WW, Li YY, Ogita S, Ashihara H (2008) Fine control of caffeine biosynthesis in tissue cultures of *Camellia sinensis*. Phytochem Lett 1:195–198

Engelmann F (1997) *In vitro* conservation research activities at the international plant genetic Resources institute (IPGRI). Plant Tiss Cult Biotech 3:46–52

Evans DA, Sharp WR (1986) Somaclonal and gametoclonal variation. In: Evans DA, Sharp WR, Ammirato PV (eds) Hand book of plant cell culture. Technique and applications, vol 4. Macmillan, New York, pp 97–132

Forrest GI (1969) Studies on the polyphenol metabolism of tissue culture derived from the tea plant (*C. sinensis* L.). Biochem J 113:765–772

Frisch CH, Camper ND (1987) Effect of synthetic auxins on callus induction from tea stem tissue. Plant Cell Tiss Org Cult 8:207–213

Furuya T, Orihara T, Tsuda Y (1990) Caffeine and theanine from cultured cells of *Camellia sinensis*. Phytochemistry 29:2539–2547

Gamborg O, Miller R, Ojima K (1968) Nutrient requirements of suspension cultures of soyabean root cells. Exp Cell Res 50:157–158

Ge G, Jiao W, Cui C, Liao G, Sun J, Hou R (2019) Thiamethoxam metabolism and metabolic effects in cell suspension culture of tea (*Camellia sinensis* L.). J Agric Food Chem 67:7538–7546

Ghanati F, Ishka MR (2009) Investigation of the interaction between abscisic acid (ABA) and excess benzyladenine (BA) on the formation of shoot in tissue culture of tea (*Camellia sinensis* L.). Int J Plant Prod 3:7–14

Ghosh B, Sen S (1994) Plant regeneration from alginate encapsulated somatic embryos of *Asparagus cooperi* baker. Plant Cell Rep 13:381–385

Hao C, Wang Y, Yang S (1994) Effects of macroelements on the growth of tea callus and the accumulation of catechins. J Tea Sci (China) 14:31–36

Haridas V, Balasaravanan T, Rajkumar R, Marimuthu S (2000) Factor influencing somatic embryogenesis in *Camellia sinensis* (L.) O. Kuntze. In: Muraleedharan N, Raj Kumar R (eds) Recent advances in plantation crops research. Allied Publishers, New Delhi, pp 31–35

Hazarika RR, Chaturvedi R (2013) Establishment of dedifferentiated callus of haploid origin from unfertilized ovaries of tea (*Camellia sinensis (L.)*O. Kuntze) as a potential source of total phenolics and antioxidant activity. In Vitro Cell Dev Biol Plant 49:60–69

Hu H (1983) Genetic stability and variability of pollen derived plants. In: Sen SK, Giles KL (eds) Plant cell culture in crop improvement. Plenum, New York, pp 145–157

Hua LD, Dai ZD, Hui X (1999) Studies on somatic embryo and adventitious bud differentiation rate among different tissues of *Camellia sinensis* L. Acta Agron Sin 25:291–295

Hwang YJ, Okubo H, Fujieda K (1992) Pollen tube growth, fertilizer and embryo development of *Camellia japonica* L. x *C. chrysantha* (Hu) Tuyama. J Jap Soc Hort Sci 60:955–961

Ikekawa N (1991) In: Patterson GW, Nes WD (eds) Physiology and biochemistry of sterols. American Oil Chemists Society, Champaign, pp 347–360

Jain SM, Newton RJ (1990) Prospects of biotechnology for tea improvement. Proc Indian Natl Sci Acad 6:441–448

Janeiro LV (1996) Almacenamiento en frio de especies lenosas propagadas in vitro. Aplicación de las tecnologías de semilla artifical y criopreservación en el género *Camellia*, Doctoral Thesis, University of Santiago de Compostela, Santiago de Compostela

Janeiro LV, Ballester A, Vieitez AM (1995) Effect of cold storage on somatic embryogenesis systems of *Camellia*. J Hort Sci 70:665–667

Janeiro LV, Ballestter A, Vieitez AM (1996) Cryopreservation of somatic embryos and embryonic axes of *Camellia japonica* L. Plant Cell Rep 15:699–703

Janeiro LV, Ballester A, Vieitez AM (1997) *In vitro* response of encapsulated somatic embryo of *Camellia*. Plant Cell Tiss Org Cult 51:119–125

Jha TB, Jha S, Sen SK (1992) Somatic embryogenesis from immature cotyledon of an elite Darjeeling tea clone. Plant Sci 84:209–213

Jiao W, Ge G, Hua R, Sun J, Li Y, Hou R (2019) Study on the metabolism of six systemic insecticides in a newly established cell suspension culture derived from tea (*Camellia Sinensis* L.) leaves. J Vis Exp 148:e59312. https://doi.org/10.3791/59312

Kartha KK (1985) Meristem culture and germplasm preservation. In: Kartha KK (ed) Cryopreservation of plant cells and organs. CRC, Boca Raton, pp 115–134

Kato M (1982) Results of organ culture on *Camellia japonica* and *C. sinensis*. Jpn J Breed 32:267–277

Kato M (1986a) Micropropagation through cotyledon culture in *Camellia japonica* L. and *Camellia sinensis* L. Jap J Breed 36:31–38

Kato M (1986b) Micropropagation through cotyledon culture in *Camellia sasanqua*. Jap J Breed 36:82–83

Kato M (1989a) *Camellia sinensis* L. (tea): *In vitro* regeneration. In: Bajaj YSP (ed) Biotechnology in agriculture and forestry. Medicinal and aromatic plants II, vol 7. Springer, Berlin, Heidelberg, pp 83–98

Kato M (1989b) Polyploids of *Camellia* through culture of somatic embryos. Hort Sci 24:1023–1025

Kato K (1996) Somatic embryogenesis from immature leaves of *in vitro* grown tea shoots. Plant Cell Rep 15:920–923

Katsuo K (1969) Anther culture in tea plant (a preliminary report). Study Tea 4:31

Koretskaya TF, Zaprometov MN (1975) Phenolic compounds in the tissue culture of *Camellia sinensis* and effect of light on their formation. Fiziol Rast 22:941–946

Kuboi T, Suda M, Konishi S (1991) Preparation of protoplasts from the leaves. In: Proceedings of the International Symposium on Tea Science, Shizuoka, Japan, pp 427–431

Kuranuki Y, Yoshida S (1991) Cryopreservation of tea seeds and excised embryonic axes in liquid nitrogen. In: Proceedings of the International Symposium on Tea Science, Shizuoka, Japan, pp 419–420

Laskar MA, Das SC, Deka PC (1993) Tissue culture of tea: anther culture for haploid plant production. In: Proceedings of the International Symposium on Tea Science and Human Health, Tea Research Association, Calcutta, India, 11–14 January

Lecouteux CG, Lai FM, Mckersie BD (1993) Maturation of alfalfa (*Medicago sativa* L.) somatic embryos by abscisic acid, sucrose and chilling stress. Plant Sci 94:207–213

Liang H, Zhang Z, Zhang X (1986) Investigation of the sexual process in interspecific crosses between *Camellia pitardii* var. *yunnanensis* and *C chrysantha*. Acta Bot Yunnan 8:147–152

Liu YJ, Gao LP, Xia T, Gao KJ (2009) Study on cryopreservation of suspension culture cells by vitrification in *Camellia Sinensis*. J Tea Sci 29:120–126

Llyod G, McCown B (1980) Commercially feasible micropropagation of mountain laurel, *Kalmia latifolia* by use of shoot tip culture. Comb Proc Int Plan Prop Soc 30:421–427

Lü J, Chen R, Zhang M, da Silva JA, Ma G (2013) Plant regeneration via somatic embryogenesis and shoot organogenesis from immature cotyledons of Camellia nitidissima Chi. J Plant Physiol 170:1202–1211

Mariya John KM, RajKumar R (2006) Factors influencing synthetic seed germination in tea (*Camellia sinensis* (L.) O. Kuntze). J Plant Crops 34:40–42

Matsuura T, Kakuda T, Kinoshita T, Takeuchi N, Sasaki K (1991) Production of theanine by callus culture of tea. In: Proceedings of the International Symposium on Tea Science, Shizuoka, Japan, pp 432–436

Maugh TH (1981) The natural occurring brassionoide in the plant species. Science 212:33–34

Mondal TK (2002) Camellia biotechnology: a bibliographic search. Int J Tea Sci 1:28–37

Mondal TK (2007) "Tea". In: biotechnology in agricultural and forestry. In: Devey MR, Pua P (eds) Transgenic crops V, vol 60. Springer, Berlin. (ISBN no. 13978-3-540-49160-6), pp 519–536

Mondal TK (2008) Tea. In: Kole C, Hall TC (eds) A compendium of transgenic crop plants: plantation crops, ornamentals and turf grasses. Blackwell, Oxford. (ISBN no. 978-1-405-16924-0), pp 99–112

Mondal TK (2009) Tea. In: Prydarsini M, Jain SM (eds) Breeding plantation tree crops tropical species. Springer, Berlin. (ISBN: 978-0-387-71199-7), pp 545–587

Mondal TK (2011) Camellia. In: Kole C (ed) Wild crop relatives: genomics and breeding resources planation and ornamental crops. Springer, Berlin. (ISBN no. 978-3-642-21200-0), pp 15–40

Mondal TK, Bhattachraya A, Sood A, Ahuja PS (1999) An efficient protocol for somatic embryogenesis and its use in developing transgenic tea (*Camellia sinensis* (L) O. Kuntze) for field transfer. In: Altman A, Ziv M, Izhar S (eds) Plant biotechnology and *in vitro* biology in 21st century. Kluwer Academic, Dordrecht, pp 101–104

Mondal TK, Bhattacharya A, Sood A, Ahuja PS (2000) Factor effecting induction and storage of encapsulated tea (*Camellia sinensis* L (O) Kuntze) somatic embryos. Tea 21:92–100

Mondal TK, Bhattacharya A, Ahuja PS (2001a) Induction of synchronous secondary embryogenesis of tea (*Camellia sinensis*). J Plant Physiol 158:945–951

Mondal TK, Bhattacharya A, Ahuja PS, Chand PK (2001b) Factor effecting *Agrobacterium tumefaciens* mediated transformation of tea (*Camellia sinensis* (L). O. Kuntze). Plant Cell Rep 20:712–720

Mondal TK, Bhattachray A, Sharma M, Ahuja PS (2001c) Induction of *in vivo* somatic embryogenesis in tea (*Camellia sinensis*) cotyledons. Curr Sci 81:101–104

Mondal TK, Bhattachrya A, Sood A, Ahuja PS (2002) Factors affecting germination and conversion frequency of somatic embryos of tea. J Plant Physiol 59:1317–1321

Mondal TK, Bhattacharya A, Laxmikumaran M, Ahuja PS (2004) Recent advance in tea biotechnology. Plant Cell Tiss Org Cult 75:795–856

Mukhopadhyay M, Mondal TK (2016) Biotechnology of tea. In: Bag N, Bag A, Palni LMS (eds) Tea: technological initiatives: some initiatives. NIPA, New Delhi, ISBN: 978-93-85163-37, pp 301–328

Mukhopadhyay M, Sarkar B, Mondal TK (2013) Omics advances in tea (*Camellia sinensis*). In: Bhar D (ed) Omics applications in crop science. CRC, Taylor and Francis Group. (ISBN:978-1-4665-8582), Boca Raton, pp 347–366

Muralidharan EM, Mascarenhas AF (1995) Somatic embryogenesis in *Eucalyptus*. In: Jain SM, Gupta PK, Newton R (eds) Somatic embryogenesis in woody plants. Angiosperms, vol 2. Kluwer Academic, Dordrecht, pp 23–40

Murashige T, Skoog F (1962) A revised medium for rapid growth and bioassay with tobacco tissue cultures. Physiol Plant 15:473–497

Muslihatin W, Jadid N, Saputro TB, Purwani KI, Himayani CES, Calandry AW (2018) Characteristic of synthetic seeds from two medicinal plants (*Moringa oleifera* and *Camellia sinensis*). J Phys Conf Ser 1040:012005

Muthaiya MJ, Nagella P, Thiruvengadam M, Mandal AKA (2013) Enhancement of the productivity of tea (*Camellia sinensis*) secondary metabolites in cell suspension cultures using pathway inducers. J Crop Sci Biotech 16:143–149

Nadamitsu S, Andoh Y, Kondo K, Segawa M (1986) Interspecific hybrids between *Camellia vietnamensis* and *C. chrysantha* by cotyledon culture. Jap J Breed 36:309–313

Nagar PK, Sood S (2006) Changes in endogenous auxins during winter dormancy in tea (*Camellia sinensis* L.) O. Kuntze. Acta Physiol Plant 28:165–169

Nakamura Y (1983) Isolation of protoplasts from tea plant. Tea Res J 58:36–37

Nakamura Y (1985) Effect of origin of explants on differentiation of root and its varietal difference in tissue culture of tea plant. Tea Res J 62:1–8

Nakamura Y (1988a) Efficient differentiation of adventitious embryos from cotyledon culture of *Camellia sinensis* and other *Camellia* species. Tea Res J 67:1–12

Nakamura Y (1988b) Effects of the kinds of auxins on callus induction and root differentiation from stem segment culture of *Camellia sinensis* (L.) O. Kuntze. Tea Res J 68:1–7

Nguyen VT (2012) Regeneration plantlets from somatic embryos of tea plant (*Camellia sinensis* L.). J Agric Tech 8:1821–1827

Nitsch JP, Nitsch C (1969) Haploid plants from pollen grains. Science 163:185

Ogutuga DBA, Northcote DH (1970) Caffeine formation in tea callus tissue. J Exp Bot 21:258–273

Okano N, Fuchinone Y (1970) Production of haploid plants by anther culture of tea *in vitro*. Jap J Breed 20:63–64

Orihara Y, Furuya T (1990) Production of theanine and other γ-glutamine derivatives by *Camellia sinensis* cultured cells. Plant Cell Rep 9:1215–1224

Paratasilpin T (1990) Comparative studies on somatic embryogenesis in *Camellia sinensis* var. *sinensis* and *C. sinensis* var. *assamica* (mast.). Pierre J Sci Soc Thailand 16:23–41

Park YG, AHN IS, Bozhkov P (1977) Effect of exogenous plant growth regulator on morphogenetic response *in vitro* by embryo and leaf cultures of *Camellia sinensis (L.)* O. Kuntze. Korean J Plant Tiss Cult 24:129–135

Pedroso MC, Pais MS (1993) Direct embryo formation in leaves of *C. japonica* L. Plant Cell Rep 12:639–643

Pedroso MC, Pais MS (1994) Induction of microspore embryogenesis in *Camellia japonica* cv. Elegans. Plant Cell Tiss Org Cult 37:129–136

Pedroso MC, Pais MS (1995a) Factor controlling somatic embryogenesis: cell wall changes as an in vivo marker of embryogenic competence. Plant Cell Tiss Org Cult 43:147–154

Pedroso MC, Pais MS (1995b) Plant regeneration from embryogenic suspension cultures of *Camellia japonica*. In Vitro Cell Dev Biol Plant 31:31–35

Pedroso-Ubach MC (1991) Contribuicao para a preservacao e o melhoramento de *Camellia japonica* L. Master's Thesis (English abstract). Faculdade de Ciencias da Universidade de Lisboa, Lisboa, Portugal, pp 23–50

Pedroso-Ubach MC (1994) Somatic embryogenesis in *Camellia japonica* L. A search for markers. PhD Thesis, Faculdade de Ciencias da Universidade de Lisboa, Lisbon, Portugal

Pence VC (1995) Cryopreservation of recalcitrant seeds. In: Bajaj YSP (ed) Biotechnology in agriculture and forestry, Cryopreservation of plant germplasm I, vol 32. Springer, Berlin, pp 29–50

Plata E (1993) Morphogenesis *in vitro* de *Camellia reticulata*: Proceson de embryogenesis somatica regeneration de plants. Doctoral Thesis, University of Santiago de Compostela, Santiago de Compostela

Plata E, Vieitez AM (1990) *In vitro* regeneration of *Camellia reticulata* by somatic embryogenesis. J Hort Sci 65:707–714

Plata E, Ballester A, Vieitez AM (1991) An anatomical study of secondary embryogenesis in *Camellia reticulata*. In Vitro Cell Dev Biol Plant 27:183–189

Ponsamuel J, Samson NP, Sathyprakash V, Abraham GC (1996) Somatic embryogenesis and plant regeneration from the immature cotyledonary tissues of cultivated tea (*Camellia sinensis* (L.) O. Kuntze). Plant Cell Rep 16:210–214

Preeti S, Pandey S, Bhattacharya A, Nagar PK, Ahuja PS (2004) ABA associated biochemical changes during somatic embryo development in *Camellia sinensis* (L.) O *Kuntze*. Plant Physiol 161:1269–1276

Purakayastha A, Das SC (1994) Isolation of tea protoplast and their culture. In: Proceedings of 32nd Tocklai conference, Tocklai, Jorhat, Assam, 24–26 February, p 34

Raina SK, Iyer RD (1992) Multicell pollen proembryoid and callus formation in tea. J Plant Crop 9:100–104

RajKumar R, Ayyappan P (1992) Somatic embryogenesis from cotyledonary explants of *Camellia sinensis* (L.) O. Kuntze. Plant Chron May:227–229

Rajkumar R, Balasusaravanam S, Jayakumar D, Haridas V, Marimuthu S (2001) Physiological and biochemical feathers of field grown somaclonal variants of tea. UPASI Tea Res Found Bull 54:73–81

Rana MM, Ali M, Wei S (2016) Effects of explant type and plant growth regulators on callogenesis in seedling derived tea explants. Tea J Bangladesh 45:66–77

Redenbaugh K, Slade D, Viss P, Fujii JA (1987) Encapsulation of somatic embryos in synthetic seed coats. Hort Sci 22:803–809

Redenbaugh K, Fujii JA, Slade D (1991) Synthetic seed technology. In: Vasil KI (ed) Scale-up and automation in plant propagation. Cell culture and somatic cell genetics of plants, vol 8. Academic, New York, pp 35–74

Roberts DR, Sutton BCS, Flinn BS (1990) Synchronous and high frequency germination of interior spruce somatic embryo following partial drying at high relative humidity. Can J Bot 68:1086–1093

Roberts PM, Weaver CD, Oh SH (1992) Intracellular receptor proteins for calcium signals in plants. In: Gresshoff PM (ed) Plant biotechnology and development: current topics in plant molecular biology. CRC Press Inc., Florida, pp 129–113

Saha SK, Bhattacharya NM (1992) Stimulating effect of elevated temperatures on callus production of meristemmoids from pollen culture of tea (*Camellia sinensis (L.) O.* Kuntze). Indian J Exp Biol 30:83–86

Salinero MC, Silva-Pando FJ (1986) In: Peon C (ed) La multiplication delas camellias. Diputacion provincial de Pontevedra La Camellia, Pontevedra, Spain, pp 175–184

Sandal I, Kumar A, Bhattacharya A, Sharma M, Shanker A, Ahuja PS (2005) Gradual depletion of 2,4-D in the culture medium for indirect shoot regeneration from leaf explants of *Camellia sinensis* (L.) O. Kuntze. Plant Growth Regul 47:121–127

San-Jose MC, Vieitez AM (1993) Regeneration of *Camellia* plantlets from leaf explant cultures by embryogenesis and caulogenesis. Sci Hortic 54:303–315

Sarathchandra TM, Upali PD, Wijeweardena RGA (1988) Studies on the tissue culture of tea (*Camellia sinensis* (L.) O. Kuntze) 4. Somatic embryogenesis in stem and leaf callus cultures. Sri Lanka J Tea Sci 52:50–54

Seran TH, Hirimburegama K, Hirimburegama WK, Shanmugarajha V (1999) Callus formation in anther culture of tea clones, *Camellia sinensis* (L) O Kuntze. J Nat Sci Found Sri Lanka 27:165–175

Sharma P, Pandey S, Bhattacharya A, Nagar PK, Ahuja PS (2004) ABA associated biochemical changes during somatic embryo development in *Camellia sinensis (L.) O. Kuntze*. J Plant Physiol 161:1269–1276

Shervington A, Shervington LA, Afifi F, El-omari MA (1998) Caffeine and theobromine formation by tissue cultures of *Camellia sinensis*. Phytochemistry 47:1535–1536

Shimokado TT, Murata, Miyaji Y (1986) Formation of embryoid by anther culture of tea. Jap J Breed 36:282–283

Sood A, Palni LMS, Sharma M, Rao DV, Chand G, Jain NK (1993) Micropropagation of tea using cotyledon culture and encapsulated somatic embryos. J Plant Crop 21:295–300

Spedding DJ, Wilson AT (1964) Caffeine metabolism. Nature 204:73

Sriyadi B (1987) Tissue culture a method to overcome hand-pollinated difficulties in tea plant. Warta Balai Peneli The Dan Kina 13:105–111

Suganthi M, Arvinth S, Raj Kumar R (2012) Impact of osmotica and abscisic acid on direct somatic embryogenesis in tea. Int J Plant Res 2:22–27

Tahardi JS, Raisawati T, Riyadi I, Dodd WA (2000) Direct somatic embryogenesis and plant regeneration in tea by temporary liquid immersion. Menara Perkebunan 68:1–9

Tahardi JS, Riyadi I, Dodd WA (2003) Enhancement of somatic embryo development and plant recovery in *Camellia sinensis* by temporary immersion system. J Bioteck Perta 8:1–7

Takemoto M, Tanaka K (2001) Synthesis of optically active α-phenylpyridylmethanols by *Camellia sinensis* cell culture. J Mol Catal B: Enzym 15:173–176

Takemoto M, Aoshima Y, Stoynov N, Kutney JP (2002a) Establishment of *Camellia sinensis* cell culture with high peroxidase activity and oxidative coupling reaction of dibenzylbutanolides. Tetrahedron Lett 43:6915–6917

Takemoto M, Suzuki Y, Tanaka K (2002b) Highly selective oxidative coupling of 2-naphthol derivatives catalyzed by *Camellia sinensis* cell culture. Tetrahedron Lett 43:8499–8501

Thomas J, Kumar RR, Mandal AKA (2006) Metabolite profiling and characterization of somaclonal variants in tea (*Camellia* spp.) for identifying productive and quality accession. Phytochemistry 67:1136–1142

Tian LN, Brown DCW (2000) Improvement of soybean somatic embryo development and maturation by abscisic acid treatment. Can J Plant Sci 80:721–276

Tremblay L, Trembly FM (1995) Maturation of black spruce somatic embryos: sucrose hydrolysis and resulting osmotic pressure of the medium. Plant Cell Tiss Org Cult 42:39–46

Tripetch P, Kerdchoechuen O, Laohakunjit N, Pitchayangkura R (2009) Induction of callus in Assam tea leaves (*Camellia sinensis var. Assamica*). Agric Sci J 40:209–212

Tulecke W (1987) Somatic embryogenesis in woody perennials. In: Bonga JM, Durzan DJ (eds) Cell and tissue culture in forestry, vol 2. Martinus Nijhoff, Dordrecht, pp 61–69

Vieitez AM (1994) Somatic embryogenesis in *Camellia* spp. In: Jain S, Gupta P, Newton R (eds) Somatic embryogenesis in woody plants. Kluwer Academic, Dordrecht, pp 235–276

Vieitez AM, Barciela J (1990) Somatic embryogenesis and plant regeneration from embryonic tissues of *Camellia japonica* L. Plant Cell Tiss Org Cult 21:267–274

Vieitez AM, San-Jose C, Vieitez J, Ballester A (1991) Somatic embryogenesis from roots of *Camellia japonica* plantlets cultured *in vitro*. J Am Soc Hortic Sci 116:753–757

Wachira F, Ogado J (1995) *In vitro* regeneration of *Camellia sinensis* (L.) O. Kuntze by somatic embryo. Plant Cell Rep 14:463–466

Webster FB, Roberts DR, McInnis SM, Sutton BCS (1990) Propagation of interior spruce by somatic embryogenesis. Can J For Res 20:1757–1762

Wu CT, Huang TK, Chen GR, Chen SY (1981) A review on the tissue culture of tea plants and on the utilization of callus derived plantlets. In: Rao AN (ed) Proceedings COSTED symposium on tissue culture of economically important plants, Singapore, pp 104–106

Yamaguchi S, Kunitake T, Hisatomi S (1987) Interspecific hybrid between *Camellia japonica* cv. Choclidori and *C. chrysantha* produced by embryo culture. Jap J Breed 37:203–206

Yan MQ, Ping C (1983) Studies on development of embryoids from the culture cotyledons of *Thea sinensis* L. Sci Silv Sin 19:25–29

Yan MQ, Ping C, Wei M, Wang YH (1984) Tissue culture and transplanting of *Camellia oleifera*. Sci Silv Sin 20:341–350

Yoshikawa K, Yoshikawa N (1990) Inter-specific hybridization of *Camellia*. Bull Seibu Maizuru. Bot Inst 5:56–75

Zagoskina NV, Dubravina GA, Alyavina AK, Goncharuk EA (2003) Effect of ultraviolet (UV-B) radiation on the formation and localization of phenolic compounds in tea plant callus cultures. Russ J Plant Physiol 50:270–275

Zagoskina NV, Goncharuk EA, Alyavina AK (2007) Effect of cadmium on the phenolic compounds formation in the callus cultures derived from various organs of the tea plant. Russ J Plant Physiol 54:237–243

Zaprometov MN, Zagoskina MV (1979) One more evidence for chloroplast involvement in the biosynthesis of phenolic compounds. Plant Physiol (Russian) 34:165–172

Zaprometov MN, Zagoskina MV (1987) Regulation of phenolic compounds formation in cultured cells of tea plant (*Camellia sinensis*). In: Proceedings of International Tea Quality Human Health Symposium China, pp 62–65

Zhuang C, Liang H (1985a) *In vitro* embryoid formation of *Camellia reticulata* L. Acta Biol Exp Sin 18:275–281

Zhuang C, Liang H (1985b) Somatic embryogenesis and plantlet formation in cotyledon culture of *Camellia chrysantha*. Acta Bot Yunnan 7:446–450

Zhuang C, Duan J, Zhou J (1988) Somatic embryogenesis and plantlets regeneration of *Camellia sasanqua*. Acta Bot Yunnan 10:241–244

Genetic Transformation

5

5.1 Introduction

Due to the non-availability of new land for expansion of cultivation along with the objectives of reducing the cost of cultivation, tea breeders need to develop the improved tea cultivars as per the requirements of the industry (Mukhopadhyay et al. 2016). On the other hand, natural variability of important traits is not available in tea. In this context, transgenic technology offers several advantages for perennial plants like tea. They are: (1) any origin of gene can be introduced, (2) greater speed for developing transgenic material, (3) free of epistasis interaction for transgene expression and (4) being targeted in nature, it is free from linkage drag (Bhatacharya et al. 2004; Mondal 2007, 2008). Therefore, gene transfer through biotechnological means appear to be not only a time effective but also an advantageous better alternative. A summary of research work on transgenic tea is listed in Table 5.1 and discussed below.

5.2 *Agrobacterium tumefaciens*

The first protocol for production of transgenic tea, cv. Kangra jat, was developed (Fig. 5.1) via *Agrobacterium*-mediated genetic transformation (Mondal et al. 1999, 2001a). Two disarmed *A. tumefaciens* strains, EHA 105 and LBA 4404, both carrying the binary plasmid p35SGUS-INT with the *nptII* (neomycin phosphotransferase II) gene and *gus* (*β*-glucuronidase) intron were evaluated as vector systems. Several parameters were evaluated to maximize the transformation efficiency. While preculture, wounding, and acetosyringone (AS) treatment in different concentrations were inhibitory, the bacterial growth phase (OD 600 = 0.6), cell density (10^9/mL), co-cultivation period (5 d), and pH of the co-cultivation medium (5.6) had positive effects on transformation efficiency. AS is a low-molecular-weight phenolic compound naturally released by wounded plant cells and acts as an inducer of the virulence genes. Following co-cultivation, globular

T. K. Mondal, *Tea: Genome and Genetics*,
https://doi.org/10.1007/978-981-15-8868-6_5

Table 5.1 Summary of transgenic tea research

Techniques	Remarks	Trans-gene	References
A. tumefaciens	Antibiotic selection for *Camellia* species was reported	–	Tosca et al. (1996)
A. rhizogenes	First attempt to induce hairy root formation	*rolB*	Zehra et al. (1996)
A. rhizogenes	Transformation protocol was exploited to promote roots to facilitate hardening of the micropropagated tea plants	*rolB*	Konwar et al. (1998)
A. tumefaciens	Preliminary study for gene transfer to tea plants	*gus-intron*	Matsumoto and Fukui (1998)
A. tumefaciens	Standardization of somatic embryogenesis and transient expression of *gus* gene	*npt-II*	Mondal et al. (1999)
Particle bombardment	First attempt for standardization of the biolistic-mediated transformation protocol	*npt-II*	Akula and Akula (1999)
A. tumefaciens	Transgenic calli were produced utilizing the phenolic inducer AS at an effective range (20–100 mg/L)	*npt-II*	Matsumoto and Fukui (1999)
A. tumefaciens	Detailed study on *Bt* gene transformation was reported	*Bt*	Luo and Liang (2000)
A. tumefaciens	Standardization of transformation	*npt-II*	Aoshima et al. (2001)
A. tumefaciens	Development of selection system for putative transformants	*npt-II*	Mondal et al. (2001a)
A. tumefaciens	Production of transgenic plants from transformed somatic embryos	*npt-II* and *gus-intron*	Mondal et al. (2001b)
A. tumefaciens and particle bombardment	Attempt was made for standardization of the protocol	*npt-II*	Wu et al. (2003)
A. tumefaciens	Tea leaves with glabrous surface having lower phenol and wax content were identified to be more suitable for infection	*npt-II*	Kumar et al. (2004)
A.tumefaciens and particle bombardment	Green florescence protein gene was transferred with organelle target signals	*gfp*	Kato et al. (2004)
A. tumefaciens	Attempts to overcome the bactericidal effect of tea leaf polyphenol during transformation	*npt-II*	Sandal et al. (2007)
A. tumefaciens	Silencing of glutathione synthetase gene in callus	*gs*	Mohanpuria et al. (2008)
A. rhizogenes	High production of catechin through hairy root culture	*rolb*	John et al. (2009)
A. tumefaciens	Caffeine-free tea plant production	*Cs*	Mohanpuria et al. (2011)
Particle bombardment	Optimization of the protocol with the introduction of osmotin gene	*osmotin*	Saini et al. (2012)

(continued)

Table 5.1 (continued)

Techniques	Remarks	Trans-gene	References
Particle bombardment	Transgenic line had higher amount of flavan-3-ols and caffeine	*osmotin gene*	Bhattacharya et al. (2013)
A. tumefaciens	Optimization of transformation protocol	*nptII, gusA*	Singh et al. (2014)
A. tumefaciens and *A. rhizogenes*	Catechin present in tea tissue inhibit the transformation efficiency	*gusA, nptII*	Song et al. (2014)
A. rhizogenes	Glutamine (0.1 g/L) and PVP (5 g/L) as co-cultivation and post co-cultivation media improve of hairy root generation efficiency	*vir, gusA, cfp*	Rana et al. (2016)
A. tumeficiens	Optimization of CRISPR/CAS-9 protocol	*Cs*	Tang et al. (2016)
A. tumefaciens	Optimization of the protocol with the *PS1aG-3 vector*	*gusA*	Qianru et al. (2017)
A. rhizogenes	Optimization for high efficiency *in-planta* transformation	*Rola, b, c, d, aux1, orf13a, orf14*	Alagarsamy et al. (2018)

rolB rooting-locus gene B, *gus-intron* beta-glucuronidase, *npt-II* neomycin phosphotransferase II, *gfp* Green florescence protein, *gs* glutathione synthetase, *cs* caffeine synthase

somatic embryos (Mondal et al. 2001c, 2002) were placed on a multiplication medium and stressed with kanamycin (50 mg/L). Further selection occurred in the maturation and germination medium at an elevated kanamycin level (75 mg/L). An average of 40% transient expression was registered based on the GUS histological assay. Then the kanamycin-resistant, GUS-positive embryos were germinated, and the resulting microshoots were multiplied in vitro on MS medium fortified with TDZ (1.1 mg/L) and NAA (2 mg/L). Later, they were micrografted onto seed-grown rootstocks (Mondal et al. 2005) of cv. Kangra jat and eventually hardened in a walk-in poly-house. Integration of the transgenes into the tea nuclear genome was confirmed by PCR analysis using *nptII*- and *gus*-specific primers (Fig. 5.2) and by southern hybridization using an *nptII*-specific probe. However, use of phenolic inducer, AS, did not enhance the efficiency of transformation (Mondal et al. 2005). In several other experiments on woody plants, it had been found that AS could not increase transformation efficiency (Confalonieri et al. 1997). The inability of AS to improve the transformation efficiency could be due to the inherent prevalence of high amounts of phenolics in tea. Contrary to this, Matsumoto and Fukui (1999) found that acetosyringone has a positive role in tea transformation as this plant is considered to be one of the recalcitrant plants for *Agrobacterium* infection. Naturally occurring crown galls are hardly observed on tea plant, and this bacterium is not cited as an economically important pathogen in Japan. Therefore, they concluded that the low efficiency of the *Agrobacterium* infection was overcome by AS. Similar observation was also made by Qianru et al. (2017) who also found that 3 d co-cultivation 150 μmol/L AS increased the efficiency of transformation.

Surface disinfection of mature sinker seeds

↓

Germination on ½ MS + sucrose (30 g/l) +Agar (0.7%)

↓ 2 weeks

Induction of somatic embryos from cotyledon slices on ½ MS+ sucrose (20 g/l) + NAA (2.5 g/l) + BAP (0.2 g/l)

↓ 6-8 weeks

Multiplication of primary embryos on (EMM) MS+ BAP (2 g/l) + IBA (0.2 g/l) + L-glutamine (1 g/l)

↓ 6 weeks

Infection of globular embryos with *Agrobacterium* suspension (O.D. 600=0.6 X 10^9 cells/ml) 20 min

↓

Co-cultivation in EMM medium (pH=5.6) in darkness

↓ 5 d

Washing 2 times with sterile water and 2 times with liquid EMM containing sporidex (400 mg/l) and blot drying

↓

Transfer of embryos to EMM + sporidex (250 mg/l) + carbenicillin (250mg/l)

↓ 2 weeks

Selection on above medium + kanamycin (50 mg/l)

↓ 8-12 weeks

Transfer to MS + maltose (40 g/l) + trans-cinamic acid (3 g/l) + kanamycin (75 mg/l)

↓ 4 weeks

Transfer to MS + 1.5 mg/l GA_3 + kanamycin (75 mg/l)

↓ 7 weeks

Multiplication of transgenic shoots on MS+ TDZ (0.001 mg/l) + NAA (2 mg/l)

↓ 8 weeks

Micrografting on seedling rootstocks and growth in green house

Fig. 5.1 Protocol for *Agrobacterium* mediated genetic transformation of tea

Resistant calli emerged only on the explants treated with a higher concentration of acetosyringone. Thus, they concluded that application of AS (20 mg/L) to a co-culture medium was effective for tea transformation. In tea, AS probably played an important role in T-DNA transfer, as reported in a wide range of plant species (Godwin et al. 1992). Another important factor that affects transformation efficiency is the selection of transgenic tissue on an antibiotic medium. In order to select transformed explants, selective agents are added into the medium to produce selective pressure. In general for tea, among the different antibiotics, hygromycin was more effectively used at a low concentration (20 mg/L) for somatic embryos followed by kanamycin with a range from 50 to 200 mg/L as a selection dose. In a wild species of tea, Tosca et al. (1996) found that 75 mg/L kanamycin was lethal to the tissues. Similarly, Mondal et al. (2001b) reported that 50 mg/L kanamycin followed by an elevated dose of 75 mg/L were the levels to be used for the effective selection of transgenic somatic embryos of tea. On the other hand, Matsumoto and Fukui (1998, 1999) found that 200 mg/L kanamycin was effective for tea when leaves were used as explants. Another prerequisite for genetic transformation is the molecular characterization of putative transformants. While Matsumoto and Fukui

Fig. 5.2 *Agrobacterium*-mediated genetic transformation of tea somatic embryos: (**a**) blue spot indicating *gus* activity after 48 h of infection, (**b**) a *gus*-positive secondary somatic embryo of 4 months old, (**c**) a hand section of leaf tissue of transformed shoot of 12 months old, (**d**) kanamycin selection, (**e**) PCR amplification of a 693 bp fragment of *nptII* gene (lane M: DNA double digested with EcoR1 and HindIII; lane C: DNA from untransformed tea plant (control); lanes 1–5: DNA from independently transformed plants, (**f**) the greenhouse-grown transgenic tea plants. (Adapted from Mondal 2008)

(1998, 1999) reported stable transformations in callus on the basis of PCR analysis and southern hybridization, yet Mondal et al. (2001b) used various tools such as gus staining, kanamycin selection, RT-PCR, and southern hybridization to confirm the stable transformation. This indicated the presence of the marker kanamycin (*nptII*) linked to the *gus* gene as a single "transfer-DNA (T-DNA) strand" in the genomic DNA of the transformed plants. The failure of the rest of the lines in yielding PCR-amplification products may be attributed to the presence of "false positives" during the antibiotics selection. Further, when the leaves of these five independent transgenic plants were subjected to southern blot analysis, PstI-digested genomic DNA from each of four putative transgenic lines generated an internal transgene fragment of 1.6 kb that hybridized to the *nptII* probe. Additional shorter fragments produced in some transgenic lines further indicated a deletion of a part of the T-DNA containing *nptII*. The deletion perhaps occurred during transformation or regeneration. Different banding patterns observed in the southern hybridization could be due to multiple insertions, rearrangement, and/or deletions of the integrated transgenes in the regenerated plants as is common to *A. tumefaciens*-mediated transformation (Mercuri et al. 2000). Interestingly, efficiency of transformation was increased when the leaf explants of tea were precultured on medium containing polyvidone at 16 g/L for 2–3 d before infiltrated by *A. tumefaciens* (Wu et al. 2003). Although 80–90% survival of transgenic plants was achieved under greenhouse conditions, the stability of the transgene remains to be elucidated as tea plants take years to flower and set seeds. The host range specificity between the bacterium and five different tea cultivars as well as an unrelated plant, *Artemisia parviflora*, having extreme surface characteristics was evaluated (Kumar et al. 2004). The degree of *Agrobacterium* infection in the five cultivars of tea was affected by leaf wetness, micro-morphology, and surface chemistry. Wet leaf surfaces of TV-1, UPASI-9, and Kangra jat showed a higher rate (75%) of infection. This indicated that the leaves with a glabrous surface having lower Ø (large surface area covered by water droplet) and higher phenol and wax content were suitable for *Agrobacterium* infection. Caffeine fraction of tea promoted *Agrobacterium* infection even in leaves containing less wax (UPASI-10), whereas caffeine-free wax inhibited both *Agrobacterium* growth and infection. Thus, this study suggested the importance of leaf surface features in influencing the *Agrobacterium* infection in tea leaf explants. Considering the fact that lepidopteran and coleopteran insects account for 31.5% and 18.8% crop loss in tea, respectively, in China, Luo and Liang (2000) made an attempt to produce transgenic tea plants with *Bt* (*Bacillus thuringiensis*) toxin gene. The vector pGA471 containing *Bt* gene was digested with HindIII and BglII and inserted into the vector pCAMBIA2301. The engineered plasmid with *Bt* gene (*cryIAc*), *gus* intron gene, and *nptII* was transformed into *Escherichia coli* and introduced into *Agrobacterium* strains LBA 4404 and EHA 105 through a triparental cross. They detected the transient expression of *gus* gene in calli and leaves of putative transgenic tea plants. However, no transgenic plants were recovered. Three genes including *rolb*, *Bt*, and *chitinase* had been transferred to tea cv. TTL-1 at the Research and Development Department, Tata Tea Ltd., Kerala, India, for developing a tea plant with better yield and resistance to pest as well as to blister bight disease. Preliminary study indicated that there was no difference in quality of made tea in the transgenic plant, but transgenic plant with *rolb* gene produced more root and shoot biomass in 2-year-old young transgenic plants (data not shown) (Mondal et al. 2004). Later, Lopez et al. (2004) also produced transgenic tea plants. Cotyledon-derived embryogenic callus were co-cultivated with *A. tumefaciens* harboring a binary vector carrying the hygromycin phosphotransferase (*hptII*), glucuronidase (*uidA*), and green fluorescent protein (*gfp*) genes. Following co-cultivation, embryogenic calli were cultured in a medium containing carbenicillin (500 mg/L) for a week and then transferred on an antibiotic selection medium containing hygromycin (75 mg/L) for 8–10 weeks. Hygromycin-resistant somatic embryos were selected. Hygromycin-resistant calli were achieved when the same were co-cultivated for 6–7 d in the presence of AS (80 mg/L). Hygromycin-resistant somatic embryos developed into complete plantlets in a regeneration medium containing half-strength MS salts with BAP (1 mg/L) and GA_3 (9 mg/L). Transformants were subjected to *gfp* expression analysis, GUS histochemical assay, PCR analysis, and southern hybridization to confirm transgene integration. The GFP is a stable, cell-autonomous fluorescent protein derived from jellyfish (*Aequorea victoria*). It has been extensively used as a non-destructive reporter gene for plant transformation (Chiu et al. 1996). After 30 d of co-cultivation, *gfp* expression was evident in calli. Non-transformed tissue from original explants did not express green fluorescence, and explant growth was suppressed. Leaves of transformed plantlets also expressed *gfp*, and the green fluorescence was easily detectable with blue light illumination of calli. Production of transgenic tea tissue was also reported by Jayaraman and Nithya (2005). Embryogenic tissues of tea were co-cultivated with *A. tumefaciens* strain LBA4404. The plasmid pBi121, with *nptII* gene providing kanamycin resistance as a selectable marker and the β-glucuronidase (*uidA*) reporter gene, was used as binary vector. The highest transformation frequency was obtained with 5-day-old tissues grown in a liquid medium and co-cultivated with *Agrobacterium* for 2 d in the same medium but containing AS (10 mg/L). There was improvement in the recovery of kanamycin-resistant tissues when tissues were first grown for 10 d on a medium containing timentin (350 mg/L) to prevent bacterial overgrowth, before application of the selection pressure. Resistant tissues obtained after 6 weeks on the kanamycin selection medium showed a stable *uidA* expression. PCR analysis demonstrated the presence of the transgenes, while southern hybridization confirmed their integration into the genome. Following this protocol, they could produce transgenic plants within 4 months after co-culture. As reported for other species (Bergmann and Stomp 1992), the physiological status of the tissue was considered to be an important factor for production of transgenic plants. Several attempts have been made time to time to study the factors that affect the transformation efficiency (Zhang et al. 1994; Zhao et al. 2001). For an example, Singh et al. (2014) reported that the transformation frequency of *A. tumefaciens* (LBA 4404) with the binary vector pCAMBIA 2301 and pCAMBIA 1301 could made 2.5% and 3.3% transformation efficiency, respectively. Catechin is an important component of the tea leaf. To understand the effect of catechin on genetic transformation, Song et al. (2014) used tea leaf extract in the co-cultivation medium that found a 73–36% reduction of expression for the six virulence (*vir*) genes and an absence of transient or stable transformation events. The reduction may be due to the inhibition effect of catechin.

Bhattacharya et al. (2013) reported that transgenic tea with *osmotin* gene significantly improves the growth under in vitro conditions. The transgenic lines not only have higher reactive oxygen species-related enzyme but also contained significantly higher levels of flavan-3-ols and caffeine. Thus they are potential to meet the demands of the tea industry for stress-tolerant plants with higher yield and quality.

5.3 *Agrobacterium rhizogenes*

Several groups also transformed the tea plant with *A. rhizogenes* (Mukhopadhyay and Mondal 2018). Zehra et al. (1996) infected in vitro raised tea leaves with *A. rhizogenes* A_4 strain. A cell density of 10^8/mL for 2 min infection followed by blotting on sterile filter papers and co-cultivation in dark for 2 d induced hairy roots after 35 d. The isolated hairy roots grew healthy in the liquid medium. Mannopine from these roots were analyzed through paper electrophoresis, which confirmed stable integration of the gene. Later, Konwar et al. (1998) also transformed 4- to 6-month-old in vitro-grown tea shoots by infecting at the basal portion followed by co-cultivation in a liquid MS medium supplemented with IBA (5 mg/L) and rifampicin (100 mg/L). Root initiation from the basal portion of 66% explants after 32–45 d of in vitro culture enabled convenient hardening of the shoots in pots or nursery beds.

John et al. (2009) transformed the in vitro-raised leaves of UPASI-9 cultivars with *A. rhizogenes* strain MTCC 532. It had been found that use of AS (60 mg/L) had maximum positive effects which increased the transformation efficiency up to 70%. MS supplemented with maltose (30 mg/L) along with IAA (5 mg/L) was found suitable for hairy root culture and accumulation of phenolic compounds. They also reported the higher accumulation of catechin in transgenic cell lines. However no transgenic plants were produced.

Callus and hairy roots were obtained in vitro from the roots of five tea cultivars. The hairy roots were characterized by rapid growth, absence of geotropism, growth of dense white fibrils, and strong capacity for diversification. Two types of branching of hairy roots were observed: direct branching and branching around the stromae that initially appeared on the hairy roots. These results suggested the significant effect of the composition of the culture medium on the differentiation and multiplication of hairy roots. The level of callus induction significantly varied among the cultivars (Peng et al. 2004). Effect of media supplement for transformation efficiency was studied. Rana et al. (2016) reported that MS basal salts medium supplemented with (30 g/L) sucrose, (0.1 g/L) L-glutamine, and (5 g/L) PVPP as co-cultivation and post co-cultivation media can led to significant improvement of hairy root generation efficiency.

5.4 Biolistic-Mediated Transformation

Although no transgenic tea plants have produced via particle gun, a preliminary study on transient expression was reported by Akula and Akula (1999). Somatic embryos were bombarded with plasmid DNA (p2k7)-coated gold particles (1.5–3 μm diameter). The transformation vector p2k7, which was used in their study, was originally derived from binary vector pBI 221 with *nptII* gene and *gus* gene. Both genes were driven individually by the cauliflower mosaic virus 35S promoter. The gold particles were coated with DNA (1 μg/μL) by precipitation with $CaCl_2$ (111 mg/mL) and spermidine (14.52 mg/mL). Optimization of various factors such as the distance between the site of delivery of the microprojectile and the target tissue, helium pressure, and the state of target tissue to obtain transient expression were evaluated on the basis of GUS assay after 30–40 h of bombardment. Following bombardment, the highest transient expression levels (up to 1085 blue spots/shot) were obtained in the somatic embryos using a helium pressure of 550 kPa with target tissue at a distance of 9.5 cm from the site of delivery of DNA. Mannitol pretreatment did not influence the transient expression as both control and treated cultures gave the same level of expression. However, further details of regeneration of transgenic somatic embryo were not mentioned.

The *gfp* gene with organelle target signals was introduced by particle bombardment or co-cultivation with *A. tumefaciens* using embryogenic callus in tea plant. Putative transformed embryos were obtained from the embryogenic callus grown on the medium with kanamycin. The *gfp* expression was observed by a spectral imaging system to eliminate the yellow-green auto-fluorescence in tea plant and Q-PCR analysis had been done to confirm the *gfp* gene integration into the tea genome and its expression (Kato et al. 2004).

5.5 *In Planta* Transformation

This method of genetic transformation overcome some of the challenges of in vitro-based techniques particularly to avoid the labor intensive disinfection of bacterium in post-infection stages, high cost of anitibiotic selection (e.g., hygromycine), etc. However, this technique is better suitable for the plant having small seed size. Nevertheless in tea a protocol of *in planta* transformation has been standardized with *A. rhizogenes* for the first time (Alagarsamy et al. 2018). Briefly, the hypocotyl region of 2 months old seedlings were punctured using a needle carrying a drop of *Agrobacterium* paste ($OD_{600} = 0.1$) and then injured region was smeared using a bend glass rod with *Agrobacterium* paste which was essential to enhance infection. The infected plants were then kept to a humid chamber with a regular watering with 10% bacterium suspension for 2 weeks. After 2-weeks, the plants were grown growth chamber where they were alternately watered with 1% MS medium and water for 3 months. Later the roots were taken for various transgenic confirmations. Following this protocol, they could achieved 90% transgenic root. However a similar protocol with *A. tumaficiens* is not available.

5.6 Applications

Scientific efforts are there to silence the undesired genes (Mondal et al. 1997). Apart from developing the transgenic tea, the techniques have been also attempted to silence the gene. Glutathione is an important antioxidant compound that restores redox balance inside the living cell which is produced by glutathione synthetase (*gs*). Gene silencing of *gs*, that reduced glutathione content in the somatic embryos of tea, had been reported by Mohanpuria et al. (2008). A 457 bp *gs* fragment had been cloned to RNAi construct which was utilized for tea somatic embryo transformation via *A. tumefaciens*. In order to alleviate the constraints related to transgenic tea development, Mohanpuria et al. (2011) developed a rapid *Agrobacterium*-mediated root transformation system for tea. Later that was used to develop the caffeine free transgenic tea plants. They cloned 376 bp of caffeine synthase cDNA (*cs*) fragment in a RNAi construct (pFGC1008-CS), infected by wounding at the root elongation zone. The suppressed expression of *cs* gene and a marked reduction in caffeine and theobromine contents in young tea shoots were obtained after root transformation through *Agrobacterium*.

Genome editing is an emerging area for gene manipulation. Though considering the fact that manipulation does not involve any foreign gene, it is considered to be a non- transgenic, yet it deserves a merit to discuss here due to the involvement of gene delivery system. Among the several genome editing techniques, CRIPR/CAS-9 is on the recent edition which has wide application for trait manipulation particularly immensely useful for tea not only as a woody perennial but also due to the heterozygous in nature. In tea, attempt for standardization of CRISPR/CAS-9 has been reported recently (Tang et al. 2016) where a vector has been constructed to target the caffeine synthase gene by using general PCR, overlapping PCR and golden gate cloning technology. Although it is a very preliminary attempt, yet this technology will be useful for developing improved genotype with desired trait of tea. However, in tea once transgenic are developed, it can be multiplied through well-established micropropagation protocol (Mondal et al. 1998).

5.7 Conclusion

Despite the fact that the transgenic technology has tremendous scope for tea, no transgenic plants have been released commercially so far. Among the many reasons, a few could be as follows: (1) tea plant is not easily amenable for *Agrobacterium* infection; (2) non-availability of generic protocol, which can be applied for a wide range of varieties of tea plant; and (3) being woody perennial, production of transgenic tea is time consuming. However, it is evident now that among the different techniques *A. tumefaciens*-mediated transformation has been attempted with different groups but transgenic tea plant is yet to be commercialized. On the other hand, though *A. rhizogenes* transformation in tea has been demonstrated, the technique has not been exploited commercially to produce the secondary metabolites so far, which will be immensely useful for a crop like tea. Contrary to the above techniques, work on particle bombardment is at very initial stage. It is noteworthy to mention here that perhaps transgenic tea has a unique advantage as, during processing, green tea leaves are exposed to a temperature of 120 °C at the drying step during which foreign gene product such as *Bt* toxins will be destroyed, which is critical for genetically modified foods globally for some agricultural crops.

Referencess

Akula A, Akula C (1999) Somatic embryogenesis in tea (*Camellia sinensis* L) O Kuntze. In: Jain SM, Gupta PK, Newton RJ (eds) Somatic embryogenesis in woody plants, vol 5. Kluwer Academic, Dordrecht, pp 239–259

Alagarsamy K, Shamala LF, Wei S (2018) Protocol: high-efficiency *in-planta Agrobacterium*-mediated transgenic hairy root induction of *Camellia sinensis* var. *sinensis*. Plant Methods 14:17

Aoshima Y, Ugaki M, Niwa Y (2001) Investigation of gene delivery condition in tea callus by *Agrobacterium*-mediated transformation using high level expressing reporter gene. Tech Bull Shizuoka Tea Exp Stat 23:29–36

Bergmann BA, Stomp AM (1992) Effect of host plant genotype and growth rate on *Agrobacterium tumefaciens* mediated gall formation in *Pinus radiata*. Phytopathology 82:1457–1462

Bhatacharya A, Mondal TK, Sandal I, Prakash O, Kumar S, Ahuja PS (2004) In: Curtis IS (ed) Genetic transformation of tea. Kluwer Academic, Dordrecht, pp 245–255

Bhattacharya A, Saini U, Joshi R, Kaur D, Pal AK, Kumar N, Gulati A, Mohanpuria P, Yadav SK, Ahuja PS (2013) Osmotin-expressing transgenic tea plants have improved stress tolerance and are of higher quality. Transgenic Res 23:211–223

Chiu W, Niwa Y, Zeng W, Hirano T, Kobayashi H, Sheen J (1996) Engineered GFP as a vital reporter in plants. Curr Biol 6:325–330

Confalonieri M, Balestrazzi A, Cella R (1997) Genetic transformation of *Populus deltoides* and *Populus* x *euramericana* clones using *Agrobacterium tumefaciens*. Plant Cell Tiss Org Cult 48:53–61

Godwin ID, Ford-Lloyd BV, Newbury HJ (1992) *In vitro* approaches to extending the host-range of *Agrobacterium* for plant transformation. Aust J Bot 40:751–763

Jayaraman PR, Nithya MS (2005) *Agrobacterium tumefaciens*-mediated transformation of embryogenic tissues of tea (*Camellia sinensis* (L.) O. Kuntze). Plant Mol Biol Rep 23:299–302

John KMM, Joshi DS, Mondal AKA, Kumar SR, Rajkumar R (2009) *Agrobacterium rhizogenes* mediated hairy root production in tea leaves [*Camellia sinensis (L) O. Kuntze*]. Indian J Biotechnol 8:430–434

Kato M, Uematu K, Niwa Y (2004) Transformation of green fluorescent protein in tea plant. In: Proceedings of 2004 International Conference O-cha (tea) culture and science, pp 219–221

Konwar BK, Das SC, Bordoloi BJ, Dutta RK (1998) Hairy root development in tea through *Agrobacterium rhizogenes*-mediated genetic transformation. Two Bud 45:19–20

Kumar N, Pandey S, Bhattacharya A, Ahuja PS (2004) Do leaf surface characteristics affect *Agrobacterium* infection in tea (*Camellia sinensis* (L.) O. Kuntze). J Biosci 29:309–317

Lopez JS, Raj Kumar R, Pius RK, Muraleedharan N (2004) *Agrobacterium tumefaciens*-mediated genetic transformation in tea (*Camellia sinensis* [L.] O. Kuntze). Plant Mol Biol Rep 22:201–204

Luo YY, Liang YR (2000) Studies on the construction of *Bt* gene expression vector and its transformation in tea plant. J Tea Sci 20:141–147

Matsumoto S, Fukui M (1998) *Agrobacterium tumefaciens* mediated gene transfer in tea plant (*Camellia sinensis*) cells. J Agric Res Q 32:287–291

Matsumoto S, Fukui M (1999) Effect of acetosyringone application on *Agrobacterium*-mediated transfer in tea plant (*Camellia sinensis*). Bull Nat Res Inst Veg Orna Tea Shizuoka 14:9–15

Mercuri A, Beneditti LD, Burchi G, Schiva T (2000) *Agrobacterium*-mediated genetic transformation of African violet. Plant Cell Tiss Org Cult 60:39–46

Mohanpuria P, Rana NK, Yadav SK (2008) Transient RNAi based gene silencing of glutathione synthetase reduces glutathione content in *Camellia sinensis* (L.) O. Kuntze somatic embryos. Biol Plant 52:381–384

Mohanpuria P, Kumar V, Ahuja PS, Yadav SK (2011) *Agrobacterium*-mediated silencing of caffeine synthesis through root transformation in *Camellia sinensis* L. Mol Biotechnol 48:235–243

Mondal TK (2007) Tea. In: Devey MR, Pua P (eds) Biotechnology in agricultural and forestry. Transgenic crops V, vol 60. Springer, Berlin, pp 519–536

Mondal TK (2008) Tea. In: Kole C, Hall TC (eds) A compendium of transgenic crop plants: plantation crops, ornamentals and turf grasses. Blackwell, Hoboken, NJ, pp 804–838

Mondal TK, Kundu PK, Ahuja PS (1997) Gene silencing: a problem in transgenic research. Curr Sci 72:699–700

Mondal TK, Bhattacharya A, Sood A, Ahuja PS (1998) Micropropagation of tea (*Camellia sinensis (L.) O. Kuntze*) using Thidiazuron. Plant Growth Regul 26:57–61

Mondal TK, Bhattachraya A, Sood A, Ahuja PS (1999) An efficient protocol for somatic embryogenesis and its use in developing transgenic tea (*Camellia sinensis* (L) O. Kuntze) for field transfer. In: Altman A, Ziv M, Izhar S (eds) Plant biotechnology and in vitro biology in 21st century. Kluwer Academic, Dordrecht, pp 101–104

Mondal TK, Bhattacharya A, Ahuja PS (2001a) Development of a selection system for *Agrobacterium*-mediated genetic transformation of tea (*Camellia sinensis*). J Plant Crops 29:45–48

Mondal TK, Bhattacharya A, Ahuja PS, Chand PK (2001b) Factor effecting *Agrobacterium tumefaciens* mediated transformation of tea (*Camellia sinensis* (L). O.Kuntze). Plant Cell Rep 20:712–720

Mondal TK, Bhattacharya A, Ahuja PS (2001c) Induction of synchronous secondary embryogenesis of tea (*Camellia sinensis*). J Plant Physiol 158:945–951

Mondal TK, Bhattachrya A, Sood A, Ahuja PS (2002) Factors affecting germination and conversion frequency of somatic embryos of tea. J Plant Physiol 159(12):1317–1321

Mondal TK, Bhattacharya A, Laxmikumaran M, Ahuja PS (2004) Recent advance in tea biotechnology. Plant Cell Tiss Org Cult 75:795–856

Mondal TK, Parathiraj S, Mohan Kumar P (2005) Micrografting-a technique to shorten the hardening time of micropropagated shoots of tea (*Camellia sinensis* (L.) *O. Kuntze*). Sri Lanka J Tea Sci 70:5–9

Mukhopadhyay M, Mondal TK (2018) In: Sharma VS, Gunasekare MTK (eds) Genetic transformation of tea. Global tea science: current status and future needs. Burleigh Dodds Science, Cambridge. ISBN-13: 978–1786761606

Mukhopadhyay M, Mondal TK, Chand PK (2016) Biotechnological advances in tea (*Camellia sinensis* [L.] O. Kuntze): a review. Plant Cell Rep 35(2):255–287

Peng ZY, Liu DH, Xiao HJ, Zhang LX, Peng ZY, Liu DH, Xiao HJ, Zhang LX (2004) On the induction frequency of callus and hairy root in roots of tea plant. J Hunan Agril Univ 30:138–141

Qianru LV, Chen C, Xu Y, Hu S, Wang L, Sun K, Chen X, Li X (2017) Optimization of *Agrobacterium tumefaciens*-mediated transformation systems in tea plant (*Camellia sinensis*). Hort Plant J 3:105–109

Rana MM, Han ZH, Song DP, Liu GF, Li DX, Wan XC, Karthikeyan A, Wei S (2016) Effect of medium supplements on *Agrobacterium rhizogenes* mediated hairy root induction from the callus tissues of *Camellia sinensis* var. *sinensis*. Int J Mol Sci 17:1132–1149

Saini U, Kaur D, Bhattacharya A, Kumar S, Singh RD, Ahuja PS (2012) Optimising parameters for biolistic gun-mediated genetic transformation of tea [*Camellia sinensis* (L.) O. Kuntze]. J Hortic Sci Biotechnol 87:605–612

Sandal I, Saini U, Lacroix B, Bhattacharya A, Ahuja PS, Citovsky V (2007) *Agrobacterium*-mediated genetic transformation of tea leaf explants: effects of counteracting bactericidity of leaf polyphenols without loss of bacterial virulence. Plant Cell Rep 26:169–176

Singh HR, Bhattacharyya N, Agarwala N, Bhagawati P, Deka M, Das S (2014) Exogenous gene transfer in Assam tea [*Camellia assamica* (masters)] by *Agrobacterium*-mediated transformation using somatic embryo. Eur J Exp Biol 4:166–175

Song DP, Feng L, Rana MM, Gao MJ, Wei S (2014) Effects of catechins on *Agrobacterium*-mediated genetic transformation of *Camellia sinensis*. Plant Cell Tiss Org Cult 19:27–37

Tang Y, Liu L, Wang R, Chen Y, Liu Z, Liu S (2016) Development of a CRISPR/Cas9 constructed for genome editing of caffeine synthase in *Camellia sinensis*. J Tea Sci 36:414–426

Tosca A, Pondofi R, Vasconi S (1996) Organogenesis in *Camellia x williamsii*: cytokinin requirement and susceptibility to antibiotics. Plant Cell Rep 15:541–544

Wu S, Liang YR, Lu JL, Kim HS, Wu Y, Wu S, Liang YR, Lu JL, Kim HS, Wu Y (2003) Optimization of *Agrobacterium*-mediated and particle bombardment-mediated transformation systems in tea plant (*Camellia sinensis*). J Tea Sci 2:5–9

Zehra M, Banerjee S, Mathur AK, Kukreja AK (1996) Induction of hairy roots in tea (*Camellia sinensis* L.) using *Agrobacterium rhizogenes*. Curr Sci 70:84–86

Zhang XW, Liu XM, Dong YY, Zhou PH (1994) Preliminary studies of callus induction and co-culture transformation of *Thea sinensis*. J Hunan Agric Coll 20:550–554

Zhao D, Liu ZS, Lu JL, Qian LS, To YY, Xi B (2001) Study on *Agrobacterium tumefaciens*-mediated transformation of tea plant. J Tea Sci 21:108–111

Molecular Markers

6

6.1 Introduction

In addition to the shortfalls described earlier in the "Chap. 2," progress of tea breeding had also been slowed down due to the lack of reliable selection criteria (Kulasegaram 1980). Various morpho-biochemical markers had been reviewed in past (Wachira 1990; Singh 1999; Ghosh-Hazra 2001; Bandyopadhyay 2011; Mukhopadhyay et al. 2016), and it had been seen that they had marginally improved the efficacy of selection for desired agronomic traits in tea. This was mainly due to the fact that most of the morphological markers defined so far were influenced greatly by the environmental factors, a fact which is known for a long time, and hence showed a continuous variation with a high degree of plasticity. Therefore, these markers could not be separated into discrete groups for identification (Wickramaratna 1981). Recently, development of the molecular biology had resulted in alternative DNA-based markers for crop improvement of tea (Bansal et al. 2014). These markers can assist the process of traditional breeding with several efficacies. The greatest advantages of molecular markers are (1) free from the environmental influences and (2) detection of polymorphisms at an early stage of growth, and (3) accuracy of detection is much higher than the morphological markers. The different markers, which have been employed for varietal improvement of tea and its wild relatives, are reviewed below.

6.2 Morphological Markers

Morphological traits can be used as genetic markers if their expression is reproducible over a range of environments (Staub et al. 1982). Tea plants had been classified into three different types based on morphological characters. Barua (1963) provided morphological and anatomical descriptions of these three types of tea, which were later elaborated by Bezbaruah (1971). Since then several morphological as well as biochemical markers had been identified and used in the tea breeding program

T. K. Mondal, *Tea: Genome and Genetics*,
https://doi.org/10.1007/978-981-15-8868-6_6

Table 6.1 Pre-genomics markers used in tea genetic improvement

Criterion	Reference
Quantity and shape of the sclereids	Barua (1958)
Bush vigor	Barua and Dutta (1971)
Leaf geometry	Banerjee (1987)
Volatile flavor compounds	Borse et al. (2002)
Leaf pose, color, serration of the margin, and angle	Eden (1976)
Chlorophyll content/photosynthesis rate	Ghosh-Hazra (2001)
Quantitative changes in chlorophyll a and chlorophyll b and carotenoids	Hazarika and Mahanta (1984)
Epicuticular waxes	Kabir et al. (1991)
Dry matter production and partitioning	Magambo and Cannell (1981)
Green leaf catechin and ratio of dihydroxylated to trihydroxylated catechins	Magoma et al. (2000)
Root lengths	Nagarajah and Ratnasurya (1981)
Catechins, caffeine, and volatile flavor compounds	Owuor and Obanda (1998) and Singh et al. (2013)
Leaf, floral biology, and growth morphology	Sealy (1958)
Chloroform test	Sanderson (1964)
Pruning litter weights	Satyanarayan and Sharma (1982)
Anthocyanin pigmentations in young leaves	Satyanarayan and Sharma (1986) and Kerio et al. (2012)
Evenness of the flash, plucking density and recovery time of pruning	Singh (1999)
Tarpon index	Takeo (1981)
Leaf pubescences	Wight and Barua (1954)
Phloem index	Wight (1954)
Isozymes	Hairong et al. (1987), Xu et al. (1987), Ikeda et al. (1991), Chengyin et al. (1992), Anderson (1994), Singh and Ravindranath (1994), Yang and Sun (1994), Borthakur et al. (1995), Chen (1996), Sen et al. (2000), Fernández-Cáceres et al. (2001), Pedro et al. (2001), Jeyaramraja et al. (2002), Magoma et al. (2003), Neog et al. (2004), Chen et al. (2005a), Otaghvari et al. (2010) and Rajanna and Ramakrishnan (2010)
Metal contents	McKenzie et al. (2010)
Stalk length, total pekoe, fresh leaf weight, stalk fresh weight, fresh pekoe weight, dry leaf weight, stalk dry weight, and pekoe dry weight	Azka et al. (2019)

(Table 6.1). While leaf architect, pollen morphology (Chen et al. 1997a) growth habits, and floral biology were important criteria used by tea taxonomists (Banerjee 1992), the bush vigor, pruning weight, periods of recovery from prune, plant height, root mass, root–shoot ratio, plucking point density, dry matter production, as well as partitioning are considered as yield indicators in tea, and organoleptic taste and cup color are main criteria for quality (Banerjee 1992). In tea, morphological markers had been used to study genetic diversity (Wickramaratna 1981; Toyao and Takeda 1999), variation (Gunasekara et al. 2001; Piyasundara et al. 2006; Su et al. 2007), phylogeny, and classification (Chen et al. 2005a; Vo 2006; Piyasundara et al. 2008; Pi et al. 2009). For an example, in Sri Lanka, 20 genotypes of tea were characterized using 13 morphological descriptors and grouped them into 4 classes based on PCA. Of these, leaf width, leaf shape, leaf pigments, and petiole pigmentation were found to be maximum contributor for morphological variation (Piyasundara et al. 2006).

Morphological diversity of three main taxa was studied, and the importance of different descriptors in categorizing accessions into distinct groups was also examined. Twelve accessions were characterized using 15 morphological descriptors of the International Plant Genetic Resource Institute, Rome, guidelines (Annon. 1997). The results of PCA on morphological characters showed that the first two principal components accounted for 44.77% of the total variance. In the evaluated quantitative characters, all three taxa had a coefficient of variation (CV) greater than 24.85%, and within the taxon, the CV was greater than 9.59%. The qualitative characters showed a wide range of variations and yielded significant differences ($p < 0.05$). Phenotypic data had high contributing component loadings from characters such as leaf area, weight of harvested shoots, stem color, leaf pubescence, and young shoot color. Cluster analysis delineated the accessions into three groups. The implications of the results hold promise for assessing genetic diversity in germplasm collections, which was a prerequisite for their utilization, effective management, and crop improvement (Rajanna et al. 2011).

Variations in leaf anatomy were compared among the diploid, triploid, and tetraploid clones of tea. There were significant variations in the different attributes within and between the different groups. The number of stomata and epidermal cells per unit leaf area decreased with increasing ploidy, reflecting larger cells in the tetraploids than in the diploids and triploids. Specifically leaf mass was also higher in the tetraploids (Ng'-Etich and Wachira 2003). Similarly pollen morphology (Chen et al. 1992; Shu and Chen 1996) and pistil type (Yamaguchi 2001) were used to find evolutionary relationship among the different genotypes of tea and species of *Camellia*.

6.3 Artificial Neural Network (ANN): A Digital Marker

The use of ANNs in identifications of plants is gaining popularity. Applications of using ANN-based morphological traits for plant identification are well-documented now (Clark and Warwick 1998). Genetic erosion assumes an alarming significance especially in those species in which genetic improvement has originated an extremely high number of cultivars, with a consequent loss or oversight of the ancient ancestors. Among these, *C. japonica* L. (Theaceae) represents an excellent example, totaling currently more than 30,000 cultivars. Thus morphological characterization of 22 ancient *C. japonica L.* cultivars of historical garden, Villa Orsi at Compito (Lucca, Italy), was done by using quantitative morphological (i.e., phyllometric) and fractal spectra traits originated from leaf image analysis and their discrimination by building a specific ANN for data modeling. Results showed that the construction of a back-propagation neural network based on phyllometric and fractal analysis could be effectively and successfully used to discriminate *C. japonica* genotypes using simple dedicated instruments, such as a personal computer and an easily available optical scanner (Mugnai et al. 2008).

Low-cost, rapid, and accurate identification of tea varieties is always challenging. Fourier transform near-infrared (FT-NIR) spectroscopy coupled with the pattern recognition was used to identify individual tea varieties as a rapid and non-invasive analytical tool. Linear discriminant analysis (LDA) and ANN were compared to construct the identification models based on PCA. The experimental results showed that the performance of ANN model was better than LDA models. The overall results demonstrated that FT-NIR spectroscopy technology with ANN pattern recognition method was successfully applied as a rapid method to identify tea varieties (Chen et al. 2008).

Leaf characters had been successfully utilized to classify *Camellia* (Theaceae) species; however, leaf characters combined with supervised pattern recognition techniques had been explored to classify tea. Lu et al. (2012) studied the leaf morphological and venation characters of 93 *Camellia* species to assess the effectiveness of several supervised pattern recognition techniques such as learning vector quantization neural network (LVQ-ANN), dynamic architecture for ANN (DAN2), and support vector machines (SVM) for classifications and compare their accuracy. Among them, SVM offered the best classification accuracy. A hierarchical dendrogram based on leaf architecture data had confirmed the morphological classification of the studied species. The overall results suggested that leaf architecture-based data analysis using supervised pattern recognition techniques, especially DAN2 and SVM discrimination methods, was excellent for identification of *Camellia* species (Lu et al. 2012). Following the similar approach, morphological diversity of 17 tea accessions belonging to Chinese, Assamese, and Shan tea (*C. sinensis* var. *pubilimba*) groups was studied. The ANN data was able to perform a correct identification for almost all the accessions using simple dedicated instruments (Pandolfi et al. 2009).

6.4 Biochemical Markers

To overcome the problems of morphological markers, attentions were given for searching alternative biochemical markers. Presence and quantity of calcium oxalate crystals in paranchymatous cells of leaf petiole of tea, known as the phloem index, had been suggested to be a suitable criterion to classify the hybrids (Wight 1958). The variations in the quantity and morphology of sclereids in the leaf lamina had also been utilized in differentiating tea taxa (Barua 1958; Barua and Dutta 1959). Caffeines, volatile compounds (Seurei 1996), green leaf pigmentation (Banerjee 1992), and leaf pubescences (Wight and Barua 1954) were some of the parameters which had been used as potential determinants of tea quality. Despite of the disadvantages due to environmental influences, these markers were mostly adopted, frequently used by tea breeders globally. Considerable success had been achieved in identification of tea biochemical indicators (Owuor et al. 1986; Takeo 1981). These indicators were also used widely to distinguish two species of tea, namely, *C. sinensis* and *C. assamica*, and their respective clones (Owuor 1989). Takeo (1983) suggested a chemotaxonomic method of classifying tea clones based on a ratio referred to as the Terpene Index (T.I), which expresses the ratio between linalools and linalools plus geraniols. Although not fully exploited, the polyphenol oxidase activity, individual polyphenol, amino acids, and chlorophyll content varied between the hybrids and used as potential parameters in tea taxonomy (Sanderson 1964).

Paper chromatography was also used to investigate the taxonomy of different species of *Camellia* under the section Thea. It had been found that species within the Thea section of the genus *Camellia* were closely similar in chemical composition, the general pattern of which did not have relationship to the chemical composition of non-Thea Camellias (Roberts et al. 1958). Though *C. saluensis* was found to cross readily with *C. japonica*, identification of their F_1 hybrid known as *C. williamsii* was a challenge for breeders at the young stage. Interestingly, on the basis of flavor and other coloring compounds using paper chromatography, Parks and Case (1968) were successful to identify the true hybrid seedlings of C. *williamsii*, indicating the suitability of this simple technique. The presence or absence of certain phenolic substances in the shoots had also been used in establishing relationship between different taxa (Roberts et al. 1958; Xu et al. 2004). Quantitative changes in chlorophyll a and chlorophyll b and four carotenoids (carotene, lutein, violaxanthine, and neoxanthine) were used for characterization of three types of tea (Hazarika and Mahanta 1984). Total green leaf catechin concentrations and ratio of dihydroxylated to trihydroxylated catechins were used to establish genetic relationship among the 102 Kenyan tea genotypes (Magoma et al. 2000). Accumulation and ratio of the various catechins were used to classify the three types of tea. Based upon the biochemical differentiation, it was found that there was potentiality for broadening the genetic base of mainly Assam teas than the putative China and Cambod teas in Kenya.

A parameter of fresh tea leaf that correlates with black tea quality is highly desired. Twenty high- and 20 low-quality tea clones were selected from the breeding program at the Tea Research Foundation (Central Africa). The flavan-3-ol profile of fresh tea leaves was analyzed by capillary electrophoresis, while total theaflavin (TF) content was determined in the black tea manufactured from the same leaves for each clone. The above parameters were correlated with total scores and valuation from two tea tasters with regression analysis (Wright et al. 2000).

Similarly, TFs and total color (TC) and high caffeine, total soluble solids (TSS), volatile compounds, and viscosity were used to carry out the fingerprint in Indian tea. It had been found that tea of different origins had characteristic chemical constituent. This fingerprint will help to trace the origin of teas (Borse et al. 2002). Catechin and its components had been widely utilized in studying the diversity of tea germplasm (Magoma et al. 2000; Gulati et al. 2009), yet it was found total catechin of tea leaves increased with increasing of sunlight exposure, suggesting that catechin biosynthesis was also environmentally dependent (Mariya et al. 2003) and geographical location of plant (Wei et al. 2011).

Diversity of amino acids variation was also studied among the different tea genotypes. Differentiation of green, black, oolong, white, and Pu-erh teas had been carried out according to their free amino acid contents. Using back-propagation multilayer perceptron ANN, 100% success in the classification was obtained. The most differentiating amino acids were found to be glutamic acid, asparagine, serine, alanine, leucine, and isoleucine (Alcázar et al. 2007).

Li et al. (2010) demonstrated that the polyphenols of tea leaves can be used as chemotaxonomic markers. They established the genetic relationship among 89 wild, hybrids, and cultivated tea trees from China and Japan with the following types of catechins: (1) (−)-epigallocatechin 3-O-gallate, (EGCG), (2) (−)-epigallocatechin, (EGC), (3) (−)-epicatechin 3-O-gallate, (ECG), (4) (−)-epicatechin, EC (5) (+)-catechin, (CA), (6) strictinin, (STR), and (7) gallic acid (GA). Of the 13 polyphenol patterns observed, PCA indicated that the structure types of the flavonoid B-rings, such as the pyrogallol–EGCG and EGC and catechol–ECG and EC types, greatly influenced the classification. A study based on morphological characteristics, geographical information, and historical information on tea trees showed good agreement with the current theory of tea tree origins and concluded that the Xishuangbanna District and Puer City of China were among the original sites of the tea tree species. The contents of gallic acid, epigallocatechin gallate, epigallocatechin, epicatechin, epicatechin gallate, catechin, caffeine, theophylline, and theobromine were determined to differentiate the 45 tea samples according to their geographical origins. Catechins, gallic acids, and tea alkaloids were adequate chemical descriptors to distinguish tea samples cultivated in different geographical areas (Pedro et al. 2001).

The Indian tea germplasm comprised with "China," "Assam," "Cambod," and their hybrids was evaluated using biochemical markers such as total catechins and their fractions, for varietal identification and characterization of various made tea samples from North India (Gulati et al. 2009) as well as South India (Saravanan et al. 2005). Morphologically indistinct large-leaved "Cambod" variety and "Assam" varieties could not be differentiated using biochemical markers, since both varietal types taxonomically belong to a single species. Clones of "China" type showed low total catechin content and catechin ratio which were distinctly grouped. The "China–Assam" and "China–Cambod" hybrids formed intermediate groups between "China" and "Cambod"/"Assam" PC groups. Tea clones which were differentially positioned in the PC group could be explained based on the genetic contribution by other varietal types as parents (Wright et al. 2002).

The variations of the main quality components such as polyphenols, catechins, amino acids, caffeine, and water extract of 596 Chinese tea accessions were analyzed. Wide range of biochemicals was determined. It had been reported that polyphenols were found to be from 13.6 to 47.8%, catechins ranged from 81.9 g/kg to 262.7 g/kg, amino acids content ranged from 1.1 to 6.5%, caffeine content was found to be from 1.2 to 5.9%, and water extract content was found to be from 24.4 to 57.0% (Chen and Zhou 2005).

The theanine content of the leaves of 27 species of *Camellia* plants was investigated. Although, theanine was present in 21 species only, but in much lower amount than the quantity detected in tea leaves. The major free amino acids in leaves of four species belonging to the genera *Schima* and *Eurya* were glutamic acid, aspartic acid, glutamine, asparagine, alanine, and proline, and content of these amino acids was similar to or higher than theanine. Accumulation of free amino acids in these plants was generally lower than in *C. sinensis* var. *sinensis*. The theanine biosynthetic activity in roots was higher than that of leaves (Deng et al. 2010).

The inheritance of anthocyanin pigmentation was studied. Based on chemical analysis, in conjunction with geographical and literary information, it was suggested that the Xinan District, China, was the site/center of origin for the red-flowered (caused by anthocyanin pigmentation) *Camellia* species of which both *C. saluenensis* and *C. reticulata* had the most primitive anthocyanin content (Li et al. 2013). The variations of EGCG and EGC were the most prominent among the different tea varieties. High EGC content was found to be a characteristic of Assam variety which was further corroborated through multivariate analysis (Sabhapondit et al. 2012). Although accuracy was higher than morphological markers, however, accumulations of chemicals were subjected to post-transcriptional modification and thus often restricted their utility as markers (Staub et al. 1982). P/11/15 clone is a green tea cultivar. Thus to understand its biochemical composition, several biochemical parameters such as polyphenol oxidase (PPO) activity, total polyphenols, catechins, amino acid content, and enzymatic antioxidants were studied. It was found that it contained high polyphenol oxidase, sufficient catechin, polyphenol, peroxidase, catalase, and super oxide dismutase due to which it has high antioxidant activity and hence suitable for green tea production (Ramkumar et al. 2016).

6.5 Metallic Markers

Tea is an infusion, made from dried leaves, and a good dietary source of essential trace metals for us. Therefore, it is necessary to consider variations in element content of tea leaves among tea cultivars. Therefore several workers attempted to classify and to establish the metal fingerprint of tea (Pedro et al. 2001; Fernández-Cáceres et al. 2001). Thus, elemental fingerprint technique, based on elemental contents (Al, B, Ca, Cd, Co, Cr, Cu, Fe, K, Mg, Mn, Na, P, Pb, and Zn) determined by inductively coupled plasma atomic emission spectrometry (ICP-AES) and multivariate statistical analysis, had been used to differentiate tea cultivars by various groups of specific geographical region. The ranges of element concentrations of the cultivars were in good agreement with those obtained in previous studies, and the level of most elements in tea leaves was significantly different among the cultivars. The classifications of tea cultivars were 100% accurate in total by PCA, hierarchical cluster analysis (HCA), LDA, and back-propagation neural network (BPNN) analysis. Each cultivar presented a distinctive element fingerprint, and the elements in tea leaves can be significant predictors in differentiating tea cultivars (Chen et al. 2009). Similar approaches were also used to fingerprint white, green, black, oolong and Pu-erh teas. Using probabilistic neural networks (PNN), it was demonstrated that overall classification performance of metals content was successful in tea (McKenzie et al. 2010).

6.6 Isozyme Markers

Genetic analysis of isozyme variation was used for cultivar identification in a wide range of plants (Ferguson and Grabe 1986; Hirai and Kozaki 1986; Mondal 2001). Similarly, in tea also, isozymes had been analyzed by several workers for studying the genetic tendencies, cultivar identifications, and implications in hybrid breeding (Table 6.1). Among the isozymes, peroxidase and esterase had been the most widely studied on a wide range of tea cultivars and different *Camellia* species (Chen 1996; Borthakur et al. 1995; Chengyin et al. 1992; Yang and Sun 1994; Singh and Ravindranath 1994; Sharma and Deka 2002; Chung et al. 2003). Isozyme analyses of tetrazolium oxidase, aspartate aminotransferase, and alpha-amylase were studied among seven different tea cultivars along with three different species (Sen et al. 2000). The electrophoretic analysis revealed both the qualitative and quantitative variations in the isozyme-banding pattern among different species of tea and their clones. The tetrazolium oxidase enzyme system showed the highest variability among all the enzymes studied. Cluster analysis using isozyme-banding pattern produced a dendrogram, which clearly differentiated characteristics of both clones and species studied. However, in general, the applications of isozyme study in tea were limited to few enzymes with inadequate polymorphisms (Wachira et al. 1995). Besides, various isozymes were also used to study the diversity of *C. japonica* (Wendel and Parks 1982; Wendel and Parks 1983; Ikeda et al. 1991). Interestingly, Wendel and Parks (1984) did a classical study to conclude that alcohol dehydrogenase isozymes in *C. japonica* were encoded by two genes *Adh*-1 and *Adh*-2. Both loci were expressed in seeds, and their products were randomly associated with intragenic and intergenic dimmers. Electrophoresis of leaf extracts produced only the products of *Adh*-2. Formal genetic analysis indicated that the two *Adh* loci were tightly linked. Most segregation fits the expected Mendelian ratios, but in some families, distorted segregation was also observed at *Adh*-1, *Adh*-2, or both loci. Further, starch gel electrophoresis was used to score allelic variation at 20 loci in *C. japonica* collected from 60 genotypes distributed throughout Japan. In comparison with other plant species, the genetic diversity within the population was very high, i.e., 66.2% of loci were polymorphic per population, which gave an average mean number of 2.16 allele per locus. They also reported genotypic proportions at most of the loci in majority of all the population and found a good fit of the Hardy–Weinberg expectation (Wendel and Parks 1985).

6.7 Cytological Markers

Cytological markers of the genus *Camellia* were elaborately studied in the early 1970s with many interesting features. Chromosome number had been established for the most available species of *Camellia* including tea (Bezbaruah 1971, which was reviewed by Kondo (1977). Generally tea chromosomes are small in size and tend to clump together due to "stickiness." Tea is diploid (2n = 30; basic chromosome number, $x = 15$), and karyotype ranges from 1.28μ to 3.44μ (Bezbaruah 1971). The *r* value (ratio of long arm to short arm) for all the 15 pairs of chromosomes ranges from 1.00 to 1.91. This consistency in diploid chromosomes number suggests a monophyletic origin of all *Camellia* species. However few higher ploidy level such as triploids, e.g., TV-29, HS-10 A, UPASI-3, and UPASI-20 (2n = 45), tetraploids (2n = 60), pentaploids (2n = 75), and aneuploids (2n ± 1 to 29) had also been identified (Singh 1980). Karyotypic data had also been accumulated in the past for the other species of this genus (Fukushima et al. 1966; Ackerman 1971; Kondo 1975, 1978a, b). In karyotype analysis, unfortunately, grouping by chromosome size was difficult in the *Camellia* taxa, since the chromosome grade imperceptibly from the largest to the smallest. Furthermore, even in the best preparation, homologous chromosome pairs could not be appeared identical in *Camellia* (Kondo 1975, 1978a, b). Relatively little intraspecific karyotypic variation had been observed in the cultivated species of *Camellia* studied (Kondo 1975). Sat-chromosomes in karyotypes within mass accessions of certain *Camellia* species were morphologically and quantitatively variable. Thus karyotypes including characteristics of sat-chromosomes were not of taxonomic significance for *Camellia* taxa. Among the diploid species of *Camellia* studied, *C. japonica* L. sensu *lato* showed the greatest karyotypic variation; many of the accessions studied indicated similar karyotypic patterns to each other (Kondo 1975). For instance, *C. japonica* L. var. *spontanea* (Makino), *C. japonica* L. var. *macrocarpa* Masamune, *C. japonica* L. subsp. *rusticana* (Honda) Kitamura, and four cultivars, namely, "Aka-Wabisuke," "Fukurin-Wabisuke," "Kuro-Wabisuke," and "Wabisuke," carried same most common standard acetoorcein-stained karyotype if the presence of satellites was not considered: 16 metacentric, 8 submetacentric, and 6 subtelocentric chromosomes. Actually, *C. japonica* L. var. *macrocarpa* (Masamune) had satellites on four submetacentric chromosomes, and the other accessions had satellites on two submetacentric chromosomes (Kondo and Parks 1980). Prior to this, it was shown that the C-banding method can be applied to the somatic mid-metaphase chromosomes in *Camellia* taxa (Kondo and Parks 1979). These differentially stained bands in somatic mid-metaphase chromosomes permitted the identification of individual chromosomes and made it possible to match the homologous pairs of chromosomes more precisely. Karyotypic variability and divergence among the seven accessions of *C. japonica* L. sensu *lato* with same acetoorcein-stained karyotype were revealed by C-banding method (Kondo and Parks 1981). This way cytological marker was used to sort and classify the vast number of cultivars. However, due to the development of more sensitive DNA-based markers, attention was shifted towards the search of molecular markers.

6.8 DNA-Based Markers

However, limited numbers of loci of morphological, biochemical, as well as cytological markers showed the lesser polymorphism, but with the advancement of sequencing techniques, such efforts were shifted towards the study of various molecular markers (Table 6.2).

6.8.1 Random Amplified Polymorphic DNA Markers (RAPD)

Since discovery, RAPD is being used for a number of areas in population genetics. Initially it was the preferred DNA markers due to greater speed, easy-to-perform, and non-requirement of radioactive materials (Bantawa et al. 2011). In tea and other species of *Camellia*, a considerable amount of work has been carried out which are summarized below.

6.8.1.1 Germplasm Characterization and Genetic Diversity

Present-day tea plantation is developed largely from the selected genotypes based on the performance of yield, quality, and tolerance to the various biotic and abiotic stresses. As a consequence, widespread cultivation of clonal tea diminished the genetic diversity. Therefore germplasm characterization at molecular level of tea will help (1) varietal improvement of tea for agronomically important characters; (2) to preserve the intellectual property right of tea breeders; (3) to identify of individual tea cultivar by making a molecular passport; (4) to prevent the duplicate entry of different genotypes in tea gene pool; (5) efficient selection of the varieties for hybridization program, graft compatibility in composite plant production, etc.; and (6) taxonomic classification of tea genotypes on the basis of molecular markers.

Wachira et al. (1995) were the first to characterize 38 different cultivars of Kenyan tea. A total of 23 primers which were used could generate 157 polymorphic bands. The maximum polymorphism of 20 bands was detected by primer SC10–56. The similarity matrix ranged from a minimum of 43% to maximum of 96%, respectively, among the clones. Based on average linkage cluster analysis, a dendrogram was constructed which clearly discriminated different genotypes of Assam, Cambod and China tea. Further, to examine the evolutionary relationships, PCA was undertaken which was able to classify into three varietal types of tea in a manner that was consistent with both the present taxonomy of tea and with the known pedigrees of some clones. In the same year, Tanaka et al. (1995) used several 10-mer and 12-mer primers to detect variation among Korean, Japanese, Chinese, Indian, and Vietnamese tea. Among all primers, OPF-2 was found to be the most polymorphic. Their study concluded that after introduction from China, Korean tea underwent little genetic diversification. On the contrary, Japanese tea showed a closer relationship with their Chinese and Indian counter parts, which revealed the fact that tea in Japan might had brought from China as well as India. Twenty-five Indian tea cultivars and two ornamental species were characterized using RAPD markers (Mondal 2000). Out of the 40 random 10-mer primers, 11 generated polymorphic

Table 6.2 Molecular markers used in tea and other *Camellia* species

Country	Marker	Objectives of the study	Population size	Reference
India	RAPD	Genetic fidelity	17 micropropagated plants	Balasaravanan et al. (2002)
	AFLP	Genetic diversity	49 genotypes	Balasaravanan et al. (2003)
	SSR	Development of SSR	21 genotypes	Bali et al. (2013)
	RAPD	Genetic diversity	14 genotypes	Bera and Saikia (2002)
	RAPD	Genetic fidelity	7 micropropagated plants	Borchetia et al. (2009)
	RFLP	Detection of adulteration in made tea	10 different tea samples	Dhiman and Singh (2003)
	CAPS	Detection of high catechin content tea	2 parent along with their 25 cultivars	Elangbam and Misra (2016)
	SSR	miRNA-SSR was used	In silico analysis	Hazra et al. (2017)
	AFLP, SSR and RAPD	QTL identification	42 genotypes	Kamunya et al. (2010)
	AFLP	Genetic diversity	14 genotypes	Karthigeyan et al. (2008)
	AFLP	Genetic diversity	29 genotypes	Mishra and Sen-Mandi (2001)
	AFLP	Marker for drought tolerant	29 genotypes	Mishra and Sen-Mandi (2004)
	AFLP	Genetic diversity	29 genotypes	Mishra et al. (2009)
	RAPD	Genetic diversity	27 genotypes	Mondal (2000)
	ISSR	Genetic diversity	25 genotypes	Mondal (2002)
	RAPD	Development of DNA isolation technique	28 genotypes	Mondal et al. (2000)
	RAPD	Genetic fidelity	18 micropropagated plants	Mondal and Chand (2002)
	AFLP	Genetic diversity	32 genotypes	Paul et al. (1997)
	AFLP	Genetic diversity	32Genotypes	Rajasekaran (1997)
	AFLP	Genetic diversity	1644 genotypes	Raina et al. (2012)
	RAPD and ISSR	Genetic diversity	10 genotypes	Roy and Chakraborty (2007)

(continued)

Table 6.2 (continued)

Country	Marker	Objectives of the study	Population size	Reference
	RAPD and ISSR	Genetic diversity	21 genotypes	Roy and Chakraborty (2009)
	SSR	Development of SSR marker	32 genotypes of tea and 2 genotypes of *C japonica*	Sharma et al. (2009)
	AFLP	Genetic diversity	123 genotype	Sharma et al. (2010)
	SSR	Cross-species validation of SSR	18 genotypes	Sharma et al. (2011)
	5S rDNA	Phylogenetic relationship	28 genotypes	Singh and Ahuja (2006)
	RAPD	Authentication of made tea sample	11 genotypes	Singh et al. (1999)
	RAPD	Genetic fidelity	16 vegetative propagated plants	Singh et al. (2004)
	RFLP	Genetic organization	1 tea and 11 non-tea species	Singh et al. (2011)
	ISSR	Genetic fidelity of somaclonal variant	20 genotypes	Thomas et al. (2006)
Indonesia	SSR	Molecular characterization	5 genotypes	Azka et al. (2019)
Japan	SSR	Identification of cultivars	41 genotypes	Kato et al. (2008)
	*mat*K	Genetic diversity	118 genotypes	Katoh et al. (2003)
	SSR	Genetic diversity	24 genotypes	Kaundun and Matsumoto (2002)
	STS-RFLP	Authentication of made tea sample	46 samples	Kaundun and Matsumoto (2003a)
	CAPS	Species-specific probe	50 genotypes	Kaundun and Matsumoto (2003b)
	SSR	Maternal inheritance of cp genome	6 genotypes	Kaundun and Matsumoto (2011)
	SSR	Genetic diversity analysis	113 lines including 68 landraces, 44 cultivars, and one wild relative	Kubo et al. (2018)

(continued)

Table 6.2 (continued)

Country	Marker	Objectives of the study	Population size	Reference
	RFLP	Genetic analysis	30 genotypes	Matsumoto et al. (1994)
	RFLP	Genetic diversity	45 genotypes	Matsumoto et al. (2000)
	RFLP	Genetic analysis of PAL gene	472 genotypes	Matsumoto et al. (2002)
	RFLP	Genetic diversity of PAL gene	297 genotypes	Matsumoto et al. (2004)
	SSR	Development and validation of SSR	157 genotypes	Ohsako et al. (2008)
	CAPS	CafLess-TCS1 for selection of caffeine-less tea plants	101 genotypes	Ogino et al. (2019)
	e-RAPD	Development of DNA isolation	16 genotypes	Tanaka and Taniguchi (2002)
	RAPD	Identification of true crosses	2 parents and 38 F_1 progenies	Tanaka and Yamaguchi (1996)
	RAPD	Maternal inheritance	14 genotypes of F_1 population	Tanaka (2006)
	Gene-specific PCR-RFLP	Phylogenic relationship	162 genotypes of 49 *Camellia* species	Tanikawa et al. (2008)
	SSR	Development of SSR	4 different species	Ueno et al. (1999)
	SSR	Genetic diversity	518 genotypes of *C japonica*	Ueno et al. (2000)
	SSR, ALP	Genetic diversity	22 genotypes of *C. japonica*	Ueno and Tsumura (2009)
	CAPS	Cultivar identification	63 genotypes	Ujihara et al. (2011)
	SNP (ddRAD)	Genetic diversity	167 genotypes	Yamashita et al. (2019)
China	RAPD	Genetic analysis	23 species	Chen and Yamaguchi (2002)
	RAPD	Discriminating of genotypes	20 species	Chen and Yamaguchi (2005)
	RAPD	Standardization of DNA isolation protocol	15 genotypes	Chen et al. (1997a)

(continued)

Table 6.2 (continued)

Country	Marker	Objectives of the study	Population size	Reference
	RAPD and RFLP	Development of technique	7 genotypes	Chen et al. (1997b)
	RAPD	Genetic analysis	7 genotypes	Chen et al. (1998a)
	RAPD	Genetic analysis	15 genotypes	Chen et al. (1998b)
	RAPD	Genetic analysis	5 genotypes	Chen et al. (1999)
	RAPD	Genetic analysis	24 genotypes	Chen et al. (2002a)
	RAPD	Molecular phylogeny	24 species	Chen et al. (2002a)
	RAPD	Genetic polymorphism	24 species	Chen et al. (2002b)
	RAPD	Genetic diversity	15 genotypes	Chen et al. (2005b)
	RAPD	Genetic diversity	45 genotypes	Fang et al. (2003)
	ISSR	Genetic diversity	36 genotypes	Fan et al. (2010)
	SSR	Genetic diversity	185 genotypes	Fang et al. (2012)
	SNP based genotyping assay	Standardization of genotyping assay	DNA was isolated fresh and processed commercial loose-leaf teas. 40 genotypes	Fang et al. (2014)
	ISSR	Genetic diversity	14 genotypes	Hou et al. (2007)
	AFLP	Linkage map	69 F_1 genotypes	Huang et al. (2005)
	AFLP	Genetic diversity	40 genotypes	Huang et al. (2006a)
	RAPD and ISSR	Genetic mapping	94 F_1 genotypes	Huang et al. (2006b)
	SNP	Identification of SNP	40 genotypes	Huang et al. (2007)
	RAPD	Genetic diversity	23 genotypes	Hui et al. (2004)
	SSR	Development of SSR markers	24 genotypes of tea and 2 genotypes of *C. japonica*	Hung et al. (2008)
	ISSR	Genetic diversity	10 genotypes	Ji et al. (2007)
	ISSR	Genetic diversity	181 genotypes	Ji et al. (2011)
	Genic SSR	Genetic diversity	20 varieties of *C. oleifera*	Jia et al. (2015)
	SSR	Development of SSR markers	10 genotypes	Jin et al. (2006)

(continued)

Table 6.2 (continued)

Country	Marker	Objectives of the study	Population size	Reference
	SSR	Genetic diversity	42 genotypes	Jin et al. (2007)
	RAPD	Genetic diversity	2 genotypes	Li et al. (2003)
	RAPD	Genetic diversity	69 genotypes	Li et al. (2005)
	RAPD	Genetic diversity	23 genotypes	Li et al. 2007
	ISSR	Genetic diversity	10 genotypes	Li et al. (2012)
	ISSR	Genetic diversity of made tea	26 genotypes	Lin et al. (2010)
	SSR	Development of BAC library for tea genome sequence	1 genotype	Lin et al. (2011)
	ISSR	Genetic diversity	25 genotypes	Liu et al. (2008)
	ISSR	Genetic diversity	134 genotypes	Liu et al. (2010)
	PAL gene, *rpl32-trnL* spacer	Genetic diversity and erosion	21 genotypes of *C. taliensis*	Liu et al. (2012a)
	ISSR	Genetic diversity	134 genotypes	Liu et al. (2012b)
	RAPD	Genetic relationship	21 genotypes	Luo et al. (2002a)
	RAPD	Parentage identification	15 genotypes	Luo et al. (2002b)
	RAPD	Genetic diversity	71 genotypes	Luo et al. (2004)
	SSR	Genetic diversity	45 genotypes of 6 different species	Ma et al. (2010)
	SSR	Validation of markers	21 genotypes	Ma et al. (2012)
	SSR	Discovery of QTL for catechin content	183 individuals	Ma et al. (2014)
	SNP (GBS)	Core germplasm development	415 tea accessions	Niu et al. (2019)
	RAPD	Genetic diversity	15 genotypes	Shao et al. (2003a)
	RAPD	Genetic diversity	25 genotypes	Shao et al. (2003b)
	RAPD	Genetic diversity	240 genotypes	Shen et al. (2007)
	SRAP	Genetic diversity	25 genotypes	Shen et al. (2009)
	RAPD	Genetic polymorphism	27 genotypes	Shen et al. (2002)

(continued)

Table 6.2 (continued)

Country	Marker	Objectives of the study	Population size	Reference
	RAPD	Phylogenic relationship	11 different species	Shi et al. (1998)
	SSR	Development of SSR markers	12 genotypes	Shi et al. (2013)
	RAPD, AFLP	Genetic diversity	120 genotypes of *C. nitidissima*	Tang et al. (2006)
	Genic SSR	Development of linkage map	F1 population consisting of more than 300	Tan et al. (2013)
	SSR	Genetic heterozygosity test	24 cultivars	Tan et al. (2015)
	IST	Phylogenic relationship	30 species of *Camellia*	Tian et al. (2008)
	RAPD, AFLP	Genetic analysis	48 genotypes	Wachira et al. (2001)
	nSSR, cpSSR	Origin of African tea	439 genotypes of 11 different African countries	Wambulwa et al. (2016)
	nSSR, cpSSR	Genetic diversity analysis	280 African tea accessions	Wambulwa et al. (2017)
	ISSR	Genetic diversity	114 genotypes of *C. reticulata*	Wang and Ruan (2012)
	RAPD	Genetic diversity	36 genotypes	Wang et al. (2007)
	RAPD	Genetic diversity	15 genotypes	Wang et al. (2010)
	RAPD	Genetic diversity	15 genotypes	Wang et al. (2011)
	SSR	Long-core SSR was used to study the genetic diversity	66 clonal tea	Wang et al. (2016)
	ISSR	Genetic diversity	84 genotypes of *C. euphlebia*	Wei et al. (2005)
	ISSR	Genetic diversity	250 genotypes of C *nitidissima*	Wei et al. (2008)
	SSR	Development of SSR markers	25 genotypes of *C. nitidissima*	Wei et al. (2010)
	RAPD	Genetic diversity	27 genotypes	Wen et al. (2002)
	SCoT, SCAR	Genetic diversity	18 green tea cultivar	Xu et al. (2019)
	SSR	Genetic diversity	587 genotype of *C. taliensis*	Zhao et al. (2014)
	SNP	Characterization of SNP for genetic diversity study	16 genotype	Zhang et al. (2014)

(continued)

Table 6.2 (continued)

Country	Marker	Objectives of the study	Population size	Reference
	EST-SSR, SRAP, SCOT	Genetic diversity	50 genotypes	Zhang et al. (2018)
	SSR	Development of SSR markers	150 genotypes of *C. chekiangoleosa*	Wen et al. (2012)
	RAPD	Genetic diversity	31 genotypes	Wu et al. (2002a)
	RAPD	Parentage identification	3 hybrid	Wu et al. (2002b)
	EST-SSR	Development of SSRs	20 genotypes	Wu et al. (2013)
	AFLP	Genetic diversity	34 genotypes	Xiao et al. (2007)
	SSR	Development of SSR markers	24 genotypes of *C. taliensis*	Yang et al. (2009)
	RAPD	Germplasm evaluation	200 genotypes	Yang et al. (2003)
	ISSR	Parent selection	48 genotypes	Yao et al. (2008)
	SSR	Genetic diversity	450 genotypes	Yao et al. (2012)
	RAPD	Genetic classification	45 genotypes	Zhang and Li (2003a)
	RAPD	Genetic classification	45 genotypes	Zhang and Li (2003b)
	ISSR	Identification of clones	10 genotypes of *C. oleifera*	Zhang et al. (2007)
	AFLP	Genetic diversity	23 genotypes	Zhao et al. (2006)
	SSR	Genetic diversity	40 genotypes	Zhao et al. (2008)
	SSR	Genetic diversity	41 genotype of *C. Japonica*, 9 genotypes of *C. oleifera*	Zhao et al. (2017)
Kenya	RAPD, AFLP	Genetic linkage map	90 genotypes of F1 progenies	Hackett et al. (2000)
	RAPD, ISSR	Assessment of mating system	180 progenies of 6 genotypes	Muoki et al. (2007)
	RAPD	Genetic analysis	38 genotypes	Wachira et al. (1995)
	RAPD	Diagnosis of gene introgression	28 genotypes	Wachira et al. (1997)
	RAPD and AFLP	Genetic diversity	24 genotypes	Wachira et al. (2001)

(continued)

Table 6.2 (continued)

Country	Marker	Objectives of the study	Population size	Reference
Italy	SSR	Genetic relationships	132 accessions of *Camellia* species	Caser et al. (2010)
	Barcode markers such as rbcL, matK	Development barcode using and process tea sample	32 samples	De Castro et al. (2017)
Taiwan	STS, CAPS	Identification of genotypes	12 cultivar	Hu et al. (2014b)
	RAPD, ISSR	Genetic analysis	37 genotypes	Lai et al. (2001)
	RAPD	Genetic diversity	96 genotypes	Lou et al. (2004)
	Intronic sequence of *RPB2* gene	Barcoding of species	25 genotypes of 4 species	Su et al. (2009)
	ITS of nrDNA	Phylogenic relationship	7 species	Vijayan and Tsou (2008)
	Nuclear ITS	Molecular taxonomy	112 species of *Camellia*	Vijayan et al. (2009)
Turkey	AFLP	Genetic diversity	32 genotypes	Kafkas et al. (2009)
Sri Lanka	RAPD	Genetic diversity	39 genotypes	Mewan et al. (2005)
	SSR	Genetic map	3 genotypes	Mewan et al. (2007)
	SSR	Genetic diversity	27 genotypes	Ariyaratne et al. (2009)
	RAPD	Genetic diversity	46 genotypes	Goonetilleke et al. (2009)
South Africa	RAPD	Genetic diversity	5 genotypes	Wright et al. (1996)
	RAPD	Development of trait specific marker	18 genotypes	Mphangwe et al. (2013)
South Korea	RAPD	Genetic analysis	6 genotypes	Kaundun and Park (2002)
	RAPD	Phylogenic relationship	27 genotypes	Kaundun et al. (2000)
	AFLP	Genetic analysis	37 genotypes	Lee et al. (2003)
	SSR	Genetic diversity analysis	410 genotype	Lee et al. (2019)
	RAPD	Genetic diversity	20 genotypes	Park et al. (2002)

(continued)

Table 6.2 (continued)

Country	Marker	Objectives of the study	Population size	Reference
Malaysia	RAM	Genetic variation	6 genotypes	Latip et al. (2010)
Pakistan	RAPD	Genetic diversity	75 genotypes	Afridi et al. (2011)
	RAPD	Genetic diversity	24 genotypes	Gul et al. (2007)
Portuguese	RAPD	Genetic analysis	71 genotypes	Jorge et al. (2003)
USA	SBA	Evolutionary studies	30 species	Prince and Parks (1997)
	SBA	Phylogenic relationship	19 species	Prince and Parks (2000)
	SBA	Evolutionary studies	35 species	Prince and Parks (2001)
UK	SSR	Development of SSR markers	15 genotypes	Freeman et al. (2004)
Bangladesh	RAPD	Genetic diversity	18 tea genotypes	Boonerjee et al. (2013)
Russia	RAPD	Genetic fidelity of tissue culture plant	15 micropropagated plants	Samarina et al. (2019)

SBA sequence-based analysis, *ALP* amplicon length polymorphism, *RAM* random amplified microsatellite (RAMs) markers

loci which were ranged from 7 to 21 per genotype. Out of the total 154 alleles, 138 were polymorphic resulting a high genetic variability of 95.2%. Shannon's index of diversity was used to partition the total phenotypic variations into intra and inter-population components. On an average 57% within and 43% between populations variability was revealed. A dendrogram was constructed on the basis of band sharing which separated the population into three clusters, i.e., China, Assam, and Ornamental type. The PCA revealed that the Chinary clones were more dispersed than Assam clones, indicating a greater genetic diversity among Chinary clones. The results showed that RAPD markers could be used effectively to distinguish and characterize Indian tea germplasm. Subsequently, several workers across the tea-growing countries used RAPD markers to study the genetic diversity as well as to make the fingerprints of their own clones. This was primarily due to simplicity and easy-to-perform nature of RAPD markers (Table 6.2).

6.8.1.2 Detection of Genetic Fidelity Among In Vitro-Raised Plants

The most important part of any in vitro propagation system is mass multiplication of plantlets which are phenotypically uniform and genetically akin to the mother/donor plant; otherwise the advantage of having desirable characters of elite supreme clones will not be achieved. Several approaches had been applied for identifying variants among micropropagated plants. These are phenotypic variation (Vuylsteke et al. 1988), karyotypic analysis of metaphase chromosomes (Jha and Sen 1992), and biochemical analysis (Damasco et al. 1996). Importantly, a major disadvantage of these techniques was the limited number of informative markers and the influence of environmental conditions or developmental process (Rani et al. 1995). Due to these limitations, the above approaches were not fully suitable for detecting DNA sequence polymorphisms of in vitro-raised plants. On the other hand, RAPD had been used very advantageously for number of crop species to detect genetic diversity among micropropagated plants (Isabel et al. 1993; Rani et al. 1995; Damasco et al. 1996). Mondal and Chand (2002) did RAPD analysis with 50 primers to investigate genetic variability of in vitro-raised tea plants of cultivar T-78. Among them, 39 primers developed 197 monomorphic bands in all the concerned plants. The primer SC 10–12 produced maximum eight bands and whereas primer SC 10–57 produced single monomorphic band. Twenty-four, out of 221 bands produced by the remaining 11 primers (SC 10–15, SC 10–56, SC 10–71, SC 10–83, OPA-12, OPA-15, OPA-18, OPM-06, OPM-09, OPM-12, and OPM-18), were polymorphic for four micropropagated tea plants. These 24 loci were monomorphic for the remaining plants. The polymorphic amplification products of 11 primers were found to be present in the 4 plants were compared to remaining 14 plants by conducting a second DNA extraction from respective plants. In a separate experiment, survey with these 11 primers revealed monomorphism of the amplified DNA fragments among 5 vegetatively propagated plants of the same cultivar (T-78). Therefore, it was concluded that variation occurred as a result of mutation during micropropagation. Interestingly, the marker profiles among these 4 plants were identical with all 11 primers suggesting a complete homogeneity among them. Various studies indicated that in vitro culture produced abundant cytological anomalies affecting both structural and numerical chromosome constitutions, which can be correlated with many phenotypic abnormalities (Evans and Sharp 1986). Subsequently several workers used RAPD markers to detect the genetic fidelity of in vitro as well as vegetatively derived tea plants (Table 6.2).

Morphological variation along with genotypic stability of tea cv. "Kolkhida" plantlets after 7 years of propagation was studied between 15 randomly chosen individuals from a set of 100 plantlets regenerated from callus and 10 individuals, randomly chosen from a set of 100 micropropagated plants. It has showed higher morphological variability among callus-derived regenerants than in micropropagated plants. Flow cytometry analysis showed no significant variability of nuclear DNA content among micropropagated plants; however changes in DNA indicated aneuploidy in three out of ten callus-derived regenerants. ISSR analysis showed lesser genetic distances among micropropagated plantlets and maternal cultivar "Kolkhida" compared with callus-derived plantlets. Yet, there is some degree of genetic instability after long-term micropropagation. This study clearly showed the genetic instability among the micropropagated plants of tea (Samarina et al. 2019). Various reasons had been assigned to the occurrence of variability among in vitro-raised plants. Smith (1998) reviewed the factors contributing to this variation and divided them into two intrinsic factors, which largely depend on the genetic stability of the explants, and extrinsic factors such as culture media, particular growth regulators, etc. Different types of cultured tissue produced different genetic instability. They had been ranked in the order of low to high from micropropagation of isolated shoot tips and meristems, adventitious shoots, somatic embryos, organogenesis from callus, and protoplast-derived cells (Scowcroft 1984). Such variants at times had a negative effect, affecting the high commercial value of the crop such as tea. Importantly, Hammerschlag (1992) suggested that somaclonal variations among tissue-cultured plants could be either eliminated or minimized if special efforts are made to distinguish between axillary and adventitious shoots that are produced during in vitro propagation and roughing out the adventitious shoots. However, these operations are not always possible to monitor fully in case of commercial tissue culture laboratory where wages of the laboratory workers are given on the basis of number of plants subcultured per day. Therefore, polymorphic ISSR and RAPD primers can be used for testing the genetic fidelity of micropropagated tea plant.

6.8.1.3 Cultivar Identification

RAPD markers were also employed for identification of the true-crossing progenies in tea breeding program and to determine relationship between parents and their hybrids (Saha-Roy et al. 2010; Bantawa et al. 2012). For example, two Japanese tea cultivars, Yutakamidori and Meiryoku for which parentage identification for registration documents were identified using this marker (Tanaka and Yamaguchi 1996). Wright et al. (1996) used the same technique to characterize five different South African tea cultivars, namely, SFS 150, SFS 204, PC1, PC81, and MFS87. Of the 20 arbitrary primers tested, only 1 (ABI-17) yielded a unique set of fingerprints for each cultivar, which allowed cultivar discrimination. Singh et al. (1999) isolated DNA from ten different processed dried commercial black and green tea samples. The isolated DNA was subjected for PCR amplification by random primers. Thus they demonstrated that this method had tremendous potential for testing the originality of commercial tea and for the identification of cultivars used by a tea manufacturer for a particular brand. Mondal et al. (2000) described a simple method of DNA isolation from 8 different polyphenol-rich genus as well as from 20 commercially important tea cultivars. The method did not require liquid nitrogen or phenol purification step. The DNA was successfully used as a template for PCR amplification, which indicated the wide applicability of the marker. Liang et al. (2000) investigated the possibility of classification and identification of tea as well as closely related species. The results showed that the RAPD markers could specifically discriminate between species and varieties. While both Assam and China tea had a specific band, Japanese tea was closer to Chinese tea than others. Some of the tea varieties from Vietnam were the hybrids of Assam and China. Tanaka et al. (2001) used RAPDs to identify the pollen parent of popular Japanese green tea cultivar "Sayamakaori." They screened 78 putative male plants, most of which were introduced from China and concluded that pollen parent of "Sayamakaori" was not present among the putative population. The genetic diversity and molecular phylogeny of 24 ornamental *Camellia* species and varieties were investigated by

RAPD analysis. Fifteen decamer oligonucleotide primers were selected from the 61 screened, which generated a total of 95.3% polymorphism of the amplified bands. The molecular phylogenetic dendrogram of 24 species was constructed using UPGMA that generated two groups, corresponding to 3- and 5-locular ovary in morphology. The genetic relationship and the molecular phylogeny among section Thea were discussed by Chen and Yamaguchi (2002). Maternal inheritance of chloroplast DNA (cpDNA) in some cross progenies between *C. vernalis* and *C. japonica* was investigated using the polymorphism of *atpH-atpI* region by RAPDs. The cpDNAs of all *C. vernalis* cultivars showed the same type as those of *C. sasanqua*, and all the progenies from *C. vernalis*, either open-pollinated or crossed, had the same cpDNA type as their maternal plants (Tateishi et al. 2007).

Internal transcribed spacer (ITS) of nrDNA had been widely employed for reconstructing phylogenetic relationships in plants, especially at the species level. In order to assess the efficacy of nrITS in elucidating the interspecific relationships of *Camellia*, Vijayan and Tsou (2008) conducted an experiment with seven closely or distantly related species. Extensive study of *Camellia*, based on Pfu polymerase-amplified ITS sequences, showed well-resolved interspecies relationships. Thus, the potential of nrITS in deducing the phylogenetic relationships in *Camellia* was demonstrated. RAPD markers were used for identification of *C. japonica* and related species as well as their hybrids. A wide range of markers such as random 10-mer to chloroplast-specific sequences were used and checked with the previously published monogram on *Camellia* for phylogenetic relationship. Even the taxonomic classification as mentioned in the Chang's manual for different *Camellia* species was confirmed using RAPD markers (Prince and Parks 1997, 2000; Thakor 1997; Tiao and Parks 1997, 2003; Yoshikawa and Parks 2001; George and Adam 2006; Orel et al. 2007). However, being dominant marker and limited degree of polymorphism, attention was shifted for alternative advance markers.

Bangladesh traditionally a back tea producing country. In maiden attempt to understand the genetic diversity of tea, 20 RAPD were used for 18 tea (*Camellia sinensis* L.) clones which generated an average of 37.75 RAPD bands per primer. Among all the bands, 97.41% were polymorphic in nature. Although a very preliminary work but still this diversity information was used for further breeding work of tea (Boonerjee et al. 2013).

6.8.2 Inter-Simple Sequence Repeat Markers (ISSR)

ISSR had been used for genetic characterization of various plant species (Tsumura et al. 1996). Because of greater length of ISSR primers, they show greater repeatability and stability of map position in the genome comparing genotypes of closely related individual (Zietkiewicz et al. 1994).

Twenty-five diverse tea cultivars were analyzed using the simple sequence repeat anchored polymerase chain reaction (SSR-anchored PCR) or inter-SSR-PCR (ISSR). Based upon the polymorphism, 12 out of 45 primers were chosen for final study. These 12 primers amplified a total of 130 bands, of which 108 (84%) were polymorphic. A dendrogram was constructed using UPGMA method and revealed three distinct clusters of Cambod, Assam, and China type, which concurred with the known taxonomical classification of tea. These results suggested that the ISSR-PCR method can be used potentially for genetic fingerprinting and taxonomic classification of tea genotypes (Mondal 2002). Subsequently, it had been noticed that ISSR markers had been used for number of applications such as to study the genetic diversity of tea germplasm as well as different species of *Camellia*, genetic fidelity of tissue cultured raised plants and for formulating the conservation strategies (Table 6.2).

6.8.3 Restriction Fragment Length Polymorphism Markers (RFLP)

RFLP had been used to investigate genetic diversity of cultivated plants and wild relatives (Tanksley et al. 1989). In tea, Matsumoto et al. (1994) cloned the *pal* gene using the rice cDNA as heterologous probe and used for studying the genetic variations of Japanese green tea using RFLP marker. On the basis of the numbers and molecular weight of the detected fragments, they classified Japanese green tea cultivars into five groups that had different routes of origin. All the Assam hybrids used in their study could be distinguished from Japanese green tea cultivars on the basis of the grouping. Furthermore, the inheritance of *pal* gene in tea was investigated. It was concluded that *pal* gene was a single copy per haploid genome and was inherited as a single gene according to the Mendelian ratio of 1:2:1. The work was further extended to the Japanese tea with green tea cultivars using *pal* as DNA marker. The DNA fragments detected by RFLP markers were named as A, B, and D which were inherited as multiple allelic genes at one locus. They concluded that most of the cultivars belonging to the AA genotype group had been selected from local tea plants about 50 years ago. On the other hand, all the cultivars in the BD genotype group were either from cultivar Yabukita or its hybrids. The other genotypes (AD, AB, and DD) included cultivars selected from local tea plants and their hybrids. To investigate the reason for the absence of any BB genotypes, allelic frequency of all genotypes were studied and found that B was the least among the three with a frequency of 0.08. Hence, the chance of occurrence of BB genotype was low, i.e., 0.0064 (0.08×0.08) (Matsumoto et al. 2002). RFLP technique was also applied to identify the processed tea sample. In Japan, tea produced from "Yabukita" fetched better realization in the market, and hence there was a trade tendency to adulterate the low-grade tea with Yabukita, which was difficult to detect either visually or through tester tong. To solve the problem, Kaundun and Matsumoto (2003a) employed the sequence tag site (STS)-RFLP using the sequence information of three genes, namely, phenyl-ammonia lyase (*PAL*), chalcone synthase (*CHS*), and dihydroflavonol 4-reductase (*DFR*). The restriction digestion of the specific amplification authenticated 46 tea samples.

6.8.4 Simple Sequence Repeat Markers (SSR)

SSRs, known also as microsatellites, are tandemly repeated DNA sequence motifs (usually 2–5 bp long) that are highly polymorphic in plant genomes (Wu and Tanksley 1993; Kuttubuddin et al. 2015). Due to their availability, hypervariability, relative ease of scoring by PCR, co-dominant nature, and high reproducibility, they are considered to be one of the most reliable genetic markers (Ganie et al. 2014). Since its inception during 1999, initial effort was made by several countries to generate the SSR markers in absence of much sequencing data (Bali et al. 2013; Sharma et al. 2009; Jin et al. 2006; Lin et al. (2011); Shi et al. (2013); Wei et al. (2010); Wen et al. (2012) Yang et al. (2009); Freeman et al. (2004) and their validation among the different tea genotypes (Ma et al. 2012; Ohsako et al. 2008). Later, it has become a very popular choice of DNA marker, (Table 6.2) for diverse purpose in tea and its wild species. They are for identification of cultivars, genetic diversity analysis, maternal inheritance of cp genome, linkage map construction, development of BAC library for tea genome sequence, genetic relationships, etc. There are several variations of SSRs in tea. They are nuclear SSRs, genic SSRs, chloroplast-specific SSRs, miRNA-based SSRs, and long-core SSRs.

Ueno et al. (1999) were pioneer to develop the SSRs from *C. japonica*, a closely related species of tea. Out of the total 339 amplifications by RAPD markers, 21 were found to contain microsatellite repeats. Finally, four primer pairs were designed, which yielded single locus polymorphic amplification products. Using these primer pairs, 53 *C. japonica* ecotypes were genotyped, and population genetic parameters were calculated. The following year, Ueno et al. (2000) investigated the spatial genetic structure of *C. japonica* using four of these microsatellite primers. Spatial distribution of individuals was also assessed to obtain an insight into spatial relationships between individuals and alleles. Morisita's index of dispersion plotted 518 individuals of *C. japonica* in a single clump, and Moran's I spatial autocorrelation coefficient revealed weak genetic structure, indicating a low level of allele clustering among the individuals. Subsequently, SSR markers had been used by several workers for cultivar identifications, study of genetic diversity of tea as well as wild *Camellia* species (Table 6.2). In certain occasions, SSRs had been mined in silico and validated either on diverse genotypes of tea or tested for their cross transferability on different species of *Camellia* (Yang et al. 2009).

C. taliensis is another species which is used to make tea in China. Fourteen nuclear microsatellite loci were used to determine the genetic diversity and domestication origin of *C. taliensis*, which were represented by 587 individuals from 25 wild, planted and recently domesticated populations. *C. taliensis* showed a moderate high level of overall genetic diversity. The greater reduction of genetic diversity and stronger genetic drift were detected in the wild group than in the recently domesticated group, indicating the loss of genetic diversity of wild populations due to overexploitation and habitat fragmentation. A little and non-significant reduction in genetic diversity was found during domestication. The long life cycle, selection for leaf traits, and gene flow between populations delayed the emergence of bottleneck in planted trees (Zhao et al. 2014).

6.8.4.1 miRNA-SSR

Micro RNAs are short, non-coding RNA molecules that are involved in post-transcriptional mode of gene regulation and thus effects on related phenotype. Thus development of microsatellite markers from the conserved non-coding genomic regions is being worthwhile for reviewing the genetic diversity of closely related species or self-pollinated species (Mondal and Ganie 2014; Guha et al. 2019). Although several SSR markers have been reported, in tea, the trait-specific SSR markers, leading to be useful in marker-assisted breeding technique, are yet to be identified. In this study, microsatellite motifs within the reported and predicted miRNA precursors were identified that are effectively followed by designing of primers from SSR flanking regions. Furthermore, 18 SSR motifs are found to be in 13 of all 33 predicted miRNAs. Trinucleotide motifs are most abundant among all followed by dinucleotides. Since miRNA-based SSR markers are evidenced to have significant role on genetic fingerprinting study, these outcomes would pave the way in developing novel markers for tagging tea-specific agronomic traits as well as substantiating non-conventional breeding program (Hazra et al. 2017).

6.8.4.2 Genetic SSR

One way to utilize publicly available expressed sequence such as ESTs or RNAseq data is to develop the SSRs. Sharma et al. (2009) used unigene that was available in NCBI during 2009 to develop the SSR markers, following that they validated the same and tested the cross species transferability. In wild species such as *C. oleifera*, RNAseq data from seeds generated 69,798 unigenes. A total of 6949 putative microsatellites were discovered among 6042 SSR-containing unigenes. Then, 150 simple sequence repeats (SSRs) were evaluated in 20 varieties of *C. oleifera* to study their effectiveness for revealing the genetic diversity.

6.8.4.3 Long-Core SSR

Long-core motifs are efficient for revealing the better variability. In tea, Wang et al. (2016) analyzed 66 elite clonal tea cultivars in China with 33 initially chosen long-core motif SSR markers covering all the 15 linkage groups of tea plant genome. A set of six SSR markers were selected as core SSR markers after further selection. The polymorphic information content (PIC) of the core SSR markers was >0.5, with ≤ 5 alleles in each marker containing 10 or fewer genotypes. Phylogenetic analysis revealed that the core SSR markers were not strongly correlated with the trait "cultivar processing-property." The combined probability of identity (PID) between two random cultivars for the whole set of six SSR markers was estimated to be 2.22×10^{-5}, which was quite low, and confirmed the usefulness of the proposed SSR markers for fingerprinting analyses in tea. Moreover, for the sake of quickly discriminating the clonal tea cultivars, a cultivar identification diagram (CID) was subsequently established using these core markers, which fully reflected the identification process and provided the immediate information about which SSR markers were needed to identify a cultivar chosen among the tested ones. The results suggested that long-core motif SSR markers used in the investigation contributed to the accurate and efficient identification of the clonal tea cultivars and enabled the protection of intellectual property (Wang et al. 2016).

6.8.5 Amplified Fragment Length Polymorphism Markers (AFLP)

AFLP being a reliable and the most robust DNA marker (Vos et al. 1995) can detect more number of polymorphisms than RFLPs or RAPDs. Thus AFLP markers offer an opportunity to perform detailed genetic studies in closely related population (Meksen et al. 1995). In tea, Paul et al. (1997) were the first to employ AFLP markers to detect diversity and genetic differentiation of 32 tea clones comprising Indian and Kenyan origin. Five enzyme-primer combinations revealed total 73 unambiguous polymorphic bands. The size of polymorphic bands ranged from 106 to 218 bp. Genetic diversity within population showed that the Chinary types were more variable than Assam or Cambod type. The similarity matrix coefficient varied from 35% to 96%. The dendrogram constructed on the basis of shared fragment into three known types, i.e., Assam, China, and Cambod, was generally consistent with the existing knowledge on the biosystematics of tea. According to the PCA, Assam clones from India and Kenya clustered closely indicating a common ancestry. In the same year, Rajasekaran (1997) reported the AFLP analysis of 42 tea clones that comprised 23 UPASI, 17 popular South Indian estate clones, and 2 Kenyan tea clones. Therefore, it was concluded that 90% of the UPASI clones were inbred and thus unsuitable for commercial cultivation. Importantly, three clones, namely, SMP-1, UPASI-15, as well as TRI-2025, clustered together, all of which were resistant to blister blight disease of tea whereas highly susceptible clones of the same diseases such as UPASI-4, UPASI-7 and UAPASI-12 grouped together. Therefore, he concluded that AFLP could be utilized successfully for developing the markers linked to blister blight diseases tolerance. Later several workers, in several countries, used AFLP markers for study the genetic diversity of tea as well as other *Camellia* species (Table 6.2). Interestingly, it had been also applied to identify the markers linked with drought tolerance in tea (Mishra and Sen-Mandi 2004).

6.8.6 Single Nucleotide Polymorphism (SNP)

Detection of allelic differences or variations in the PGRs is an important application of genomic resources which can be achieved by highly robust DNA-based marker such as SNP or its haplotype (i.e., group of SNPs that are inherited together and may link to a particular trait). Due to higher availability, predominantly biallelic nature, and stability during inheritance as compared to other markers, such as SSRs, SNPs provide enhanced possibilities for studying PGRs management in several ways such as cultivar identification, construction of genetic maps, detection of genetic diversity, detection of genotype vs. phenotype associations, and marker-assisted breeding. Although large-scale SNPs had been discovered in several crops using various sources of sequences, in tea not few SNPs have been identified till 2007. Huang et al. (2007) were first to identify the SNPs present in the coding region of polyphenol oxidase from different tea genotypes. Later, there was an attempt to scale up the throughput of discovery of SNPs. For an example, Zhang et al. (2014) identified 818 putative SNPs from expressed sequence tag (EST) databases for the tea plant, which produced a frequency of 1 SNP/170 bp. A direct sequencing method was then used to verify 253 putative SNPs in genome DNA of 17 tea varieties. Fifty (20%) candidates and 299 new SNPs were identified. The results indicate that *Camellia sinensis*-derived ESTs provide a valuable resource for SNP discovery. Furthermore, the abundance of SNPs in tea varieties is anticipated to generate the development of associated genetic studies, in addition to enhancing tea plant breeding programs (Zhang et al. 2014). Simultaneously, using a set of SNP markers developed from the EST database of tea, they genotyped diverse group of tea varieties, including both fresh and processed commercial loose-leaf teas. The validation led to the designation of 60 SNPs that unambiguously identified all 40 tested tea varieties. This method provides a powerful tool for variety authentication and quality control for the tea industry. It is also highly useful for the management of tea genetic resources and breeding, where accurate and efficient genotype identification is essential (Fang et al. 2014).

Due to several advantages, there was a tremendous progress for developing the SNP from the reduced representation of the genome as tea genome is quite large. Thus several protocols have been developed to generate the SNPs which were used for population structure analysis, linkage map, or genome-wide association study for marker–trait relationships. For an example, SNP markers were generated for using specific locus amplified fragment sequencing (SLAF-seq) to develop linkage map, double-digest restriction site-associated DNA sequencing (ddRAD-seq), or genotyping by sequencing (GBS) for developing the linkage map of tea.

To obtain genetic information about the germplasm of tea in Japan, 167 accessions including 138 var. *sinensis* (96 Japanese var. *sinensis* and 42 exotic var. *sinensis*) and 29 Assam hybrids were analyzed using SNPs markers identified by ddRAD-seq analysis. Approximately 10,000 SNPs were identified by ddRAD-seq and were mapped across the whole genome. The 167 tea accessions were classified into three genetic subgroups: (1) Japanese var. *sinensis*, (2) Japanese and exotic var. *sinensis*, and (3) Assam hybrids and exotic var. *sinensis*. Leaf morphology varied widely within each genetic subgroup. The 96 Japanese var. *sinensis* were classified into four genetic subgroups as follows: two subgroups of Shizuoka (the largest tea production region) landraces, Uji (most ancient tea production region) landraces, and the pedigree of "Yabukita," the leading green tea cultivar in Japan. These results indicated that the SNP markers obtained from ddRADseq are a useful tool to investigate the geographical background and breeding history of Japanese tea. This genetic information revealed the ancestral admixture situation of the "Yabukita" pedigree and showed that the genome structure of "Yabukita" is clearly different from those of other Japanese accessions (Yamashita et al. 2019).

6.8.7 Sequence Tagged Microsatellite Site Markers (STMS)

Recently, Caser et al. (2010) used STMS markers to study the genetic variations with 132 accessions of *Camellia* spp., which included 24 accessions representing 22 different species as well as 63 cultivars of *C. japonica*, 33 cultivars of *C. sasanqua*, 7 cultivars of *C. vernalis*, 3 cultivars of *C. hiemalis*, and 2 cultivars of *C. hybrida*. Cross transferability of 96 alleles was obtained that fruitfully amplified polymorphic alleles in all analyzed species which were scored. PCA could differentiate the *C. japonica* cultivars from the winter Camellias. The distribution of genetic variation, attributed by AMOVA, particularly highlighted genetic overlap among *C. sasanqua* cultivars and those belonging to *C. vernalis*, *C. hiemalis*, and *C. hybrida*.

Similarly, an initial study of STMS markers was undertaken by Matteo et al. (2010) with 132 accessions of *Camellia* species, which included 24 accessions representing 22 different species or varieties as well as 63 cultivars of *C. japonica*, 33 cultivars of *C. sasanqua*, 7 cultivars of *C. vernalis*, 3 cultivars of *C. hiemalis*, and 2 cultivars of *C. hybrida*. The four primer sets used (MSCJAF37, MSCJAH46, MSCJAF25, and MSCJAH38) successfully amplified polymorphic alleles in all the species analyzed, showing cross transferability. Overall, 96 alleles were scored. MSCJAH38 primer produced the highest number of alleles (30), while MSCJAH46 primer yielded the lowest number of alleles (15). Again the PCA plot revealed a main differentiation between the *C. japonica* cultivars and the winter Camellias. The distribution of the genetic variation, attributed by AMOVA, particularly highlighted genetic overlap among *C. sasanqua* cultivars and the cultivars belonging to *C. vernalis*, *C. hiemalis*, and *C. hybrida*. Thus both the studies demonstrated that STMS markers offered a suitable method for detection of genetic variability and molecular study of *Camellia* genotypes.

6.8.8 Single-Strand Conformation Polymorphism (SSCP)

Wachira et al. (1997) analyzed species introgression into cultivated gene pool of tea using 5 different organelle-specific primers in 19 species as well as 9 tea cultivars. Out of the five, three non-coding chloroplast regions as well as one mitochondrial region that amplified with universal primers did not reveal any polymorphisms. Remaining one cp DNA-specific PCR product revealed a SSCP. This SSCP in the intergenic spacer between the *trnL* (UAA) 3′ exon and *trn* F (GAA) indicated that four species, namely, *C. furfuracea*, *C. assimilis*, *C. nokoensis*, and *C. tsaii* shared a common haplotype. Thus the study indicated a possible hybridization between species of the sections involved.

6.8.9 Cleaved Amplified Polymorphic Sequence (CAPS)

CAPSs are alternative molecular markers, which combine both PCR and RFLP techniques. The techniques require minute amount of DNA and simple electrophoresis system to reveal polymorphisms. CAPS markers were successfully applied to a number of crops and forest species for which extensive nucleotide information are available. The advantages of these markers are development of mapped cDNA clones that represent expressed genes. The genetic diversity of tea in both taxa was investigated based with CAPS analysis of *PAL*, *CHS*, and *DFR* gene (Kaundun and Matsumoto 2003b). CAPS-based markers for *PAL* and *CHS* are also used to identify the high catechin content of tea (Elangbam and Misra 2016). These genes are involved in catechin and tannin biosynthesis, hence directly responsible for tea aroma. The critical factor in development of CAPS markers is the right choice of PCR primers targeting a unique sequence in a genome at a time, in order to ensure reliable results. This was facilitated here because both *PAL* and *DFR* exist in one and two copies, respectively, in tea. The *CHS* gene is present in three copies but with sufficient sequence difference among them, allowing specific amplification of *CHS2* only. Based on their analysis, it was revealed that China type was more diverse than Assam type and that a higher proportion of overall diversity resided within genotypes as compared to between genotypes. Even though no specific DNA profile was found for either tea varieties following any single PCR-RFLP analysis, a factorial correspondence analysis was carried out on all genotypes and markers separated the tea sample into two distinct groups according to their origin. This reflects the larger differences between the two taxa in their polyphenolic profiles. Thus, STS-based markers developed will be very useful in future mapping, population genetics, and fingerprinting studies of this important species and other *Camellia* species. This work also opens the way for correlative analysis between newly established tea genotypes based on phenylpropanoid genes and chemotypes, mainly polyphenolics.

CAPS markers are also used for the study of genetic diversity of tea. For an example, Ujihara et al. (2011) developed EST-based CAPS markers and used them to identify different tea cultivar in Japan. In another study, 37 CAPS markers were derived from genome and ESTs of tea to distinguish the tea germplasm. For identifying 12 prevailing tea cultivars in Taiwan, five core markers, including each one of mitochondria and chloroplast and three nuclear markers, were developed. Based on principal coordinate analysis and cluster analysis, 55 tea germplasm in Taiwan were divided into three groups: China type (*C. sinensis* var. sinensis), Assam type (*C. sinensis* var. assamica), and Taiwan wild species (*C. formosensis*). The result of genetic diversity analysis revealed that both sinensis and assamica types had higher genetic diversity than wild species. Thus CAPS markers are also found to be an efficient tool for tea germplasm discrimination and genetic diversity analysis (Hu et al. 2014a).

To develop the functional marker for identifying caffeine-less tea cultivars, *caffeine synthase* (*TCS*), gene was chosen. Attempt was made to determine the sequence of the six *tea caffeine synthase* (*TCS*) genes to search for polymorphisms and to prepare one of the *TCS* genes as a selection marker. Six *TCS* genes and the caffeine-less trait were mapped on the reference linkage map of tea. Strong linkage between the caffeine-less phenotype and *TCS1* indicated that it was a promising candidate as a causative gene of the caffeine-less trait. A three nucleotides insertion in *TCS1* was used as a selection marker named "CafLess-TCS1" that can be distinguished by sequencing. Caffeine-less individuals appeared in the progeny population of caffeine-less heterozygous individuals selected using "CafLess-TCS1." These results confirmed that the developed "CafLess-TCS1" will be an effective selection marker for breeding of caffeine-less tea cultivars (Ogino et al. 2019).

6.8.10 Start Codon Targeted (SCoT) and Sequence-Characterized Amplified Region (SCAR)

In recent years, many green tea cultivars have been released globally to meet the demand of green tea consumption. Therefore varietal improvement of green tea cultivars is also becoming the focus area of tea breeding. In order to assess the genetic diversity of the 18 popular green tea cultivars of China, SCoT and SCAR markers were used as a diagnosis tool to identify the cultivar. Thirty-one SCoT primers produced 264 loci, 226 of which were polymorphic indicating that they are highly useful. The genetic similarity coefficients among these green tea cultivars found to be high. The partitioning of groups using the UPGMA and PCoA were similar due to their robustness, and much of the clustering was highly consistent with the classification of tea cultivars according to their genetic backgrounds. Further, a unique SCoT band, SCoT4–1649, specific to the tea cultivar "Yingshuang," was transformed into a SCAR marker. This SCAR marker is highly useful for the identification and germplasm conservation of green tea cultivars in China (Xu et al. 2018).

Qinba area has a long history of tea planting and is a Northern most region of China where tea is grown. In order to provide basic data for selection and optimization of molecular markers of tea plants, 40 SRAP along with some other EST and SCoT markers were used to evaluate the genetic diversity of 50 tea plant genotypes collected from Qinba tea germplasm. These SRAP 40 markers generated 338 alleles (8.45 per primer). Furthermore, 320 alleles have been detected using 40 EST and SCoT primers (8.00 per primer). These results indicated that SCoT markers had higher efficiency. Further it was shown that the correlation between the genetic distance matrix based on EST-SSR and that based on SRAP was very small ($r = 0.01$), followed by SCoT and SRAP ($r = 0.17$), then by SCoT and EST-SSR ($r = 0.19$). The 50 tea samples were divided into two sub-populations. Furthermore, there is no obvious relationship between the results produced using sub-populational and geographical data.

Correlation between genetic distances produced by three different molecular markers was very small; thus it is not recommended to use a single marker to evaluate genetic diversity and population structure. It is hence suggested that combining of different types of molecular markers should be used to evaluate the genetic diversity and population structure (Zhang et al. 2018).

6.9 InDel Markers

Insertion deletion or InDel markers are particularly useful for those laboratories with limited resources, which also showed reliable transferability between distinct populations (Hu et al. 2014b). In a study, genome-wide genetic variations between tea "Shuchazao" and "Yunkang 10" were identified; 7,511,731 SNPs and 255,218 InDels markers based on their whole genome sequences generated their distinct types and distribution patterns. A total of 48 InDel markers that yielded polymorphic and unambiguous fragments were developed when screening six tea cultivars. These markers were further deployed on 46 tea cultivars for transferability and genetic diversity analysis which showed that the phylogenetic relationships among these tea cultivars that are highly consistent with their genetic backgrounds or original places. Interestingly, they observed that the catechin/caffeine contents between "Shuchazao" and "Yunkang 10" were significantly different, and a large number of SNPs/InDels were identified within catechin/caffeine biosynthesis-related genes. The identified genome-wide genetic variations and newly developed InDel markers will provide a valuable resource for tea plant genetic and genomic studies, especially the SNPs/InDels within catechin/caffeine biosynthesis-related genes, which may serve as pivotal candidates for elucidating the molecular mechanism governing catechin/caffeine biosynthesis (Liu et al. 2019).

6.10 Organelle DNA-Based Markers

Because of the relative resistance to evolutionary changes of organelle genome than nuclear genome, chloroplast (cp) and mitochondrial (mt) sequences had been widely used to investigate interspecific relationships (Jorgensen and Cluster 1989; Waugh et al. 1990). Chloroplast genome encodes for many agronomically important genes including large subunit of RUBP carboxylase oxygenase, a 32 kd thylakoid membrane protein (Bedrook and Kolodner 1979), and some other coupling factors (Nelson et al. 1980). Non-coding regions displayed higher rates of evolution than coding regions, hence former were desirable targets for phylogenic studies. The resolutions of many such non-coding regions had been amplified by the universal PCR primers (Taberlet et al. 1991). However, the relatively high frequency of insertion/deletions may even, in some cases, made it possible to use the size of PCR product as a genetic marker. The choice of cp and mt DNA sequences that maximized phylogenic information, however, depended upon the evolutionary time scale of the plant system. Borthakur et al. (1998) made a first attempt to study the cp DNA of tea. Good quality of cp DNA using a discontinuous sucrose gradient was isolated from ten different tea cultivars. To overcome the phenolic problem, 0.1% polyvinylpyrrolidone along with 0.1% BSA was used during homogenization. The clear greenish yellow bands of intact chloroplasts were collected from the interface of 20–45% sucrose gradient from where cp DNA was isolated. Thakor (1997) re-examined taxonomic relationship among the genus *Camellia* with the help of five cp DNA sequences from 25 *Camellia* species covering four subgenera of tea family. Out of which, four cp DNA sequences showed a low degree of variability (less than 2%). Remaining one revealed a much higher degree of variability (3.8–20%). Interestingly, the Phylogenetic Analysis Using Parsimony (PAUP) analysis of these sequence contradicted the sectional or subgeneric grouping of either Sealy's or Chang's monograph (Chang 1981). Prince and Parks (1997) analyzed the evolutionary relationship in tea subfamily Theoideae based upon 2 cp DNA regions, namely, *rbcL* and *matK* sequence, for 4 species of subfamily Ternstroemioideae, and 24 species from Theoideae. Later on, the same workers also examined the same cp DNA region (*rbcL* and *matK*) to confirm the family Theaceae, a natural group, as well as to evaluate the validity of circumscription of tribes and genus of its subfamily Theoideae (Prince and Parks 2000). The nucleotide sequences of *rbcL* gene in chloroplast DNA were determined on the native tea varieties of Japan, Korea, China, Southeast Asia, Sri Lanka, and India. Direct sequencing of the amplified cp DNA products was carried out. Alignments were obtained by assuming two substitutions, at nucleotide position 40 (adenine in China tea) and 948 (guanine in Assam tea). The nucleotide sequences of the *rbcL* gene in China and Assam were 99.8% similar. On the other hand, the 1370 nucleotide sequences of *rbcL* gene among *C. irrawadiensis*, *C. taliensis*, and Assam tea were the same except a different base at position 627. At this position in *C. irrawadiensis* and *C. taliensis*, thymine and adenine were observed, respectively, as specific bases (Kato 2001). The two major molecular phylogenetic investigations of the genus *Camellia* (one by Vijayan et al. 2009 with *nrITS* sequence and the other with *nDNA RPB2* sequence by Xiao and Parks 2003) had provided considerable insight into the interspecies relationships, which could not be provided by many previous attempts with the use of cpDNA sequences (Yang et al. 2006). These two molecular phylogenetic investigations shared many important findings. Both studies revealed the need to revise the existing classifications and supported the monophyly of sections Thea as well as Furfuracea. Also they suggested that the species *C. hongkongensis* should be shifted from section *Camellia* to section Furfuracea, and both studies revealed that sections Eriandra and Theopsis were closely related, not separable, and that species of sections Tuberculata, Chrysantha, as well as *C. szechuanensis* from section Pseudocamellia were closely related. Finally, results of both studies equally supported the section Camellia as polyphyletic. The species from the section Camellia grouped based on geographical origin and distribution and species in this section distributed in the south-eastern and eastern China, Korea, and Japan were well-separated from those in southern and south-western China. Nevertheless, both studies disagreed on many points. The most notable disagreement was the monophyly of the section Paracamellia defined by Ming (2000) and Sealy (1958), which was bifurcated of the section Paracamellia (Xiao and Parks 2003), and supported Chang's (1981) creation of the section Oleifera from the section Paracamellia. A recent study of leaf anatomical characters also supported the separation of Oleifera from the section Paracamellia (Lin et al. 2008). Other important differences were, first, species of the section Eriandra and Theopsis which formed a monophyletic clade in our tree but mixed together with species from the section Camellia and divided into two well-separated clades in the study by Xiao and Parks (2003). Second, the positions of some species in small sections and isolates differed; for an example, *C. amplexicaulis* of section Longipedicellata was isolated and was a sister to the clade of Eriandra and Theopsis but was associated with clades of species in sections Camellia, Oleifera, and Paracamellia in the dendrogram tree of Xiao and Parks (2003). Also, *C. yunnanensis* of section Stereocarpus was embedded in the clade consisted of sections Chrysantha and Tuberculata but was allied to section Furfuracea in the dendrogram tree of Xiao and Parks (2003). These types of conflicts in results are not uncommon in molecular phylogeny and can arise from both analytical and biological factors (Rokas et al. 2003a). Analytical factors that generally affect phylogenetic reconstruction were choice of optimality criterion (Huelsenbeck 1995), data availability (Cummings et al. 1995), taxon sampling (Graybeal 1998), and specific assumptions in the modeling of sequence evolution (Yang et al. 1994). The major biological factor that affects phylogenetic reconstruction was the evolutionary dynamics that may cause the history of the genes under analysis to obscure the history of the species (Rokas et al. 2003b).

6.11 Genetic Linkage Map

Construction of genetic linkage map is very important to know about the genome structure of an organism. It has several uses in plant breeding (Mondal 2004). However, the pre-requisite for any such work is a mapping population from where segregation pattern of the marker can be tagged. This is difficult in a woody species like tea due to the several problems as discussed in earlier Chap. 2.

A preliminary linkage map for tea plant was constructed with RAPD markers along with phenotypic characters such as theanine content, date of bud sprouting, resistance to anthracnose, and tolerance to cold (Tanaka 1996). However, in tea, a population thought to be derived from two known non-inbred parents was scored for RAPD and AFLP markers in order to develop a linkage map (Hackett et al. 2000). A very high proportion of the markers exhibited unexpected segregation ratio in the light of their configurations in parents, and an exploratory statistical analysis revealed patterns in the marker scores, which could most easily be explained by the hypothesis of three male parents contributing pollen to this cross. The map had 15 linkage groups with three or more markers, agreeing with the haploid chromosome number of tea. The statistical methods that revealed the sub-population were easy to apply routinely and may prove to be useful diagnostic tools for the analysis of non-inbred mapping populations. There were 126 markers, covering 1349.7 cM, with an average distance of 11.7 cM between loci on the map. Recently, an AFLP linkage map for tea plant was also constructed in China. The map of a female parent included 17 linkage groups and contained 208 markers, covering a total length of 2457.7 cM. The average distance between markers was 11.9 cM. A map from male parent included 16 linkage groups and located 200 markers, covering a total length of 2545.3 cM, and the average distance between markers was 12.8 cM (Huang et al. 2005). Most recently, a partial genetic map of backcross F_1 generation between "Zhenong 129" (selected from the open pollination of "Fuding Dabaicha" X "Yunnna Dayecha" parents) and "Fuding Dabaicha" was also generated using RAPD and ISSR markers (Huang et al. 2006a). However, in previous studies, the number of individuals for mapping was limited to 94 only, and the density was not high enough to meet the demand of precise mapping. These constructed maps were still limited to locate quantitative trait loci (QTLs) linked with some important traits due to their low distribution of molecular markers.

Further, a population of 54 F_1 clones derived from reciprocal crosses between "Sayamakaori" and "Kana-Ck17" was used for the linkage analysis. Maps of both parents were constructed from the F_1 population that was taken for BC_1 population. The order of most of the dominant markers in the parental maps was consistent. Thus a core map was constructed by merging the linkage data for markers that detected polymorphisms in both the parents. The core map contained 15 linkage groups, which corresponds to the basic chromosome number of tea. The total length of the core map was 1218 cM. Thus, the reference map as a central core map sandwiched between the parental maps for each linkage group; the combined maps contain 441 SSR, 7 CAPS, 2 STS, and 674 RAPD markers which were developed. This newly constructed linkage map can be used as a basic reference linkage map of tea (Taniguchi et al. 2007, 2012).

Another integrated genetic map of tea using a segregating F_1 population derived from a cross between two commercial cultivars ("TTES 19" and "TTES 8") was developed. Of the 505 mapped markers, there were 265 paternal markers (52.5%), 163 maternal markers (32.3%), 65 doubly heterozygous dominant markers (12.9%), and 12 co-dominant markers (2.4%). The co-dominant markers and doubly heterozygous dominant markers were used as bridge loci for the integration of the paternal and maternal maps. The integrated map comprised 367 linked markers, including 36 SSR, 3 CAPS, 1 STS, 250 AFLP, 13 ISSR, and 64 RAPD markers that were assigned to 18 linkage groups. The linkage groups represented a total map length of 4482.9 cM with a map density of 12.2 cM. The average density of the integrated genetic map constructed in this study was 12.2 cM (based on Haldane function) or 10.4 cM (based on Kosambi function), which was similar to the result of Huang et al. (2005) by using 208 and 200 markers to construct maternal and paternal map with density of 11.9 and 12.8 cM, respectively. This genetic map had the highest genetic coverage so far, which could be applied to comparative mapping, QTL mapping, and marker-assisted selection in the future (Hu et al. 2013). A consensus genic SSR-based linkage map was constructed that covered 1156.9 cM with 237 SSR markers distributed in 15 linkage groups (Tan et al. 2013).

Ma et al. (2015) developed a total of 6042 valid SNP markers using specific locus amplified fragment sequencing (SLAF-seq) and subsequently mapped them into the previous framework map. The final map contained 6448 molecular markers, distributing on 15 linkage groups corresponding to the number of tea plant chromosomes. The total map length was 3965 cM, with an average inter-locus distance of 1.0 cM. This map is the first SNP-based reference map of tea plant, as well as the most saturated one developed to date. The SNP markers and map resources generated in this study provide a wealth of genetic information that can serve as a foundation for downstream genetic analyses, such as the fine mapping of QTL, map-based cloning, marker-assisted selection, and anchoring of scaffolds to facilitate the process of whole genome sequencing projects for tea plant.

6.12 QTL Discovery

The application of genetic linkage map is many but primarily to identify the QTL and their subsequent transfer to elite variety. However, the prerequisite for identification of QTL is either developing the mapping population or identifying an association panel (Ganie et al. 2016; Mazumder et al. 2020). Being a woody plant, development of bi-parental mapping population is extremely difficult, and hence like any other woody plant, pseudo-testcross are widely used in tea for this purpose.

The history of discovery of QTL of tea went back to 2010 when Kamunya et al. (2010) identified 23 minor QTLs, yet approximately 250 polymorphic SSR markers validated these QTLs by various groups (Kato et al. 2008; Sharma et al. 2009). They explored using progeny from a cross between clones TRFCA SFS150 and AHP S15/10. The 42 clones were tested in two distinct tea-growing regions in Kenya. Bulk segregant analysis was performed by SSR genotyping. Out of 260 informative markers, 100 markers that showed 1:1 segregation were used to construct a linkage map. Much later, to characterize QTLs for catechin content in the tender shoots of tea plant, a moderately saturated genetic map was constructed using 406 SSR markers, based on a pseudo-testcross population of 183 individuals derived from an intraspecific cross of two tea varieties with diverse catechins composition. The map consisted of 15 linkage groups (LGs), corresponding to the haploid chromosome number of tea. The total map length was 1143.5 cM, with an average locus spacing of 2.9 cM. A total of 25 QTLs associated with catechins content were identified over 2 measurement years. The population variability explained by each QTL was predominantly at moderate-to-high levels and ranged from 2.4% to 71.0%, with an average of 17.7%. This is the first report on the identification of QTL for catechins content in tea plant. The results of this study provide a foundation for further cloning and functional characterization of catechin QTLs for utilization in improvement of tea plant (Ma et al. 2014). Understanding the genetic basis of theobromine and caffeine accumulation in the tea plant is important due to their contribution to tea flavor. Thus QTL analyses were carried out to identify genetic variants associated with theobromine and caffeine contents and ratio using a pseudo-testcross population derived from an intervarietal cross between two varieties of tea. A total of ten QTLs controlling caffeine content (CAF), theobromine content (TBR), sum of caffeine and theobromine (SCT), and caffeine-to-theobromine ratio (CTR) were identified over four measurement years. The major QTL controlling CAF, qCAF1, was mapped onto LG01 and validated across years, explaining an average of 20.1% of the phenotypic variance. The other QTL was detected in 1 or 2 years, and of them, there were four, two, and three for TBR, SCT, and CTR, respectively. It provides valuable information for further fine mapping and cloning functional genes and for genetic improvement in tea plant (Ma et al. 2018). To identify the flavonoids governing QTLs, a F_1 population (LJ43 × BHZ) was genotyped using 2d-RAD sequencing. A total of 13,446 polymorphic SNP markers was developed 1678.52-cM high-density map at an average interval of 0.40 cM which generated a total of 27 QTLs related to flavonoids or caffeine content (CAF) were mapped to 8 different linkage groups, LG01, LG03, LG06, LG08, LG10, LG11, LG12, and LG13, with an LOD from 3.14 to 39.54, constituting 7.5% to 42.8% of the phenotypic variation. Highest-density genetic map was constructed since the large mapping population of tea plants was adopted in present study. Moreover, novel QTLs related to flavonoids and CAF were identified based on the new high-density genetic map (Xu et al. 2018). To identify the QTLs related to purple coloration, a F_1 mapping population of a green (LJ43) and purple Zijuan (ZJ) cultivars was used which identified the transcription factor CsMYB75 and phi (F) class glutathione transferase CsGSTF1 as being associated with anthocyanin hyperaccumulation. Further it was found CsMYB75 promoted the expression of CsGSTF1 in transgenic tobacco. Although CsMYB75 elevates the biosynthesis of both catechins and anthocyanins, only anthocyanins accumulate in purple tea, indicating selective downstream regulation. In tea, anthocyanins accumulate in multiple vesicles, with the expression of CsGSTF1 correlated with BLC, but not with catechin content, in diverse germplasm (Wei et al. 2019). Another genetic analysis was done with "Ziyan," an anthocyanin-rich tea cultivar, and 176 of its naturally pollinated offspring. For two consecutive years, anthocyanins and catechins of "Ziyan" and the offspring population were quantified. While >60% of the offspring accumulated less than half of the amount of anthocyanins of "Ziyan," 17 (in first year) and 15 (in second year) individuals exceeded "Ziyan" in anthocyanin content. A negative correlation between anthocyanin and total catechin content ($r = -0.59$, $P < 0.001$) was observed. The population was genotyped with 131 SSR markers spanning all linkage groups of tea which identified 10 markers significantly associated with anthocyanins, catechins, and their ratios in both years. QTL analyses using the interval mapping method detected 13 QTLs, suggesting the dark purple trait of "Ziyan" is because of the pyramiding of anthocyanin-promoting alleles on at least five linkage groups. Two genetic loci reversely related to anthocyanin and total catechin contents were identified. This study provides valuable information for genetic improvement of purple tea cultivars and for fine-mapping related genes (Tan et al. 2020).

To identify putative QTL controlling, black tea quality and percent relative water traits in two tea genotypes and their F_1 progeny were used. A total of 1421 DArTseq markers derived from the linkage map identified 53 DArTseq markers to be linked to black tea quality and % RWC. All 53 DArTseq markers with unique best hits were identified in the tea genome. A total of 84 unigenes in 15 LGs were assigned to 25 different KEGG database pathways based on categories of secondary metabolite biosynthesis. The three major enzymes identified were transferases (38.9%), hydrolases (29%), and oxidoreductases (18.3%). The functional annotation of putative QTL identified in this current study may be associated with caffeine and catechins biosynthesis and % RWC. This study may help breeders in selection of parents with desirable DArTseq markers for development of new tea cultivars with desirable traits (Koech et al. 2018, 2019).

6.13 Population Genetics, Linkage Disequilibrium (LD), and Genome-Wide Association Studies (GWAS)

Association mapping is an important approach particularly for woody species such as tea. In order to characterize the LD pattern to facilitate GWAS and marker-assisted selection, Niu et al. (2019) did genetic diversity analysis of 415 tea accessions mainly from China and analyzed population structure, and LD pattern using GBS approach. A total of 79,016 high-quality SNPs were identified; PIC and GD based on these SNPs showed a higher level of genetic diversity in cultivated type than in wild type. Further analyses identified that 415 accessions were clustered into 4 groups, the Pure Wild Type, Admixed Wild Type, ancient landraces and modern landraces using structure, and the results were confirmed by PCA and UPGMA tree method. A higher level of genetic diversity was detected in ancient landraces and Admixed Wild Type than that in the Pure Wild Type and modern landraces. A relatively fast LD decay with a short range (kb) was observed, and the LD decays of four inferred populations were found to be different (Niu et al. 2019). The timing of spring bud flush (TBF) is of economic importance for tea plant breeding. GWAS was conducted to identify favorable SNP allelic variations as well as candidate genes that control TBF of tea using SLAFseq in a diversity panel comprising 151 tea plant germplasm resources. GWAS analysis revealed 26 SNPs associated with TBF in 3 years, and they identified a final significant SNP for TBF. To identify candidate genes possibly related to TBF, they screened seven candidate genes within 100 kb regions surrounding the trait-related SNP loci. Furthermore, the favorable allelic variation, the "TT" genotype in the SNP loci, was discovered, and a derived cleaved amplified polymorphism (dCAPS) marker was designed that cosegregated with TBF, which could be used for marker-assisted selection (MAS) breeding in tea. The results obtained from this study can provide a theoretical and applied basis for the MAS of early breeding in tea plants in the future (Wang et al. 2019).

An accurate depiction of the genetic relationship, the development of core collection, and GWAS are key for the effective exploitation and utilization of genetic resources. Here, GBS was used to characterize 415 tea accessions of China. A total of 30,282 high-quality SNPs was used to estimate the genetic relationships, develop core collections, and perform GWAS. Finally 198 and 148 accessions were found to represent the core set and mini-core set, which consisted of 47% and 37% of the whole collection, respectively, and contain 93 and 95% of the total SNPs. Furthermore, the frequencies of all alleles and genotypes in the whole set were very well-retained in the core set and mini-core set. The 415 accessions were clustered into 14 groups; the core and the mini-core collections contain accessions from each group, species, cultivation status, and growth habit. By analyzing the significant SNP markers associated with multiple traits, nine SNPs were found to be significantly associated with four leaf size traits. This study characterized the genetic distance and relationship of tea collections, suggested the core collections, and established an efficient GWAS analysis of GBS result (Niu et al. 2020).

6.14 Genomic Resource

Genomic resources are important component for plants where sequenced-driven research such as marker-assisted breeding, association mapping, cloning of genes, and mapping of QTLs are extremely important to develop the elite genotypes within a short period of time. In *Camellia*, 516,896 expressed sequence tags (as on 30 June 2020) are available in NCBI which are mainly associated with quality and few with abiotic as well as biotic stress (Table 6.3). Lin et al. (2011) constructed bacterial artificial chromosome library of tea and generated 4,01,280 clones with an average insert size of 135 kb. Among the other genomic resources, while Kamunya et al. (2010) identified 137 QTLs, yet, approximately 935,547 SSR markers for various *Camellia* species including tea are available (Kato et al. 2008; Sharma et al. 2009), 33,400 lncRNA (Varshney et al. 2019), 5070 miRNAs (Table 6.3). Apart from that, several SNPs (Huang et al. 2007) and microRNAs of tea (Das and Mondal 2010; Prabu and Mandal 2010; Mohanpuria and Yadav 2012; Zhu and Luo 2013) had also been identified.

C. reticulata was found as diploid, tetraploid, as well as hexaploid and hence expected to have a complex genome. To resolve the polyploid complexity, physical maps of the 18S–26S rDNA ribosomal RNA genes (rDNA) were generated by fluorescent in situ hybridization (FISH) for *C. reticulata*, including three types of ploidy of *C. reticulata* as well as with its related species such as *C. japonica*, *C. yunnanensis*, *C. pitardii*, and *C. saluenensis*. An advanced method was used for preparing chromosome spreads. Eight, twelve and eighteen rDNA sites were observed on the genomes of diploid, tetraploid, and hexaploid *C. reticulata*, respectively. Eight, four, five, and four rDNA loci were located on the chromosomes of

Table 6.3 Different genomic resources of tea

Name	Quantity	Reference
Popular cultivars	600	Mondal et al. (2004)
Different species of *Camellia*	325	Mondal et al. (2004)
BAC clones	401,280	Lin et al. (2011)
QTLs	137	Kamunya et al. (2010) and Ma et al. (2014, 2018)
Cloned genes	175	NCBI
lncRNA	20,000	Varshney et al. (2019)
miRNAs	5070	Das and Mondal (2010), Prabu and Mandal (2010), Mohanpuria and Yadav (2012) and Zhu and Luo (2013)
ESTs	516,896	NCBI
SSR markers	935,547	Kato et al. (2008), Sharma et al. (2009), Shi et al. (2013) and Dubey et al. (2020)

C. pitardii, *C. japonica*, *C. saluenensis*, and *C. yunnanensis*, respectively. The number and position of rDNA sites in these species were compared for analysis. The results supported some of the earlier phylogenetic speculations about this complex genome and suggested the occurrence of some structural rearrangements in chromosome (Gu and Xiao 2003).

6.15 Conclusion

In general, genetic improvement of tea was mainly contributed by conventional breeding; however, in the last two decades, great attempts had been made to intervene some of the problems of conventional breeding using modern techniques such molecular markers. Till now, several DNA markers have been used to make fingerprints, which need to be documented systematically and should be made available for public use to preserve the intellectual property right of tea breeders. Priority in the area of molecular breeding should be given: (1) to undertake a massive characterization of tea germplasm across the world through a common "Tea germplasm characterization consortium," which already exist for several similar crops; (2) DNA markers need to be identified to do early selection at nursery stage for various biotic (e.g., flower blight) and abiotic stresses (e.g., cold hardiness), which will revolutionize *Camellia* breeding where works suffer due to the lack of selection criteria and long gestation periods; (3) to develop the molecular markers for hybrid identification; (4) to generate large-scale SSRs as well as SNPs which have high potential for marker-assisted breeding; (5) to develop the trait-specific "core" germplasm; (6) allele mining for important genes and their utilization either for developing markers or for transgenic plant production; and (7) to develop a haplotype mapping using large-scale sequencing data of many genotypes. Finally, being a woody perennial, highly out-breeding in nature, tea possesses a good potential for association mapping than the simple bi-parental mapping which is totally unexplored in this genus.

References

Ackerman WL (1971) Genetic and cytological studies with *Camellia* and related genera. Technical Bull. No. 1427, USDA. U.S. Govt Print Office, Washington, DC, p 115

Afridi SG, Ahmad H, Alam M, Khan IA, Hassan M (2011) DNA landmarks for genetic diversity assessment in tea genotypes using RAPD markers. Afr J Biotechnol 10:15477–15482

Alcázar A, Ballesteros O, Jurado JM, Pablos F, Martín MJ, Vilches JL, Navalón A (2007) Differentiation of green, white, black, oolong, and Pu-erh teas according to their free amino acids content. J Agric Food Chem 55:5960–5965

Anderson S (1994) Isozyme analysis to differentiate between tea clones. Inligtingsbulletin Institut vir Tropiese en Subtropiese Gewasse 266:15

Annon (1997) IPGRI. Descriptors for tea (*Camellia sinensis*). International Plant Genetic Resources Institute, Rome

Ariyaratne PNK, Mewan KM, Goonetilleke WASNST, Attanayake DPSTG (2009) Study of genetic relationships of tea (*Camellia sinensis* (L.) O. Kuntze) using SSR markers and pedigree analysis. In: Proceedings of 9th agricultural research symposium, vol 1, pp 293–298

Azka NA, Widhianata H, Taryono (2019) Morphological and molecular characterization of 5 accessions of tea (*Camellia sinensis* (L.) O. Kuntze) exploited to develop high quality and quantity yield. In: AIP Conference Proceedings 020003. https://doi.org/10.1063/1.5098408

Balasaravanan T, Pius PK, Kumar RR (2002) Assessment of genetic fidelity among the in vitro propagated culture lines of *Camellia sinensis* (L.) O Kuntze using RAPD markers. In: Proceedings of the 15th plantation crops symposium PLACROSYM XV Mysore, 10–-13 December, pp 181–184

Balasaravanan T, Pius PK, Kumar RR, Muraleedharan N, Shasany AK (2003) Genetic diversity among south Indian tea germplasm (*Camellia sinensis, C. assamica* and *C. assamica* spp. Lasiocalyx) using AFLP markers. Plant Sci 165:365–372

Bali S, Raina SN, Bhat V, Aggarwal RK, Goel S (2013) Development of a set of genomic microsatellite markers in tea (*Camellia L.*) (Camelliaceae). Mol Breed 32:735–741

Bandyopadhyay T (2011) Molecular marker technology in genetic improvement of tea. Int J Plant Breed Genet 5:23–33

Banerjee B (1987) Can leaf aspect affect herbivory? A case study with tea. Ecology 68:839–834

Banerjee B (1992) Botanical classification of tea. In: Wilson KC, Clifford MN (eds) Tea cultivation to consumption. Chapman and Hall, London, pp 25–51

Bansal KC, Lenka SK, Mondal TK (2014) Genomic resources for breeding crops with enhanced abiotic stress tolerance. Plant Breed 133:1–11

Bantawa P, Das A, Ghosh PD, Mondal TK (2011) Detection of genetic diversity among the *Gaultheria fragrantissima* landraces by RAPDs: An endangered aromatic plants of Indo-China Himalayas. Indian J Biotechnol 10(July):294–300

Bantawa P, Das A, Ghosh PD, Mondal TK (2012) Genetic variation of extremely threatened medicinal plant Nepalese Kutki (*Picrorhiza scrophulariiflora*). Indian J Genet Plant Breed 72:103–106

Barua DN (1958) Leaf sclereids in the taxonomy of the *Thea Camellias*. I Wilson's and related *Camellias*. Phytomorphology 8:257–264

Barua PK (1963) Classification of tea plant. Two Bud 10:3–11

Barua DN, Dutta AC (1959) Leaf sclereids in taxonomy of *Thea camellias* II. *Camellia sinensis* L. Phytomorphology 9:372–382

Barua DN, Dutta KN (1971) Distribution of shoots on the plucking surface of a tea bush and its relation to spacing. Two Bud 18:8–11

Bedrook JR, Kolodner R (1979) The structure of chloroplast DNA. Annu Rev Plant Physiol 30:593–620

Bera B, Saikia H (2002) Randomly amplified polymorphic DNA (RAPD) marker analysis in tea (*Camellia sinensis L*) generative clones. In: Proceedings of the 15th plantation crops symposium PLACROSYM XV 10–13 December, pp 235–238

Bezbaruah HP (1971) Cytological investigation in the family theaceae-I. Chromosome numbers in some *Camellia* species and allied genera. Caryologia 24:421–426

Boonerjee S, Nurul Islam M, Hoque MI, Sarker RH (2013) Genetic diversity analysis of eighteen tea (*Camellia sinensis* L.) clones of Bangladesh through RAPD. Plant Tiss Cult Biotech 23 (2):189–199

Borchetia S, Das SC, Handique PJ, Das S (2009) High multiplication frequency and genetic stability for commercialization of the three varieties of micropropagated tea (*Camellia* spp.). Sci Hortic 120:544–550

Borse BB, Rao LJM, Nagalakshmi S, Krishnamurthy N (2002) Fingerprint of black teas from India: identification of the region-specific characteristics. Food Chem 79:419–424

Borthakur S, Mondal TK, Borthakur A, Deka PC (1995) Variation in peroxidase and esterase isoenzymes in tea leaves. Two Bud 42:20–23

Borthakur S, Mondal TK, Parveen SS, Guha A, Sen P, Borthakur A, Deka PC (1998) Isolation of chloroplast DNA from tea, *Camellia sp*. Indian J Exp Biol 36:1165–1167

Caser M, Torello Marinoni D, Scariot V (2010) Microsatellite-based genetic relationships in the genus camellia: potential for improving cultivars. Genome 53:384–399

Chang HT (1981) A taxonomy of the genus *Camellia*. In: Acta Scientarum Naturalium Universitatis, Sunyatseni, Monographic series, vol 1, pp 1–180

Chen C (1996) Analysis on the isozymes of tea plants F<sub(1)> hybrids. J Tea Sci 16:31–33

Chen L, Yamaguchi L (2002) Genetic diversity and phylogeny of tea plant (*Camellia sinensis*) and its related species and varieties in the section Thea genus *Camellia* determined by randomly amplified polymorphic DNA analysis. J Hortic Sci Biotech 77:729–732

Chen L, Yamaguchi S (2005) RAPD markers for discriminating tea germplasms at the inter-specific level in China. Plant Breed 124:404–409

Chen L, Zhou Z (2005) Variations of main quality components of tea genetic resources [*Camellia sinensis* (L.) O. Kuntze] preserved in the China National Germplasm Tea Repository. Plant Food Human Nutr 60:31–35

Chen L, Tong Q, Zhuang W (1992) Studies on pollen morphology and fuzzy clustering analysis of tea. Acta Agric Univ Zhejiang (Chinese) 18:29–36

Chen L, Qiqing T, Qikang G, Jilin S, Fulian Y (1997a) Observation on pollen morphology of 8 species and 1 variety in genus *Camellia*. J Tea Sci 17:183–188

Chen L, Chen D, Gao Q, Yang Y, Yu F (1997b) Isolation and appraisal of genomic DNA from tea plant (*Camellia sinensis* (L.) O. Kuntze). J Tea Sci 17:177–181

Chen L, Gao Q, Yang Y, Yu F, Chen D (1998a) Optimum amplification procedure and reaction system for RAPD analysis of tea plant (*Camellia sinensis* (L.) O. Kuntze). J Tea Sci 18:16–20

Chen L, Yang Y, Yu F, Gao Q, Chen D (1998b) Genetic diversity of 15 tea (*Camellia sinensis* (L.) O. Kuntze) cultivars using RAPD markers. J Tea Sci 18:21–27

Chen L, Yu F, Yang Y, Gao Q, Chen D, Xu C (1999) A study on genetic stability of excellent germplasm (*Camellia sinensis* (L.) O. Kuntze) using RAPD markers. J Tea Sci 19:13–16

Chen L, Wang PS, Yamaguchi S (2002a) Discrimination of wild tea germplasm resources (*Camellia* sp.) using RAPD markers. Agric Sci China 1:1105–1110

Chen L, Yamaghuchi S, Wang PS, Xu M, Song WX, Tong QQ (2002b) Genetic polymorphism and molecular phylogeny analysis of section Thea based on RAPD markers. J Tea Sci 22:19–24

Chen J, Wang PS, Xia YM, Xu M, Pei SJ (2005a) Genetic diversity and differentiation of *Camellia sinensis* L. (cultivated tea) and its wild relatives in Yunnan province of China, revealed by morphology, biochemistry and allozyme studies. Genet Res Crop Evol 52:41–52

Chen L, Gao QK, Chen DM, Xu CJ (2005b) The use of RAPD markers for detecting genetic diversity, relationship and molecular identification of Chinese elite tea genetic resources [*Camellia sinensis (L.) O. Kuntze*] preserved in a tea germplasm repository. Biodivers Conserv 14:1433–1444

Chen Q, Zhao J, Liu M, Cai J (2008) Nondestructive identification of tea (*Camellia sinensis* L.) varieties using FT-NIR spectroscopy and pattern recognition. Czech J Food Sci 26:360–367

Chen Y, Yu M, Xu J, Chen X, Shi J (2009) Differentiation of eight tea (*Camellia sinensis*) cultivars in China by elemental fingerprint of their leaves. J Sci Food Agric 89:2350–2355

Chengyin L, Weihua L, Mingjun L (1992) Relationship between the evolutionary relatives and the variation of esterase isozymes in tea plant. J Tea Sci 12:15–20

Chung MY, Epperson BK, Chung MG (2003) Genetic structure of age classes in *Camellia japonica* (THEACEAE). Evolution 57:62–73

Clark JY, Warwick K (1998) Artificial keys for botanical identification using a multilayer perception neural network (MLP). Artif Intell Rev 12:95–115

Cummings MP, Otto SP, Wakeley J (1995) Sampling properties of DNA sequence data in phylogenetic analysis. Mol Biol Evol 12:814–822

Damasco OP, Godwin ID, Smith MK, Adkins SW (1996) Gibberellic acid detection of dwarf off-types in micropropagated Cavendish bananas. Aus J Expt Agric 36:237–241

Das A, Mondal TK (2010) *In silico* analysis of miRNA and their targets in tea. Am J Plant Sci 1:77–86

De Castro O, Comparone M, Di Maio A, Del Guacchio E, Menale B, Troisi J et al. (2017) What is in your cup of tea? DNA Verity Test to characterize black and green commercial teas. PLoS ONE 12(5):e0178262

Deng WW, Ogita S, Ashihara H (2010) Distribution and biosynthesis of theanine in Theaceae plants. Plant Physiol Biochem 48:70–72

Dhiman B, Singh M (2003) Molecular detection of cashew husk *(Anacardium occidentale)* adulteration in market samples of dry tea (*Camellia sinensis*). Planta Med 69:882–884

Dubey H, Rawal HC, Rohilla M, Lama U, Mohan Kumar P, Bandyopadhyay T, Gogoi M, Singh NK, Mondal TK (2020) TeaMiD: a comprehensive database of simple sequence repeat markers of tea. Database 2020:baaa013

Eden T (1976) Tea. Longman, London, p 236

Elangbam M, Misra AK (2016) Development of CAPS markers to identify Indian tea (*Camellia sinensis*) clones with high catechin content. Genet Mol Res 15:gmr7860

Evans DA, Sharp WR (1986) Somaclonal and gametoclonal variation. In: Evans DA, Sharp WR, Ammirato PV (eds) Hand book of plant cell culture. Technique and applications, vol 4. Macmillan, New York, pp 97–132

Fan K, Hong Y-C, Ding Z-T, Wang Y (2010) Analysis of genetic diversity among natural hybrid progenies of *Camellia sinensis* 'Huangshanzhong'. Acta Hort Sin 37:1357–1362

Fang SW, Hua PR, Sheng WP, Mei X, Xing DH, Ping ZY, Hua LJ, Shao WF, Pang RH, Wang PS, Xu M, Duan HX, Zhang YP, Li JH (2003) RAPD analysis of tea trees in Yunnan. Sci Agrire Sinica 36:1582–1587

Fang W, Cheng H, Duan Y, Jiang X, Li X (2012) Genetic diversity and relationship of clonal tea (*Camellia sinensis*) cultivars in China as revealed by SSR markers. Plant Syst Evol 298:469–483

Fang WP, Meinhardt LW, Tan HW, Zhou L, Mischke S, Zhang D (2014) Varietal identification of tea (*Camellia sinensis*) using nanofluidic array of single nucleotide polymorphism (SNP) markers. Hort Res 1:14035

Ferguson JM, Grabe DF (1986) Identification of cultivars of perennial rye grass by SDS-PAGE of seed proteins. Crop Sci 26:170–176

Fernández-Cáceres PL, Martín MJ, Pablos F, González AG (2001) Differentiation of tea (*Camellia sinensis*) varieties and their geographical origin according to their metal content. J Agric Food Chem 49:4775–4779

Freeman SJ, West CJ, Lea V, Mayes S (2004) Isolation and characterization of highly polymorphic microsatellites in tea (*Camellia sinensis*). Mol Ecol Notes 4:324–326

Fukushima E, Iwasa S, Endo N, Yoshinari T (1966) Cytogenetics studies in *Camellia.* I Chromosome survey in some *Camellia* species. Jap J Hort 35:413–421

Ganie SA, Karmakar J, Roychowdhury R, Mondal TK, Dey N (2014) Assessment of genetic diversity in salt-tolerant rice and its wild relatives for ten SSR loci and one allele mining primer of salT gene located on 1st chromosome. Plant Syst Evol 300(7):1741–1747

Ganie SA, Borgohain MJ, Kritika TAK, Pani DR, Mondal TK (2016) Assessment of genetic diversity of *Saltol* QTL among the rice (*Oryza sativa* L.) genotypes. Physiol Mol Biol Plants 22(1):61–73

George O, Adam M (2006) Investigation into the evolutionary origins of Theaceae and genus *Camellia*. Int Camellia J 38:78–89

Ghosh-Hazra N (2001) Advances in selection and breeding of tea-a review. J Plant Crop 29:1–17

Goonetilleke WASNST, Priyantha PGC, Mewan KM, Gunasekare MTK (2009) Assessment of genetic diversity of tea (*Camellia sinensis* L.O. Kuntze) as revealed by RAPD-PCR markers. J Nat Sci Found Sri Lanka 37:147–150

Graybeal A (1998) Is it better to add taxa or characters to a difficult phylogenetic problem? Syst Biol 47:9–17

Gu Z, Xiao H (2003) Physical mapping of the 18S-26S rDNA by fluorescent *in situ* hybridization (FISH) in *Camellia reticulata* polyploid complex (Theaceae). Plant Sci 164:279–285

Guha PK, Mazumder A, Das A, Pani DR, Mondal TK (2019) *In silico* identification of long non-coding RNA based simple sequence repeat markers and their application in diversity analysis in rice. Gene Rep 16:100418

Gul S, Ahmad H, Khan IA, Alam M (2007) Assessment of diversity of tea genotypes through RAPD markers. Pak J Biol Sci 10:2609–2611

Gulati A, Rajkumar S, Karthigeyan S, Sud RK, Vijayan D, Thomas J, Rajkumar R, Das SC, Tamuly P, Hazarika M, Ahuja PS (2009) Catechin and catechin fractions as biochemical markers to study the diversity of Indian tea (*Camellia sinensis (L.) O.* Kuntze) germplasm. Chem Biodivers 6:1042–1052

Gunasekara MTK, Arachchige JDK, Mudalige AK, Peiris TUS (2001) Morphological diversity of tea (*Camellia sinensis* L.) genotypes in Sri Lanka. In: Proceedings of the 57th annual session of Sri Lanka Association for the Advancement of science (SLAAS) part I, Colombo, Sri Lanka, p 83

Hackett CA, Wachira FN, Paul S, Powell W, Waugh R (2000) Construction of a genetic linkage map for *Camellia sinensis* (tea). Heredity 85:346–355

Hairong X, Qiqing T, Wanfanz Z (1987) Studies on the genetic tendency of tea plant hybrid generation using isozyme technique. In: Proceedings of the international symposium on tea quality and human health, China, 4–9 December, p 21

Hammerschlag FA (1992) Somaclonal variation. In: Hammerschlag FA, Lit RE (eds) Biotechnology of perennial fruit crops. CAB International, Wallingford, pp 35–55

Hazarika M, Mahanta PK (1984) Composition changes in chlorophylls and carotenoids during the four flushes of tea in north-East India. J Sci Food Agric 35:298–303

Hazra A, Dasgupta N, Sengupta C, Das S (2017) Extrapolative microRNA precursor based SSR mining from tea EST database in respect to agronomic traits. BMC Res Notes 10:261

Hirai M, Kozaki I (1986) Isozymes of citrus leaves. In: Kitaura K, Akihama T, Kukimura H, Nakajima H, Horie M, Kozaki I (eds) Development of new technology for identification and classification of tree crops and ornamentals fruit tree research station. Ministry of Agriculture, Forestry and Fisheries, Tokyo, pp 73–76

Hou YJ, He Q, Li ZL, Li PW, Liang GL, Xu J (2007) ISSR applied to the germplasms identification of *Camellia sinensis*. Southwest China J Agric Sci 26:1272–1276

Hu CY, Lee TC, Tsai HT, Tsai YZ, Lin SF (2013) Construction of an integrated genetic map based on maternal and paternal lineages of tea (*Camellia sinensis*). Euphytica 191(1):141–152

Hu CY, Tsai YZ, Lin SF (2014a) Development of STS and CAPS markers for variety identification and genetic diversity analysis of tea germplasm in Taiwan. Bot Stud 55:12

Hu YY, Mao BG, Peng Y, Sun YD, Pan YL, Xia YM, Sheng XB, Li YK, Tang L, Yuan LP et al (2014b) Deep re-sequencing of a widely used maintainer line of hybrid rice for discovery of DNA polymorphisms and evaluation of genetic diversity. Mol Genet Genomics 289:303–315

Huang JA, Li JX, Huang YH, Luo JW, Gong ZH, Liu ZH (2005) Construction of AFLP molecular markers linkage map in tea plant. J Tea Sci 25:7–15

Huang JN, Li J, Huang Y, Luo J, Zong Z, Liu Z (2006a) Genetic diversity of tea 〔*Camellia sinensis* (L.) O. Kuntze〕 cultivars revealed by AFLP analysis. Acta Hort Sin 33:317–322

Huang FP, Liang YR, Lu JL, Chen RB (2006b) Genetic mapping of first generation of backcross in tea by RAPD and ISSR markers. J Tea Sci 26:171–176

Huang JA, Huang YH, Luo JW, Li JX, Gong ZH, Liu ZH (2007) Identification of single nucleotide polymorphism in polyphenol oxidase gene in tea plant (*Camellia sinensis*). J Hunan Agric Univ 33:454–458

Huelsenbeck JP (1995) Performance of phylogenetic methods in simulation. Syst Biol 44:17–48

Hui LX, Lin LC, Peng SZ, Wu LJ, Wen SC, Hua GZ, Xuan C, Li XH, Liu CL, Shi ZP, Luo JW, Shen CW, Gong ZH, Chen X (2004) Analysis of genetic relationships among "Rucheng Baimao cha" plants with RAPD method. J Tea Sci 24:33–36

Hung CY, Wang KH, Huang CC, Gong X, Ge XJ, Chiang TY (2008) Isolation and characterization of 11 microsatellite loci from *Camellia sinensis* in Taiwan using PCR-based isolation of microsatellite arrays (PIMA). Conserv Genet 9:779–781

Ikeda N, Kawada M, Takeda Y (1991) Isozymic analysis of *Camellia sinensis* and its interspecific hybrids. In: Proceedings of the International Symposium of Tea Science, Shizuoka, Japan, 26–28 August, p 98

Isabel NL, Tremblay MM, Tremblay FM, Bousquet J (1993) RAPD as an aid to evaluate the genetic integrity of somatic embryogenesis derived population of *Picea mariana* (Mill) B.S.P. Theor Appl Genet 86:81–87

Jeyaramraja PR, Jayakumar D, Kumar RR, Pius PK (2002) Peroxidase isozyme—a versatile tool for screening of drought tolerant clones. Newsl UPASI 12:1–7

Jha TB, Sen SK (1992) Micropropagation of an elite Darjeeling tea clone. Plant Cell Rep 11:101–104

Ji PZ, Zhang J, Wang PS, Huang XQ, Xu M, Tang YC, Liang MZ (2007) Genetic diversity of ancient tea plant in Yunnan province of China revealed by inter-simple sequence repeat (ISSR) polymerase chain reaction. J Tea Sci 27:271–279

Ji PZ, Li H, Gao LZ, Zhang J, Chen GZQ, Huang XQ (2011) ISSR diversity and genetic differentiation of ancient tea (*Camellia sinensis* var. *assamica*) plantations from China: implications for precious tea germplasm conservation. Pak J Bot 43:281–291

Jia BG, Lin Q, Feng YZ, Hu XY, Tan XF, Shao FG, Zhang L (2015) Development and cross-species transferability of unigene-derived microsatellite markers in an edible oil woody plant, *Camellia oleifera* (Theaceae). Genet Mol Res 14(2):6906–6916

Jin JQ, Cui HR, Chen WY, Lu MZ, Yao YL, Xin Y, Gong XC (2006) Data mining for SSR in ESTs and development of EST-SSR marker in tea plant (*Camellia sinensis*). J Tea Sci 26:17–23

Jin JQ, Cui HR, Gong XC, Chen WY, Xin Y (2007) Studies on tea plants (*Camellia sinensis*) germplasms using EST-SSR marker. Yi Chuan 29:103–108

Jorge S, Pedroso MC, Neale DB, Brown G (2003) Genetic differentiation of Portuguese tea plant using RAPD markers. Hort Sci 38:1191–1197

Jorgensen RA, Cluster PD (1989) Modes and tempos in the evolution of nuclear ribosomal DNA: new character for evolutionary studies and new markers for genetic and population studies. Ann Mo Bot Gard 75:1238–1247

Kabir SE, Ghosh-Hajra N, Chaudhuri TC (1991) Performance of certain clones under the agroclimatic conditions of Darjeeling. Tea Board India Tech Bull 5:1–8

Kafkas S, Ercisxli S, Doğan Y, Ertürk Y, Haznedar A, Sekban R (2009) Polymorphism and genetic relationships among tea genotypes from Turkey revealed by amplified fragment length polymorphism markers. J Am Soc Hort Sci 134:428–434

Kamunya SM, Wachira FN, Pathak RS, Korir R, Sharma V, Kumar R, Bhardwaj P, Chalo R, Ahuja PS, Sharma RK (2010) Genome mapping and testing for quantitative trait loci in tea (*Camellia sinensis* (L.) O. Kuntze). Tree Genet Genomes 6:915–929

Karthigeyan S, Rajkumar S, Sharma RK, Gulati A, Sud RK, Ahuja PS (2008) High level of genetic diversity among the selected accessions of tea (*Camellia sinensis*) from abandoned tea gardens in Western Himalaya. Biochem Genet 46:810–819

Kato M (2001) Analysis of differentiation of tea using DNA markers in Evergreens forest. In: International Conference on O-CHA (Tea) Culture and Science, Shizuoka, Japan, p 18

Kato F, Taniguchi F, Monobe M, Ema K, Hirono H, Maeda-Yamamoto M (2008) Identification of Japanese tea (*Camellia sinensis*) cultivars using SSR marker. J Jap Soc Food Sci Tech 55:49–55

Katoh Y, Katoh M, Takeda Y, Omori M (2003) Genetic diversity within cultivated teas based on nucleotide sequence comparison of ribosomal RNA maturase in chloroplast DNA. Euphytica 134:287–295

Kaundun SS, Matsumoto S (2002) Heterologous nuclear and chloroplast microsatellite amplification and variation in tea, *Camellia sinensis*. Genome 45:1041–1048

Kaundun SS, Matsumoto S (2003a) Identification of processed Japanese green tea based on polymorphism generated by STS-RFLP analysis. J Food Chem 51:1765–1770

Kaundun SS, Matsumoto S (2003b) Development of CAPS markers based on three key genes of the phenylpropanoid pathway in tea, *Camellia sinensis* (L.) O. Kuntze, and differentiation between assamica and sinensis varieties. Theor Appl Genet 106:375–383

Kaundun SS, Matsumoto S (2011) Molecular evidence for maternal inheritance of the chloroplast genome in tea, *Camellia sinensis* (L) O Kuntze. J Sci Food Agric 91:2660–2663

Kaundun SS, Park YG (2002) Genetic structure of six Korean tea populations revealed by RAPD-PCR markers. Crop Sci 42:594–601

Kaundun SS, Zhyvoloup A, Park YG (2000) Evaluation of genetic diversity among elite tea (*Camellia sinensis var. sinensis*) accessions using RAPD markers. Euphytica 115:7–16

Kerio LC, Wachira FN, Wanyoko JK, Rotich MK (2012) Characterization of anthocyanins in Kenyan teas: extraction and identification. Food Chem 131:31–38

Koech RK, Malebe PM, Nyarukowa C, Mose R, Kamunya SM, Joubert F, Apostolides Z (2018) Functional annotation of putative QTL associated with black tea quality and drought tolerance traits. Sci Rep 9:1465

Koech RK, Malebe PM, Nyarukowa C et al (2019) Functional annotation of putative QTL associated with black tea quality and drought tolerance traits. Sci Rep 9:1465. Published 2019 Feb 6

Kondo K (1975) Cytological studies in cultivated species of *Camellia*. PhD thesis. Univ. N.C., Chapel Hill, p 260

Kondo K (1977) Chromosome number in the genus *Camellia*. Biotrapica 9:86–94

Kondo K (1978a) Cytological studies in cultivated species of *Camellia* in: Encyclopedia of *Camellia* Japan. Camellia Soc, vol 2. Kodansha, Tokyo, p 456

Kondo K (1978b) Cytological studies in cultivated species of *Camellia*. Shi-Kaki 99:41–53

Kondo K, Parks CR (1979) Giemsa C-banding and karyotype of *Camellia* (−banded karyotypes can tell more detail on inter and intra-specific relationships in *Camellia*). Am Camellia Year Book 34:42–47

Kondo K, Parks CR (1980) Giemsa C-banding and karyotype of *Camellia*. In: Proceedings of the inter camellia Cong Kyoto, pp 55–57

Kondo K, Parks CR (1981) Cytological studies in cultivated species of *Camellia*. VI Giemsa C-banded karyotypes of seven accessions of *Camellia japonica* L *sensu lato*. Jap J Breed 31:25–34

Kubo N, Mimura Y, Matsuda T, Nagano AJ, Hirai N, Higashimoto S, Yoshida H, Uemura N, Fujii T (2018) Classification of tea (*Camellia sinensis*) landraces and cultivars in Kyoto, Japan and other regions, based on simple sequence repeat markers and restriction site associated DNA sequencing analysis. Genet Resour Crop Evol 66:441–451

Kulasegaram S (1980) Technical developments in tea production. Tea Q 49:157–183

Kuttubuddin AM, Debnath AB, Ganie SA, Mondal TK (2015) Identification and analysis of novel salt responsive candidate gene based SSRs (cgSSRs) from rice (*Oryza sativa L.*). BMC Plant Biol 15:122

Lai JA, Yang WC, Hsiao JY (2001) An assessment of genetic relationships in cultivated tea clones and native wild tea in Taiwan using RAPD and ISSR markers. Bot Bull Acad Sin 42:93–100

Latip SNH, Muhamad R, Manjeri G, Tan SG (2010) Genetic variation of selected *Camellia sinensis* (cultivated tea) varieties in Malaysia based on random amplified microsatellite (RAMs) markers Pertanika J. Trop Agric Sci 33:259–267

Lee S, Kim J, Sano J, Ozaki Y, Okubo H (2003) Phylogenic relationships among tea cultivars based on AFLP analysis. J Food Agric 47:289–299

Lee KJ, Lee JR, Sebastin R, Shin MJ, Kim SH, Cho GTm Hyun DY (2019) Assessment of genetic diversity of tea germplasm for its management and sustainable use in Korea gene bank. Forests 10:780

Li B, Yin Y, Zhou Y, Deng PQ, Yang HW (2003) Genetic diversity of two sexual tea cultivars detected by RAPD markers. J Tea Sci 23:46–150

Li J, Jiang CJ, Wang ZX (2005) RAPD analysis on genetic diversity of the preconcentrated core germplasms of *Camellia sinensis* in China. Yi Chuan 27:765–771

Li XH, Zhang CZ, Liu CL, Shi ZP, Luo JW, Chen X (2007) RAPD analysis of the genetic diversity in Chinese tea germplasm. Acta Hort Sin 34:507–508

Li JH, Nesumi A, Shimizu K, Sakata Y, Liang MZ, He QY, Zhou HJ, Hashimoto F (2010) Chemosystematics of tea trees based on tea leaf polyphenols as phenetic markers. Phytochemistry 71:1342–1349

Li L, Zhong YH, Ji YZ, Zhi YZ, Sui N (2012) Analysis on genetic diversity of ten insular populations of *Camellia japonica*. Acta Hort Sin 39:1531–1539

Li JB, Hashimoto F, Shimizu K, Sakata Y (2013) Chemical taxonomy of red-flowered wild camellia species based on floral anthocyanins. Phytochemistry 85:99–106

Liang YR, Tanaka JC, Takeda YY (2000) Study on diversity of tea germplasm by RAPD method. JI Zejiang Forest Coll 17:215–218

Lin XY, Peng QF, Tang X, Hu ZH (2008) Leaf anatomy of camellia sect. Oleifera and sect. Paracamellia (Theaceae) with reference to their taxonomic significance. J Syst Evol 46:183–190

Lin YP, Hu CY, Tsai YZ, Lin SF (2010) Studies on the fast molecular identification technologies of made tea (*Camellia sinensis*) varieties and their applications. Crop Environ Bioinform 7:37–51

Lin J, Kudrna D, Wing RA (2011) Construction, characterization, and preliminary BAC-end sequence analysis of a bacterial artificial chromosome library of the tea plant (*Camellia sinensis*). J Biomed Biotechnol 2011:476723–476731

Liu BY, Wang PS, Ji PZ, Xu M, Cheng H (2008) Study on genetic diversity of peculiar sect. *Thea* (L) Dye in Yunnan by ISSR markers. J Yunn Agric Univ 23:302–308

Liu BY, Li YY, Tang YC, Wang LY, Cheng H, Wang PS (2010) Assessment of genetic diversity and relationship of tea germplasm in Yunnan as revealed by ISSR markers. Acta Agro Sin 36:391–400

Liu Y, Yang SX, Ji PZ, Gao LZ (2012a) Phylogeography of *Camellia taliensis* (Theaceae) inferred from chloroplast and nuclear DNA: insights into evolutionary history and conservation. BMC Evol Biol 12:92–117

Liu B, Sun X, Wang Y, Li Y, Cheng H, Xiong C, Wang P (2012b) Genetic diversity and molecular discrimination of wild tea plants from Yunnan Province based on inter-simple sequence repeats (ISSR) markers. Afr J Biotechnol 11:11566–11574

Liu S, An Y, Tong W, Qin X, Samarina L, Guo R, Xia X, Wei C (2019) Characterization of genome-wide genetic variations between two varieties of tea plant (*Camellia sinensis*) and development of InDel markers for genetic research. BMC Genomics 20:935

Lou ZC, Shin YL, Ying LC (2004) Genetic relationship of Taiwan tea varieties. In: Proceeding of 2004 international Conf on O-CHA culture and Sci. 4–6 Shizuoka, Japan, pp 226–228

Lu H, Jiang W, Ghiassi M, Lee S, Nitin M (2012) Classification of *Camellia* (Theaceae) species using leaf architecture variations and pattern recognition Techniques. PLoS One 7:29704–29722

Luo JW, Shi ZP, Shen CW, Liu CL, Gong ZH, Huang YH (2002a) Stuies on genetic relationships of tea cultivars (*Camellia sinensis (L.) O.* Kuntze) by RAPD analysis. J Tea Sci 22:140–146

Luo JW, Shi ZP, LiJX SCW, Huang YH, Gong ZH (2002b) Study on the application of RAPD techniques to parentage identification of tea plant. J Huanan Agric Univ 3:3–6

Luo JW, Shi ZP, Shen CW, Liu CL, Gong ZH, Huang YH, Luo JW, Shi ZP, Shen CW, Liu CL, Gong ZH, Huang YH (2004) The genetic diversity of tea germplasms [*Camellia sinensis (L.) O.* Kuntze] by RAPD analysis. Acta Agro Sin 30:266–269

Ma JQ, Zhou YH, Ma CL, Yao MZ, Jin JQ, Wang XC, Chen L (2010) Identification and characterization of 74 novel polymorphic EST-SSR markers in the tea plant, *Camellia sinensis* (Theaceae). Am J Bot 97:153–156

Ma JQ, Ma CL, Yao MZ, Jin JQ, Wang ZL, Wang XC, Chen L (2012) Microsatellite markers from tea plant expressed sequence tags (ESTs) and their applicability for cross-species/genera amplification and genetic mapping. Sci Hortic 134:167–175

Ma JQ, Yao MZ, Ma CL, Wang XC, Jin JQ et al (2014) Construction of a SSR-based genetic map and identification of QTLs for catechins content in tea plant (*Camellia sinensis*). PLoS One 9(3): e93131

Ma JQ, Huang L, Ma CL, Jin JQ, Li CF, Wang RK et al (2015) Large-scale SNP discovery and genotyping for constructing a high-density genetic map of tea plant using specific-locus amplified fragment sequencing (SLAF-seq). PLoS One 10(6):e0128798

Ma JQ, Jin JQ, Yao MZ, Ma CL, Xu YX, Hao WJ, Chen L (2018) Quantitative trait loci mapping for theobromine and caffeine contents in tea plant (*Camellia sinensis*). J Agric Food Chem 66 (50):13321–13327. Epub 2018 Dec 10

Magambo MJS, Cannell MGR (1981) Dry matter production and partition in relation to yield of tea. Exp Agric 17:33–38

Magoma GN, Wachira FN, Obanda M, Imbuga M, Agong SG (2000) The use of catechins as biochemical markers in diversity studies of tea (*Camellia sinensis*). Genet Resour Crop Evol 47:107–114

Magoma GN, Wachira FN, Imbuga MO, Agong SG (2003) Biochemical differentiation in *Camellia sinensis* and its wild relatives as revealed by isozyme and catechin patterns. Biochem Syst Ecol 31:995–1010

Mariya JKM, Sasikumar R, Balasubramanian M, Saravanan M, RajKumar R (2003) Influence of light on catechin biosynthesis in tea. Tea 24:80–86

Matsumoto S, Takeuchi A, Hayastsu M, Kondo S (1994) Molecular cloning of phenylalanine ammonia-lyase cDNA and classification of varieties and cultivars of tea plants (*Camellia sinensis*) using the tea PAL cDNA probes. Theor Appl Genet 89:671–675

Matsumoto S, Kiriwa Y, Takeda Y (2000) Analysis of genetic diversity in Chinese tea (*Camellia sinensis*) using RFLP and detection of difference on tea from Japanese. Breed Sci 2:209

Matsumoto S, Kiriwa Y, Takeda Y (2002) Differentiation of Japanese green tea as revealed by RFLP analysis of phenyl-alanine ammonia-lyase DNA. Theor Appl Genet 104:998–1002

Matsumoto S, Kiriiwa Y, Yamaguchi S (2004) The Korean tea plant (*Camellia sinensis*):RFLP analysis of genetic diversity and relationship to Japanese tea. Breed Sci 54:231–237

Matteo C, Marinoni T, Daniela VS (2010) Microsatellite-based genetic relationships in the genus camellia: potential for improving cultivars. Genome 53:384–399

Mazumder A, Rohilla M, Bisht DS, Krishnamurthy SL, Barman M, Sarma RN, Sharma TR, Mondal TK (2020) Identification and mapping of quantitative trait loci (QTL) and epistatic QTL for salinity tolerance at seedling stage in traditional aromatic short grain rice landrace Kolajoha (*Oryza sativa* L.) of Assam, India. Euphytica 216(75):1–14

McKenzie JS, Jurado JM, Pablos FD (2010) Characterization of tea leaves according to their total mineral content by means of probabilistic neural networks. Food Chem 123:859–864

Meksen K, Leister D, Peleman J, Zabeau M, Salamini F, Gebhardt C (1995) A high resolution map of the vicinity of the R1 locus on chromosome V of potato based on RFLP and AFLP markers. Mol Gen Genet 249:74–81

Mewan KM, Liyanage AC, Everard JMDT, Gunasekara MTK, Karunanayake EH (2005) Studing genetic relationships among tea (Camellia sinensis) cultivars in Sri Lanka Using RAPD Markers. SL J Tea Sci 70(1):42–53

Mewan KM, Saha MC, Konstatin C, Pang Y, Abeysinghe ISB, Dixon RA (2007) Construction of genomic and EST-SSR based genetic linkage map of tea (*Camellia sinensis*). In: The 4th international conference on O-cha (tea) culture and science. November 2–4, 2007, Shizuoka, Japan, p 52

Ming TL (2000) Monograph of the genus *Camellia*. Kunming Institute of Botany, Chinese Academy of Sciences, Yunnan Science and Technology Press, Kunming, China

Mishra RK, Sen-Mandi S (2001) DNA fingerprinting and genetic relationship study of tea plants using amplified fragment length polymorphism (AFLP) technique. Ind J Plant Genet Resour 14:148–149

Mishra RK, Sen-Mandi S (2004) Molecular profiling and development of DNA marker associated with drought tolerance in tea clones growing in Darjeeling. Curr Sci 87:60–66

Mishra RK, Chaudhary S, Ahmad A, Pradhan M, Siddiqi TO (2009) Molecular analysis of tea clones (*Camellia sinensis*) using AFLP markers. Int J Integr Biol 5:130–135

Mohanpuria P, Yadav SK (2012) Characterization of novel small RNAs from tea (*Camellia sinensis* L.). Mol Biol Rep 39:3977–3986

Mondal TK (2000) Studies on RAPD marker for detection of genetic diversity, *in vitro* regeneration and *Agrobacterium*-mediated genetic transformation of tea (*Camellia sinensis*). Ph.D thesis, Utkal University, India

Mondal TK (2001) Camellia biotechnology: a bibliographic search. Int Soc Tea Sci 1:28–37

Mondal TK (2002) Detection of genetic diversity among the Indian tea (*Camellia sinensis*) germplasm by inter-simple sequence repeats (ISSR). Euphytica 128:307–315

Mondal TK (2004) Biotechnological improvements of tea ISB news letter. Texas Tech University, Lubbock, TX, p 24

Mondal TK, Chand PK (2002) Detection of genetic instability among the micropropagated tea (*Camellia sinensis*) plants. In Vitro Cell Dev Biol Plant 37:1–5

Mondal TK, Ganie SA (2014) Identification and characterization of salt responsive miRNA-SSR markers in rice (*Oryza sativa*). Gene 535(2):204–209

Mondal TK, Singh HP, Ahuja PS (2000) Isolation of genomic DNA from tea and other phenolic rich plants. J Planta Crop 28:30–34

Mondal TK, Bhattacharya A, Laxmikumaran M, Ahuja PS (2004) Recent advance in tea biotechnology. Plant Cell Tiss Org Cult 75:795–856

Mphangwe NIK, Vorster J, Steyn JM, Nyirenda HE, Taylor NJ, Apostolides Z (2013) Screening of tea (*Camellia sinensis*) for trait-associated molecular markers. Appl Biochem Biotechnol 171 (2):437–449

Mugnai S, Pandolfi C, Azzarello E, Masi E, Mancuso S (2008) *Camellia japonica* L. genotypes identified by an artificial neural network based on phyllometric and fractal parameters. Plant Syst Evol 270:95–108

Mukhopadhyay M, Mondal TK, Chand PK (2016) Biotechnological advances in tea (*Camellia sinensis* [L.] O. Kuntze): a review. Plant Cell Rep 13:1–33

Muoki RC, Wachira FN, Pathak RS, Kamunya SM (2007) Assessment of the mating system of *Camellia sinensis* in biclonal seed orchards based on PCR markers. J Hortic Sci Biotech 82:733–738

Nagarajah S, Ratnasurya R (1981) Clonal variability in root growth and drought resistance in tea (*Camellia sinensis*). Plant and Soil 60:153–155

Nelson N, Melson H, Schatz G (1980) Biosynthesis and assembly of the protein-translocating adenosine triphosphete complex from chloroplasts. Natl Acad Sci 77:1361–1364

Neog B, Yadav RNS, Singh ID (2004) Peroxidase, polyphenol oxidase and acid phosphatase activities in the stigma-style tissue of *Camellia sinensis* (L) O. Kuntze following compatible and incompatible pollination. J Indian Inst Sci 84:47–52

Ng'-Etich WK, Wachira FN (2003) Variations in leaf anatomy and gas exchange in tea clones with different ploidy. J Hortic Sci Biotech 78:173–176

Niu S, Song Q, Koiwa H, Qiao D, Zhao D, Chen Z, Liu X, Wen X (2019) Genetic diversity, linkage disequilibrium, and population structure analysis of the tea plant (*Camellia sinensis*) from an origin center, Guizhou plateau, using genomewide SNPs developed by genotyping-by sequencing. BMC Plant Biol 19:328

Niu S, Koiwa H, Song Q, Qiao D, Chen J, Zhao D, Chen Z, Wang Y, Zhang T (2020) Development of core-collections for Guizhou tea genetic resources and GWAS of leaf size using SNP developed by genotyping-by-sequencing. Peer J 8:e8572

Ogino A, Taniguchi F, Yoshida K, Matsumoto S, Fukuoka H, Nesumi A (2019) A new DNA marker CafLess-TCS1 for selection of caffeine-less tea plants. Breed Sci 69:393–400

Ohsako T, Ohgushi T, Motosugi H, Oka K (2008) Microsatellite variability within and among local landrace populations of tea, *Camellia sinensis* (L.) O. Kuntze, in Kyoto, Japan. Genet Resour Crop Evol 55:1047–1053

Orel G, Marchant AD, Wei CF, Curry AS (2007) Molecular investigation and assessment of *C. azalea* (syn. *C. changii* Ye 1985) as potential breeding material. Int Camellia J 39:64–75

Otaghvari AM, Garehyazie B, Hassanpur M, Sehgal D (2010) Isozyme banding patterns in Iranian tea clones. Bioinfolet 7:7–12

Owuor PO (1989) Differentiation of teas by the variations of linalools and geraniol contents. Bull Chem Soc Ethip 3(1):31–35

Owuor PO, Obanda M (1998) The use of chemical parameters as criteria for selecting for quality in clonal black tea in Kenya: achievements, problems and prospects: a review. Tea 19:49–58

Owuor PO, Reeves SG, Wanyoko JK (1986) Co-relation of flavins content and valuation of Kenyan black teas. J Sci Food Agric 37:507–513

Pandolfi C, Mugnai S, Azzarello E, Bergamasco S, Masi E, Mancuso S (2009) Artificial neural networks as a tool for plant identification: a case study on Vietnamese tea accessions. Euphytica 166:411–421

Park YG, Kaundun SS, Zhyvoloup A (2002) Use of the bulked genomic DNA-based RAPD methodology to assess the genetic diversity among abandoned Korean tea plantations. Genet Res Crop Evol 49:159–165

Parks CR, Case KF (1968) Chromatographic evidence for the genetic contamination of *Camellia saluenensis* in cultivation. In: American Camellias Year book, pp 124–134

Paul S, Wachira FN, Powell W, Waugh R (1997) Diversity and genetic differentiation among population of Indian and Kenyan tea (*Camellia sinensis* (L.) O. Kuntze) revealed by AFLP markers. Theor Appl Genet 94:255–263

Pedro L, Caceres F, Martin MJ, Pablos F, Gonzalez G (2001) Differentiation of tea (*Camellia sinensis*) varieties and their geographical origin according to their metal content. J Agric Food Chem 49:4775–4779

Pi E, Peng Q, Lu H, Shen J, Du Y, Huang F, Hu H (2009) Leaf morphology and anatomy of *Camellia* section *Camellia* (Theaceae). Bot J Linn Soc 159:456–476

Piyasundara JHN, Gunasekare MTK, Peiris TUS, Wickramasinghe IP (2006) Phenotypic diversity of Sri Lankan tea (*Camellia sinensis L*) germplasm based on morphological descriptors. Trop Agric Res 18:237–243

Piyasundara JHN, Gunasekare MTK, Wikremasinghe IP (2008) Identification of discriminating morphological descriptors for characterization of tea (*Camellia sinensis* L.) germplasm in Sri Lanka. Trop Agric Res 20:193–199

Prabu GR, Mandal AK (2010) Computational identification of miRNAs and their target genes from expressed sequence tags of tea (*Camellia sinensis*). Genomics Proteomics Bioinformatics 8:113–121

Prince LM, Parks CR (1997) Evolutionary relationships in the tea subfamily Theoideae based on DNA sequence data. Int Camellia J 29:135–144

Prince LM, Parks CR (2000) Estimation on relationships of Theoideae (Theaceae) inferred from DNA data. Int Camellia J 32:79–84

Prince ML, Parks CR (2001) Phylogenic relationships of Theaceae inferred from chloroplast DNA sequence data. Am J Bot 88:2320–2320

Raina SN, Ahuja PS, Sharma RK, Das SC, Bhardwaj P, Negi R, Sharma V, Singh SS, Sud RK, Kalia RK, Pandey V, Banik J, Razdan V, Sehgal D, Dar TH, Kumar A, Bali S, Bhat V, Sharma S, Prasanna BM, Goel S, Negi MS, Vijayan P, Tripathi SB, Bera B, Hazarika M, Mandal AKA, Kumar RR, Vijayan D, Ramkumar S, Chowdhury BR, Mandi SS (2012) Genetic structure and diversity of India hybrid tea. Genet Resour Crop Evol 59:1527–1541

Rajanna L, Ramakrishnan M (2010) Isozyme studies on some selected camellia clones. Int J Eng Sci Tech 2:6918–6921

Rajanna L, Ramakrishnan M, Simon L (2011) Evaluation of morphological diversity in south Indian tea clones using statistical methods. Maejo Int J Sci Technol 5:1–12

Rajasekaran P (1997) Development of molecular markers using AFLP in tea. In: Varghese JP (ed) Molecular approaches to crop improvement. Proceedings of National Seminar on molecular approaches to crop improvement 29–31 Dec. Kottayam, Kerala, India, pp 54–58

Ramkumar S, Kumar PS, Sudhakar G, Anitha J, Geetha S, Mohankumar P, Gopalakrishnan VK (2016) Biochemical and molecular analysis of *Camellia sinensis* (L.) O. Kuntze tea from the selected P/11/15 clone. J Genet Eng Biotechnol 14:69–75

Rani V, Ajay P, Raina SN (1995) Random amplified polymorphic DNA (RAPD) markers for genetic analysis in micropropagated plants of *Populus deltoides* marsh. Plant Cell Rep 14:459–462

Roberts EAH, Wight W, Wood DJ (1958) Paper chromatography as an aid to the identification of *Thea camellias*. New Phytol 57:211–225

Rokas A, King N, Finnerty J, Carroll SB (2003a) Conflicting phylogenetic signals at the base of the metazoan tree. Evol Dev 5:346–359

Rokas A, Williams BL, King N, Carroll SB (2003b) Genome scale approaches to resolving incongruence in molecular phylogenies. Nature 425:798–804

Roy SC, Chakraborty BN (2007) Evaluation of genetic diversity in tea of the Darjeeling foot hills, India using RAPD and ISSR markers. J Hill Res 20:13–19

Roy SC, Chakraborty BN (2009) Genetic diversity and relationship among tea (*Camellia sinensis*) cultivars as revealed by RAPD and ISSR based fingerprinting. Indian J Biotechnol 8:370–376

Sabhapondit S, Karak T, Bhuyan LP, Goswami BC, Hazarika M (2012) Diversity of catechin in northeast Indian tea cultivars. Sci World J. https://doi.org/10.1100/2012/485193

Saha-Roy O, Bantawa P, Ghosh SK, Tripati SB, Ghosh PD, Mondal TK (2010) Assessment of genetic diversity in Banana (*Musa* spp.) of north-eastern India by RAPD. Bioremediat Biodivers Bioavail 4:93–98

Samarina L, Gvasaliya M, Koninskaya N, Rakhmangulov R, Efremov A, Kiselyova N, Ryndin A, Hanke MV (2019) A comparison of genetic stability in tea [*Camellia sinensis* (L.) Kuntze] plantlets derived from callus with plantlets from long term *in vitro* propagation. Plant Cell Tiss Org Cult 138:467–474

Sanderson GW (1964) The chemical composition of fresh tea flush as affected by clone and climate. Tea Q 35:101–109

Saravanan M, Maria John KM, Raj Kumar R, Pius PK, Sasikumar R (2005) Genetic diversity of UPASI tea clones (*Camellia sinensis (L.)* O. Kuntze) on the basis of total catechins and their fractions. Phytochemistry 66:561–565

Satyanarayan N, Sharma VS (1982) Biometric basis for yield prediction in tea clonal selection. In: Proc PLACROSYM IV, Dec 3–5, 1981, Mysore, India, pp 237–243

Satyanarayan N, Sharma VS (1986) Tea (*Camelli*a L. spp.) germplasm in South India. In: Srivastava HC, Vatsya B, Menon KKG (eds) Plantation crops: opportunity and constraints. Oxford IBH Publishing, New Delhi, pp 173–179

Scowcroft WR (1984) Genetic variability in tissue culture: impact on germplasm conservation and utilization. Report on the IBPGR, vol 152. IBPGR, Rome

Sealy JR (1958) A revision of the genus camellia. Royal Horticultural Society, London

Sen P, Bora U, Roy BK, Deka PC (2000) Isozyme characterization in camellia spp. Crop Res 19:519–524

Seurei P (1996) Tea improvement in Kenya: a review. Tea 17:76–81

Shao WF, Pang RH, Duan HX, Wang PS, Xu M, Zhang YP, Li JH (2003a) RAPD analysis of tea in Yunnan. Sci Agric Sin 36:1582–1587

Shao WF, Pang RH, Duan HX, Wang PS, Xu M, Zhang YP, Li JH (2003b) Use of RAPD analysis to classify tea trees in Yunnan. Agric Sci China 2:1290–1296

Sharma P, Deka PC (2002) Identification of tea hybrids by leaf and pollen isozyme analysis. Res Crop 3:411–415

Sharma RK, Bhardwaj P, Negi R, Mohapatra T, Ahuja PS (2009) Identification, characterization and utilization of unigene derived microsatellite markers in tea (*Camellia sinensis* L.). BMC Plant Biol 9:53–77

Sharma RK, Negi MS, Sharma S, Bhardwaj P, Kumar R, Bhattachrya E, Tripathi SB, Vijayan D, Baruah AR, Das SC, Bera B, Rajkumar R, Thomas J, Sud RK, Muraleedharan N, Hazarika M,

Lakshmikumaran M, Raina SN, Ahuja PS (2010) AFLP-based genetic diversity assessment of commercially important tea germplasm in India. Biochem Genet 48:549–564

Sharma RKK, Negi MS, Sharma S, Bhardwaj P, Kumar R, Bhattachrya E, Tripathi SB, Vijayan D, Baruah AR, Das SC, Bera B, Rajkumar R, Thomas J, Sud RK, Muraleedharan N, Hazarika M, Sharma H, Kumar R, Sharma V, Kumar V, Bhardwaj P, Ahuja PS, Sharma RK (2011) Identification and cross-species transferability of 112 novel unigene-derived microsatellite markers in tea (*Camellia sinensis*). Am J Bot 98:133–138

Shen CW, Luo JW, Shi ZP, Gong ZH, Tang HP, Liu FZ, Huang YH (2002) Study on genetic polymorphism of tea plants in Anhua Yuntaihan population by RAPD. J Huan Agric Univ 28:320–325

Shen CW, Huang YH, Huang JA, Luo JW, Liu CL, Liu DH (2007) RAPD analysis for genetic diversity of typical tea populations in Hunan Province. J Agric Biotechnol 15:855–860

Shen CW, Ning ZX, Huang JA, Chen D, Li JX (2009) Genetic diversity of *Camellia sinensis* germplasm in Guangdong Province based on morphological parameters and SRAP markers. Ying Yong Sheng Tai Xue Bao 20:1551–1558

Shi SH, Tang SQ, Chen YQ, Qu LH, Chang HT (1998) Phylogenic relationship among yellow flowered camellia species based on random polymorphic DNA. Acta Phytotax Sin 36:317–322

Shi J, Dai X, Chen Y, Chen J, Shi J, Yin T (2013) Discovery and experimental analysis of microsatellites in an oil woody plant *Camellia chekiangoleosa*. Plant Syst Evol 299:1387–1393

Shu J, Chen L (1996) Study on the evolution route of tea pollen morphology. J Tea Sci 16:115–118

Singh ID (1980) Non conventional approaches to the breeding of tea in north-East India. Two Bud 27:3–6

Singh ID (1999) Plant improvement. In: Jain NK (ed) Global advances in tea. Aravali Book International, New Delhi, pp 427–448

Singh D, Ahuja PS (2006) 5S rDNA gene diversity in tea (*Camellia sinensis (L.)* O. Kuntze) and its use for variety identification. Genome 49:91–96

Singh HP, Ravindranath SD (1994) Occurrence and distribution of PPO activity in floral organs of some standard and local cultivars of tea. J Sci Food Agric 64:117–120

Singh M, Bandana, Ahuja PS (1999) Isolation and PCR amplification of genomic DNA from market samples of dry tea. Plant Mol Biol Rep 17:171–178

Singh M, Saroop J, Dhiman B (2004) Detection of intra-clonal genetic variability in vegetatively propagated tea using RAPD markers. Biol Plant 48:113–115

Singh M, Dhiman B, Sharma C (2011) Characterization of a highly repetitive DNA sequence in *Camellia sinensis* (L.) O. Kuntze genome. J Biotech Res 3:78–83

Singh S, Sud RK, Gulati A, Joshi R, Yadav AK, Sharma RK (2013) Germplasm appraisal of western Himalayan tea: a breeding strategy for yield and quality improvement. Genet Resour Crop Evol 60:1501–1513

Smith MK (1998) A review of factors influencing the genetic stability of micropropagated banana fruits. Aust J Exp Agric 43:219–223

Staub JE, Kuhns LJ, May B, Grun P (1982) Stability of potato tuber isozymes under different storage regimes. J Am Soc Hortic Sci 107:405–408

Su MH, Tsou CH, Hsieh CF (2007) Morphological comparisons of Taiwan native wild tea plant (*Camellia sinensis* (L.) O. Kuntze forma formosensis Kitamura) and two closely related taxa using numerical methods. Taiwania 52:70–83

Su MH, Hsieh CF, Tsou CH (2009) The confirmation of *Camellia formosensis* (Theaceae) as an independent species based on DNA sequence analyses. Bot Stud 50:477–485

Taberlet P, Gielly L, Pauton G, Bouvet J (1991) Universal primers for amplification of three non-coding regions of chloroplast DNA. Plant Mol Biol 17:1105–1109

Takeo T (1981) Variations in amounts of linalool and geraniol produced in tea shoots by mechanical injury. Phytochemistry 30:2149–2151

Takeo T (1983) Variation in the aroma compound content of semi-fermented tea and black tea. Nippon Nogei Kagaku Kaishi 57:457–459

Tan LQ, Wang LY, Wei K, Zhang CC, Wu LY et al (2013) Floral transcriptome sequencing for SSR marker development and linkage map construction in the tea plant (*Camellia sinensis*). PLoS One 8(11):e81611

Tan LQ, Zhang CC, Qi GN, Wang LY, Wei K, Chen SX, Zou Y, Wu LY, Cheng H (2015) Heterozygosities and genetic relationship of tea cultivars revealed by simple sequence repeat markers and implications for breeding and genetic mapping programs. Genet Mol Res 14:1557–1565

Tan LQ, Yang CJ, Zhou B, Wang LB, Zou Y, Chen W, Xia T, Tang Q (2020) Inheritance and quantitative trait loci analyses of the anthocyanins and catechins of *Camellia sinensis* cultivar 'Ziyan' with dark-purple leaves. Physiol Plant 170(1):109–119

Tanaka J (1996) RAPD linkage map of tea plant and the possibility of application in tea genetics and breeding. Tea Res J 84:44–45

Tanaka J (2006) Study on the utilization of DNA markers in tea breeding. Bull Nat Inst Veg Tea Sci 5:113–155

Tanaka J, Taniguchi F (2002) Emphasized-RAPD (e-RAPD): a simple and efficient technique to make RAPD bands clearer. Breed Sci 52:225–229

Tanaka J, Yamaguchi S (1996) Use of RAPD markers for the identification of parentage of tea cultivars. Bull Nat Res Inst Veg Orna Plant Tea 9:31–36

Tanaka J, Sawai Y, Yamaguchi S (1995) Genetic analysis of RAPD markers in tea. J Jap Breed 45:198–199

Tanaka J, Yamaguchi N, Nakamura Y (2001) Pollen parent of tea cultivar 'Sayamakaori' with insect and cold resistance may not exist. Breed Res 3:43–48

Tang S, Bin X, Wang L, Zhong Y (2006) Genetic diversity and population structure of yellow camellia (*Camellia nitidissima*) in China as revealed by RAPD and AFLP markers. Biochem Genet 44:449–461

Taniguchi F, Tanaka J, Kono I (2007) Construction of genetic linkage map of tea using SSR markers. In: proceedings of the 3rd international conference on O-cha (tea) culture and science (ICOS), Shizuoka, Japan

Taniguchi F, Furukawa K, Ota-Metoku S, Yamaguchi N, Ujihara T, Kono I, Fukuoka H, Tanaka J (2012) Construction of a high-density reference linkage map of tea (*Camellia sinensis*). Breed Sci 62:263–273

Tanikawa N, Onozaki T, Nakayama M, Shibata M (2008) PCR-RFLP analysis of chloroplast DNA variations in the atpI-atpH spacer region of the genus camellia. J Jap Soc Hort Sci 77:408–417

Tanksley SD, Yong ND, Paterson AH, Bonierbals MW (1989) RFLP mapping in plant breeding: new tools for an old science. Biotechniques 7:257–264

Tateishi N, Ozaki Y, Okubo H (2007) Occurrence of ploidy variation in *Camellia vernalis*. J Fac Agric Kyushu Univ 52:11–15

Thakor BH (1997) A re-examination of the phylogenetic relationships within the genus camellia. Int Camellia J 29:130–134

Thomas J, Vijayan D, Joshi SD, Joseph Lopez S, Raj Kumar R (2006) Genetic integrity of somaclonal variants in tea (*Camellia sinensis (L.)* O Kuntze) as revealed by inter simple sequence repeats. J Biotechnol 123:149–154

Tian M, LI JY, Ni S, Li XL (2008) Phylogenetic study on section *Camellia* based on ITS sequences data. J Acta Hort Sin 35:1685–1688

Tiao JX, Parks CR (1997) Identification of closely related camellia hybrid and mutant using molecular markers. Int Camellia J 29:111–116

Tiao JX, Parks CR (2003) Research for a new classification system for the genus camellia. Int Camellia J 33:109–112

Toyao T, Takeda Y (1999) Studies on geographical diversity of floral morphology of tea plant (*Camellia sinensis* (L.) O. Kuntze) using the method of numerical taxonomy. Tea Res J 87:39–57

Tsumura Y, Ohba K, Strauss SH (1996) Diversity and inheritance of inter-simple sequence repeat polymorphism in Douglas fir (*Pseudotsuga menziesii*) and sugi (*Cryptomeria japonica*). Theor Appl Genet 92:40–45

Ueno S, Tsumura Y (2009) Development of microsatellite and amplicon length polymorphism markers for Camellia japonica L. from tea plant (*Camellia sinensis*) expressed sequence tags. Mol Ecol Resour 9:814–816

Ueno S, Yoshimaru H, Tomaru N, Yamamoto S (1999) Development and characterization of microsatellite markers in *Camellia japonica* L. Mole Eco 8:335–336

Ueno S, Tomaru N, Yoshimaru H, Manabe T, Yamamoto S (2000) Genetic structure of *Camellia japonica* L. in an old-growth evergreen forest, Tsushima, Japan. Mol Ecol 9:647–656

Ujihara T, Taniguchi F, Tanaka J, Hayashi N (2011) Development of expressed sequence tag (EST)-based cleaved amplified polymorphic sequence (CAPS) markers of tea plant and their application to cultivar identification. J Agric Food Chem 59:1557–1564

Varshney D, Rawal HC, Dubey H, Bandyopadhyay T, Bera B, Mohan Kumar P, Singh NK, Mondal TK (2019) Tissue specific long non-coding RNA landscape and their association with aroma formation of tea. Ind Crop Prod 133:79–89

Vijayan K, Tsou CH (2008) Technical report on the molecular phylogeny of *Camellia* with nrITS: the need for high quality DNA and PCR amplification with *Pfu*-DNA polymerase. Bot Stud 49:177–188

Vijayan K, Zhang WJ, Tsou CH (2009) Molecular taxonomy of *Camellia* (Theaceae) inferred from NRITS sequences. Am J Bot 96:1348–1360

Vo TD (2006) Assessing genetic diversity in Vietnam tea [*Camellia sinensis* (L.) O. Kuntze] using morphology, inter-simple sequence repeat (ISSR) and microsatellite (SSR) markers, PhD Thesis, Georg-August Goettingen University, Germany

Vos P, Hogers R, Bleeker M, Reijans M, Van de Lee T, Hornes M, Freijters A, Pot J, Peleman J, Kuiper M, Zabeau M (1995) AFLP: a new techniques for DNA fingerprinting. Nucleic Acids Res 23:4407–4414

Vuylsteke D, Swennen R, Wilson GF, Langhe ED (1988) Phenotypic variation among *in vitro* propagated plantain (*Muse* sp. cultivar 'AAB'). Sci Hortic 36:79–80

Wachira FN (1990) Desirable tea plants: an overview of a search for markers. Tea 11:42–48

Wachira FN, Waugh R, Hackett CA, Powell W (1995) Detection of genetic diversity in tea (*Camellia sinensis*) using RAPD markers. Genome 38:201–210

Wachira FN, Powell W, Waugh R (1997) An assessment of genetic diversity among *Camellia sinensis* L. (cultivated tea) and its wild relatives based on randomly amplified polymorphic DNA and organelle specific STS. Heredity 78:603–611

Wachira F, Tanaka J, Takeda Y (2001) Genetic variation and differentiation in tea *(Camellia sinensis)* germplasm revealed by RAPD and AFLP variation. J Hortic Sci Biotech 76:557–563

Wambulwa MC, Meegahakumbura MK, Kamunya S, Muchugi A, Möller M, Liu J, Xu JC, Ranjitkar S, Li DZ, Gao LM (2016) Insights in to the genetic relationships and breeding patterns of the African tea germplasm based on nSSR markers and cpDNA sequences. Front Plant Sci 7:1244

Wambulwa MC, Meegahakumbura MK, Kamunya S, Muchugi A, Möller M, Liu J, Xu JC, Li DZ, Gao LM (2017) Multiple origins and a narrow genepool characterise the African tea germplasm: concordant patterns revealed by nuclear and plastid DNA markers. Sci Rep 7:4053

Wang BY, Ruan ZY (2012) Genetic diversity and differentiation in *Camellia reticulata* (Theaceae) polyploid complex revealed by ISSR and ploidy. Genet Mol Res 11:503–511

Wang XP, Ma B, Qi GN, Tian H, Fang CU, Zhang ZC, Yin XM (2007) RAPD analysis on the genetic relationships of tea cultivars grown in Sichuan. J Acta Hort Sin 34:242–244

Wang XF, Zheng WH, Zheng HX, Xie QQ, Zheng HY, Tang H, Tao YL (2010) Optimization of RAPD-PCR reaction system for genetic relationships analysis of 15 camellia cultivars. Afr J Biotechnol 9:798–804

Wang XF, Zheng HY, Zheng WH, Ao CQ, Jin HY, Zhao LH, Li N, Jia LR (2011) RAPD-based genetic diversities and correlation with morphological traits in *Camellia* (Theaceae) cultivars in China. Genet Mol Res 10:849–859

Wang RJ, Gao XF, Kong XR, Yang J (2016) An efficient identification strategy of clonal tea cultivars using long-core motif SSR markers. SpringerPlus 5:1152

Wang RJ, Gao XF, Yang J, Kong XR (2019) Genome-wide association study to identify favorable SNP allelic variations and candidate genes that control the timing of spring bud flush of tea (*Camellia sinensis*) using SLAF-seq. J Agric Food Chem 67:10380–10391

Waugh R, Vande Ven WTG, Phillips MS, Powell W (1990) Chloroplasts DNA diversity in the genus Rubus (Rosaceae) revealed by southern hybridization. Plant Syst Evol 172:65–75

Wei X, Wei JQ, Cao HL, Li F, Ye WH (2005) Genetic diversity and differentiation of *Camellia euphlebia* (Theaceae) in Guangxi, China. Ann Bot Fennici 42:365–370

Wei X, Cao H-L, JIang YS, Ye WH, Ge XJ, LI F (2008) Population genetic structure of *Camellia nitidissima* (Theaceae) and conservation implications. Bot Stud 49:147–153

Wei JQ, Chen ZY, Wang ZF, Tang H, Jiang YS, Wei X, Li XY, Qi XX (2010) Isolation and characterization of polymorphic microsatellite loci in *Camellia nitidissima* chi (Theaceae). Am J Bot 97:89–90

Wei K, Wang L, Zhou J, He W, Zeng J, Jiang Y, Cheng H (2011) Catechin contents in tea (*Camellia sinensis*) as affected by cultivar and environment and their relation to chlorophyll contents. Food Chem 125:44–48

Wei K, Wang L, Zhang Y et al (2019) A coupled role for CsMYB75 and CsGSTF1 in anthocyanin hyperaccumulation in purple tea. Plant J 97:825–840

Wen SC, Wu LJ, Peng SJ, Hua GZ, Ping TH, Zhi LF, Huan HY, Shen CW, Luo JW, Shi ZP, Gong ZH, Tang HP, Liu FZ, Huang YH (2002) Study on genetic polymorphism of tea plants in Anhua Yuntaishan population by RAPD. J Hunan Agric Univ 28:320–325

Wen Q, Xu L, Gu Y, Huang M, Xu L (2012) Development of polymorphic microsatellite markers in *Camellia chekiangoleosa* (Theaceae) using 454-ESTs. Am J Bot 99:203–205

Wendel JF, Parks CR (1982) Genetic control of isozyme variation in *Camellia japonica* L. J Hered 73:197–204

Wendel JF, Parks CR (1983) Cultivar identification in *Camellia japonica* L. using allozyme polymorphisms. J Am Soc Hort Sci 108:290–295

Wendel JF, Parks CR (1984) Distorted segregation and linkage of alcohol dehydrogenase genes in *Camellia japonica* L. (Theaceae). Biochem Genet 22:739–748

Wendel JF, Parks CR (1985) Genetic diversity and population structure in *Camellia japonica* L. (Theaceae). Am J Bot 72:52–65

Wickramaratna MRT (1981) Variations in some leaf characteristics in tea (*Camellia sinensis* L.) and their use in identification of clones. Tea Q 50:183–198

Wight W (1954) Morphological basis of quality in tea. Nature 173:630–631

Wight W (1958) The agrotype concept in tea taxonomy. Nature 181:893–895

Wight W, Barua DN (1954) Morphological basis of quality in tea. Nature 173:630–631

Wright LP, Apostolides Z, Louw AI (1996) DNA fingerprinting of tea clones. In: Whittle AM, Khumalo FRB (eds) Proceedings of the 1st Regional Tea Research Seminar. Blantyre, Malawi 22–23rd March. 1995, pp 44–50

Wright LP, Mphangwe NIK, Nyirenda HE, Apostolides Z (2000) Analysis of caffeine and flavan-3-ol composition in the fresh leaf of *Camellia sinensis* for predicting the quality of the black tea produced in central and southern Africa. J Sci Food Agric 80:1823–1830

Wright LP, Mphangwe NIK, Nyirenda HE, Apostolides Z (2002) Analysis of the theaflavin composition in black tea (*Camellia sinensis*) for predicting the quality of tea produced in central and southern Africa. J Sci Food Agr 82:517–525

Wu KS, Tanksley SD (1993) Abundance, polymorphism and genetic mapping of microsatellite in rice. Mol Genet 241:225–235

Wu LJ, Peng SJ, Wen SC, Lin LC, Hua GZ, Huan HY, Luo JW, Shi ZP, Shen CW, Liu CL, Gong ZH, Huang YH (2002a) Studies on genetic relationships of tea cultivars [*Camellia sinensis (L.)* O. Kuntze] by RAPD analysis. J Tea Sci 22:140–146

Wu LJ, Peng SZ, Xian LJ, Wen SC, Huan HY, Hua GZ, Luo JW, Shi ZP, Li JX, Shen CW, Huang YH, Gong ZH (2002b) Studies on genetic diversity of tea cultivars [*Camellia sinensis (L.)* O. Kuntze] by RAPD analysis. J Tea Sci 12:121–127

Wu H, Chen D, Li J, Yu B, Qiao X, Huang H, He Y (2013) *De novo* characterization of leaf transcriptome using 454 sequencing and development of EST-SSR markers in tea (*Camellia sinensis*). Plant Mol Biol Rep 31:524–538

Xiao TJ, Parks CR (2003) Molecular analysis of the genus camellia. Int Camellia J 35:57–65

Xiao LZ, Yan CY, Li JX, Luo JW, He YM, Zhao CY (2007) AFLP analysis on genetic diversity of Fenghuang-Dancong tea plant germplasm. J Tea Sci 27:280–285

Xu H, Ton Q, Zhuang W (1987) Studies on genetic tendency of tea plant hybrid generation using isozyme technique. In: Proceedings of International Tea Quality and Human Health Symposium, pp 21–25

Xu ZX, Liu ZH, Wang KB, Liu F, Shi L, Gao DZ, Xu ZX, Liu ZH, Wang KB, Liu F, Shi L, Gao DZ (2004) On relationship between green tea polyphenols and catechins and green tea raw materials. J Hunan Agric Univ 30:257–260

Xu LY, Wang LY, Wei K, Tan LQ, Su JJ, Cheng H (2018) High-density SNP linkage map construction and QTL mapping for flavonoid-related traits in a tea plant (*Camellia sinensis*) using 2d-RAD sequencing. BMC Genomics 19:955

Xu YX, Shen SY, Chen W, Chen L (2019) Analysis of genetic diversity and development of a SCAR marker for green tea (*Camellia sinensis*) cultivars in Zhejiang Province: the most famous green tea-producing area in China. Biochem Genet 57(4):555–570

Yamaguchi S (2001) Summarized remarks on the origin of Japanese tea by the genetic resource study in eastern Asia and Japan. In: Proceedings of 2001 International Conference on O-cha cult science. Shizuka, Japan. Oct 5–8

Yamashita H, Katai H, Kawaguchi L, Nagano AJ, Nakamura Y, Morita A et al (2019) Analyses of single nucleotide polymorphisms identified by ddRAD-seq reveal genetic structure of tea germplasm and Japanese landraces for tea breeding. PLoS One 14(8):e0220981

Yang Y, Sun T (1994) Study on the esterase isoenzyme in tea mutagenic breeding. China Tea 16:4–9

Yang Z, Goldman N, Friday A (1994) Comparison of models for nucleotide substitution used in maximum-likelihood phylogenetic estimation. Mol Biol Evol 11:316–324

Yang YJ, Yu FL, Chen L, Zeng JM, Yang SJ, Li SF, Shu AM, Zhang ZF, Wang YS, Wang HS, Wang PS, Xu M, Song WX, Guo JC, Yang RX, Zhang WJ, Chen ZH, Yang YJ, Yu FL, Chen L, Zeng JM, Yang SJ, Li SF, Shu AM, Zhang ZF, Wang YS, Wang HS, Wang PS, Xu M, Song WX, Guo JC, Yang RX, Zhang WJ, Chen ZH (2003) Elite germplasm evaluation and genetic stability of tea plants. J Tea Sci 23:1–8

Yang JB, Li HT, Yang SX, Li DZ, Yang YY (2006) The application of four DNA sequences to studying molecular phylogeny of camellia (Theaceae). Acta Bot Yunn 28:108–114

Yang JB, Yang J, Li HT, Zhao Y, Yang SX (2009) Isolation and characterization of 15 microsatellite markers from wild tea plant (*Camellia taliensis*) using FIASCO method. Conserv Genet 10:1621–1623

Yao MZ, Chen L, Liang YR (2008) Genetic diversity among tea cultivars from China, Japan and Kenya revealed by ISSR markers and its implication for parental selection in tea breeding prog. Plant Breed 127:166–172

Yao MZ, Ma CL, Qiao TT, Jin JQ, Chen L (2012) Diversity distribution and population structure of tea germplasms in China revealed by EST-SSR markers. Tree Genet Genomes 8:205–220

Yoshikawa N, Parks CR (2001) Systematic studies of *Camellia japonica* and closely related species. Int Camellia J 33:117–121

Zhang YP, Li JH (2003a) RAPD analysis of tea trees in Yunnan. Sci Agric Sin 36:1582–1587

Zhang YP, Li JH (2003b) Use RAPD analysis to classify tea trees in Yunnan. Agric Sci China 2:1290–1296

Zhang GW, Zhong WB, Wu Y, Tan XF, Du TZ (2007) Identification of oil tea (*Camellia Oleifera*) superior clones by ISSR molecular marker. Forest Res 20:278–282

Zhang CC, Wang LY, Wei K, Cheng H (2014) Development and characterization of single nucleotide polymorphism markers in *Camellia sinensis* (Theaceae). Genet Mol Res 13 (3):5822–5831

Zhang Y, Zhang X, Chen X, Sun W, Li J (2018) Genetic diversity and structure of tea plant in Qinba area in China by three types of molecular markers. Hereditas 155:22

Zhao CY, Zhou LH, Luo JW, Huang JA, Tan HP (2006) AFLP analysis of genetic diversity of tea plant germplasm in Guangdong province. J Tea Sci 26:249–252

Zhao LP, Liu Z, Chen L, Yao MZ, Wang XC (2008) Generation and characterization of 24 novel EST derived microsatellites from tea plant (*Camellia sinensis*) and cross-species amplification in its closely related species and varieties. Conserv Genet 9:1327–1331

Zhao DW, Yang JB, Yang SX, Kato K, Luo JP (2014) Genetic diversity and domestication origin of tea plant *Camellia taliensis* (Theaceae) as revealed by microsatellite markers. BMC Plant Biol 14:14

Zhao Y, Ruan CJ, Ding GJ, Mopper S (2017) Genetic relationships in a germplasm collection of *Camellia japonica* and *Camellia oleifera* using SSR analysis. Genet Mol Res 16(1): gmr16019526

Zhu QW, Luo YP (2013) Identification of microRNA and their targets in tea (*Camellia sinensis*). J Zhejian Uni Sci B 5:1–10

Zietkiewicz E, Rafalski A, Labuda D (1994) Genome fingerprinting by simple-sequence repeat (SSR) anchored polymerase chain reaction amplification. Genomics 20:176–183

Physiology and Biochemistry 7

7.1 Introduction

Being perennial, tea plants experience multiple abiotic and biotic stresses (ABstreses) in various combinations with varying duration that adversely influence and reduce the growth as well as yield of the plants (Mondal 2003, 2009). While biotic stresses typically reduced 20% yield, abiotic stresses estimated to reduce 65% reduction of yield in tea (Waheed et al. 2012). Stress evaluation and exploration of alleviatory actions are the two most active and kinetic research fields of plant science. Tea plants encounter stresses in several ways due to their unique growth habits. They are:

1. Being perennial, with a life span more than 100 years, bushes experience all season stress throughout their life.
2. The plants are maintained in a height of 60–80 cm by continuous pruning that leads to mechanical injury throughout their lives.
3. The young leaves are continually harvested which lead to wounding stress.
4. Usually grown in higher altitude and thus experience UV radiation stress.
5. Necessary but the shade trees grown in tea plantation provide low light intensity stress.
6. Being cultivated as monoculture, they provide complimentary circumstances for diverse pests.
7. Near equator, they encounter winter dormancy.
8. Always grown in low soil pH or acidic soils (Mondal et al. 2004).

The work related to stresses are discussed here.

T. K. Mondal, *Tea: Genome and Genetics*,
https://doi.org/10.1007/978-981-15-8868-6_7

7.2 Abiotic Stress

Being a perennial bush, tea plants need to combat an array of stresses throughout their whole life to facilitate adaptation as well as survival through the counteractive mechanisms to endure stresses. Nevertheless, excessive stresses beyond tolerance level will unavoidably culminate in oxidative damage attributable to intensive production of reactive oxygen species (ROS) (Smirnoff 1993; Mondal and Saha 2013). Oxidative stress is a phenomenon that surfaces as a consequence of unevenness involving the ROS production (Apel and Hirt 2004). In this chapter, progress in the research of abiotic and biotic stresses of tea plants has been appraised.

7.2.1 Moisture Stress

The physiochemical aspects of tea plant due to low moisture stress had been studied quite details. Water stress is a major limitation for plant survival and growth (Kefei et al. 1997; Bhat et al. 2020). Tea suffers both moisture deficit and inundation (Mokgalaka et al. 2004). Tea plants under low moisture stress showed reduced water potential (ψ) and sluggish growth rate (Yang et al. 1987). Tender leaves of tea exhibited higher diffusion resistance through stomatal regulation, but tolerant clones possessed higher relative water content (RWC) than their drought-sensitive counterparts (Das et al. 2015; Sandanam et al. 1981). However, clones that minimized water loss and possessed efficient osmotic adjustments can endure moisture deficiency. Soil moisture stress reduced RWC, ψ, phosynthetic rate (*Pn*); severely affected PS-II reaction centers; and increased diffusive resistance (Wijeratne et al. 1998). A colossal 258% increase in *Pn* had been observed in fully irrigated tea compared to drought-stressed tea (Smith et al. 1994). Experimental evidences suggested that critical levels of leaf ψ and transpiration (*E*) of drought tolerant tea clones were relatively higher than the sensitive genotypes. Similarly, maximal fluorescence (Fm), variable fluorescence (Fv), and their ratio (Fv/Fm) as well as activity potential of PS-II (Fv/Fo) also shrank leading to non-photochemical quenching and reduction of *Pn* (Guo et al. 2009). It had been detected that unavailability of moisture drastically reduced shoot-root ratio, leaf area, stem diameter, stomatal conductance (*gs*), and net CO_2 assimilation and adjusted the dry matter partitioning to roots which ultimately decreased yield (Bore 2008). Soil water status influenced *gs* that, in turn, affect *Pn* and *E* and played a key role in shaping the water status in tea leaves. Decreased *gs* in response to increasing irradiance, leaf temperature (T_L), and air vapor pressure was noticed in tea (Smith et al. 1993, 1994; De Costa et al. 2007). Further, it was found that total chl content increased during rainy season and decreased during summer indicating that chlorophyll (chl) synthesis too is affected by low moisture. Studies indicated that high humidity generated maximum *Pn*, whereas low temperature along with low soil moisture eased *Pn*. Under excessive water along with high humidity, photosynthetic photon flux density (PPFD) and *Pn* were reduced (Hajra and Kumar 2002), whereas low moisture stress increased electrical conductivity (Yang et al. 1987).

Several studies indicated that low soil moisture induced the activity of antioxidant enzymes such as peroxidase (POD), superoxide dismutase (SOD), catalase (CAT), glutathione reductase (GR), polyphenol oxidase (PPO), also phenyl ammonia lyase (PAL), etc. along with antioxidants such as polyphenol and osmolites such as proline. Although the increments were rapid in tolerant cultivars with compare to sensitive cultivars which might acted as defense mechanism to combat of low moisture stress (Yang et al. 1987; Wu and Pan 1995; Jeyaramraja et al. 2002a; Chakraborty et al. 2002; Upadhyaya and Panda 2004; Upadhyaya et al. 2008). Tea seeds under desiccation accrued H_2O_2 and increased ascorbate peroxidise (APX) and SOD activities (Chen et al. 2011b), whereas boron deficiency and water stress together decreased dry weight as well as photochemical parameters, but increment in proline and antioxidant activity was also evident (Hajiboland and Bastani 2012). In another study, low moisture stress reduced PAL activity in UPASI-2, UPASI-8, and UPASI-9 tea clones along with lowering the synthesis of EGCG and ECG due to molecular rearrangement with elevated leaf temperature. Similarly, flooding (high moisture stress) deteriorated made tea quality, because gallic acid and caffeine concentrations declined leading to reduced synthesis of epitheaflavic acid, epitheaflavic acid-3′-gallate, and theaflavic acid (Jeyaramraja et al. 2003a). Drought tolerant clones had been found to translocate more photosynthates to shoots apart than sensative one (Jeyaramraja et al. 2003b). Drought along with high temperature influenced on dry matter partitioning, and more dry matter was diverted to roots than to shoots during drought. Under drought, the amount of dry matter partitioned to leaves, stems, and harvested shoots was decreased by 80–95% (Burgess and Carr 1996).

It was found that low soil moisture stress reduced the accumulation of polyphenols in the leaf of tea, but again the reduction was much higher in sensitive genotype than tolerant genotypes. Additionally tolerant clones maintained a high amount of polyphenols even at low soil moisture content. Thus polyphenols may be used as biochemical indicators for selecting drought-tolerant cultivars (Cheruiyot et al. 2007).

The underlying mechanism of post-drought stress recovery in tea had been studied. Application of $CaCl_2$ thwarts drought induced oxidative stress by increasing DW, prolines, phenols, SOD, POD, CAT, GR activities, and plummeting H_2O_2 as well as MDA (Upadhyaya et al. 2011). Accumulation of K, Ca, Mn, and B accelerated during post-drought recovery because they positively modulated enzymes like SOD, CAT, POX, PPO, and GR (Upadhyaya et al. 2012). Under drought, continuous loss of soil moisture, by evapotranspiration, led to water deficiency in plants that eventually became fatal. Hence, application of antitranspirants is a rewarding choice. Phenylmercuric acetate (PMA) was the most useful (Nagarajah and Ratnasooriya 1977) due to its efficient translocation and obliteration within 20 d of application (Nagarajah and Ratnasooriya 1977). Alongside ABA, Raliidhan and Antistress (anti-transparent) were crucial as they increased stomatal diffusion resistance, ψ, and relative turgidity and reduced E (Handique and Manivel 1990). A linear relationship between total DW and the ratio between E and mean saturation vapor pressure deficit in clonal tea plants had been established that manifested discernible significance (Anandacoomaraswamy et al. 2000).

Several reports indicated that water deficiency regulated the catechin metabolism in tea plants. Longer rainy season (sufficient moisture stress) along with high humidity altered the quality constituents, such as EGC, epicatechin, ECG, EGCG, TC, and caffeine (Wang et al. 2011). Interestingly, EC quinone (ECQ) and EGCG quinone (EGCGQ), the oxidation products of EC and EGCG, increased up to 100- and 30-fold, respectively, in tea plants exposed to 19 d of water deficit. Oxidation of EC and EGCG preceded proanthocyanidins accumulation in leaves, which increased from 35 to 53 mg/g DW after 26 d of water deficit. Under low moisture stress, formation of ECQ and EGCGQ negatively correlated with the extent of lipid peroxidation in leaves, thus supporting a protective role for these compounds in drought-stressed plants (Hernandez et al. 2006). While epigallocatechin gallate (EGCg) and epicatechin gallate (ECG) showed no significant response to changes in soil water content, the shoot contents of EC and EGC in the six tea clones showed varied responses, with a distinct pattern in the water stress-tolerant clones (TRFK 6/8 and TRFK 31/30). The results suggested a potential use for EC and EGC as indicators to identify drought-tolerant cultivars in tea (Cheruiyot et al. 2008).

Low moisture stress increased water loss rate and decreased relative water content, dry mass, chl, carotenoid, and total phenolic contents of leaf and antioxidants like ascorbate and glutathione in tea. Antioxidant enzymes such as SOD, CAT, GR, etc. showed differential activities, whereas there was an increase in ROS and lipid peroxidation with decreased POX activities with progressive stress. Drought stress altered antioxidative response with apparent decrease in mineral nutrient (Zn, Ca, Na, Fe, Mg, and K) contents of leaves suggesting that mineral deficiency mediated drought stress-induced oxidative damage in tea. Tea plants exposed to heavy metals (e.g., Cd, Cu, Al) also showed reduction in growth and antioxidative responses. Further, a post-drought recovery study in tea also reveals that drought-induced biochemical damages are not permanent, as the plant recovers on rehydration. Mineral nutrients play an important role in post-drought recovery in tea. The process of recovery was significantly influenced by foliar spray of K, Ca, Zn, etc., leading to improved antioxidant potential. Thus, drought tolerance and post-drought recovery can be improved by application of nutrients like K, Ca, Zn, B, etc. (Upadhyaya and Panda 2013).

Recently genotypic variation of 16 tea clones was studied in Iran. It was found that there was significantly variation of several biochemical and physiological parameters among the genotype under drought stress at field level. For an example, the clone 100 can tolerate drought stress better because of high protein content. Using this study, the group has identified drought-tolerant clone in Iran (Rahimi et al. 2018). It has also been found that drought stress modify cuticle wax coverage, increased thickness, and osmiophilicity of young and mature leaf of tea. Further drought stress induced new wax species and remodeled the composition of existing waxes; the chain length distributions of alkanes, esters, glycols, and terpenoids were altered in complex manners (Chen et al. 2020).

7.2.2 Temperature Stress

Low temperature is one of the key environmental stresses that tea bushes necessitate to cope with their entire life cycle. Tea plants growing beyond 16°N or S latitudes encounter winter dormancy with corresponding variations in day and night temperatures (Barua 1989; Nandi and Palni 1993). Environmental cues including low temperature forced plants to undergo oxidative stress, which was accompanied by enhanced production of reactive oxygen species (ROS). These toxic ROS were scavenged using an array of enzymes. Among them, SOD plays an important role which was characterized from tea leaf (Vyas and Kumar 2005a).

Barua (1969) induced dormant tea buds by application of GA_3 and suggested that the balance of growth-promoting and growth-inhibiting substances might determine the state of dormancy in tea shoots. High levels of free ABA and phenols were observed in dormant shoots of tea, and it was suggested that they could possibly play major role in dormancy (Nagar 1996). The possible involvements of PAs (Kakkar and Nagar 1997), GA_3 (Nagar and Kumar 2000), and brassinosteroids (Gupta et al. 2004) were established in tea shoots during dormancy. Conjugated IAA levels were increased with onset of dormancy reaching its maximum value when free IAA levels were lowest. Nagar and Sood (2006) reported that with increase of free IAA levels, the conjugated IAA levels decreased in shoots prior to dormancy suggesting metabolic inter-conversion of IAA during these periods. Effects of low temperature on the response of two cultivars of tea with contrasting periods of winter dormancy were studied. The two cultivars were "Teenali 17/154" (TNL) and "Kangra Jat" (KNJ), where the former showed shorter periods of winter dormancy than the latter. Low temperature treatment (5 °C) resulted in an increase of superoxide (O_2^-) ion and cellular damage (as measured by tetrazolium chloride reduction test) in both the clones; however, the increase was lesser in the case of TNL compared to KNJ. Activities of SOD, APX, and GR increased in both the clones in response to low temperature; however, GR activity exhibited significant differences ($P < 0.05$) between the two clones. Low temperature caused increase in the activity of various isozymes such as SOD, APX, and GR. A new isozyme of SOD (Cu/Zn type) was induced in both the clones at low temperature. Significantly, higher GR activity in both the clones suggested a role of this enzyme in imparting better protection to tea at low temperature. Also, variations for GR isozyme were observed between the clones. Therefore, it was concluded that TNL, a clone with relatively lesser period of winter dormancy, experienced lesser oxidative stress in response to low temperature compared to KNJ (Vyas and Kumar 2005b).

Frost during winter further injured young pluckable shoots which led to defoliations (Mondal 2003). During winter dormancy, low temperatures alone or in combination with low irradiation induced oxidative stress in plants. It had been found that genotypes having shorter dormancy period displayed elevated activities of antioxidant enzymes such as SOD, APX, and GR due to accrual of ROS. Efficient scavenging of toxic oxygen species lowered down accumulations during winter months, and thus it was coupled with reduced winter dormancy (Vyas et al. 2007). Photosynthesis can be a potential marker to categorize low temperature stress effects on plants. In fact, low temperature-tolerant tea genotypes possessed higher MDA, proline, and carotenoid contents along with enhanced ROS scavenging machinery than the sensitive clones (Netto et al. 2005). Clones, for instance, TTL-1, TTL-4, and UPASI-9, were found to be frost tolerant compared to clones TTL-6, SM/OM/54, as well as SMP-1, and thus former clones were suitable for a region having recurrent frost occurrence because cultivation of these clones will ensure the higher yield and quality (Netto et al. 2005). It was reported that low temperature induced albinism in tea but reverted back the albino phenomenon when the temperature was above 20 °C. In albino tea genotype, the development of chl from etioplast and the accumulation of chl a and b was blocked at 15 °C, thus culminating in albino leaf. Experimental evidences specified that chl content reduced with decrease in temperature (Du et al. 2008). However, external applications of proline and betaine on tea leaves decreased cold susceptibility, enhanced the content of methylglyoxal as well as MDA, and increased GR as well as glutathione-*S*-transferase activity (Kumar and Yadav 2009). Tea growing under the shade tree (*Grevillea robusta*) had substantially lower transpiration rates than unshaded tea because shade tree reduced *E* principally by minimizing irradiance that strike the canopy and by lowering the canopy temperature (Anandacoomaraswamy et al. 2000).

7.2.3 Ultraviolet (UV) Radiation Stress

It is well-documented that UV-B is harmful to plants (Frederick et al. 1989). It reduced the synthesis of chl a/b-binding proteins as well as the D1 polypeptide of PS-II, thus inhibited the overall growth (Stapleton 1992; Mackerness et al. 1997). In response to UV-B radiation, tea plants upregulated gene required for synthesis of protective phenolic compounds and catechins (Mamati et al. 2006). Short-term irradiation of UV-B stimulated the accumulation of total catechins. Conversely, excessive irradiation of UV-B suppressed catechin's accumulation (Zheng et al. 2008). Under in vitro culture, UV-B radiation impeded a number of callus-forming cells and promoted the accumulation of soluble phenols. Deposition of phenolic compounds on cell wall and intercellular spaces was increased (Zagoskina et al. 2003). White light supplemented with UV-B induced formation of chl-bearing cells in callus culture and increased the phenols. PS-II activity was stimulated in the phototrophic cells of the callus culture in tea (Zagoskina et al. 2005). Tea plants were accustomed to broad range of irradiance that regulated C assimilation, and experimental evidences revealed that 1200 μmol m^{-2} s^{-1} PPFD generated maximum *Pn* (Barman et al. 2008).

7.2.4 Low Light-Induced Stress

Under shaded conditions, tea plants possess photosynthetic apparatus customized to function, with utmost capacity (De Costa et al. 2007). Shade can influence *Pn* indirectly by regulating leaf temperature. Experimental evidences indicated that photosynthetic apparatus was tailored to function with maximum capacity under shade (Mohotti and Lawlor 2002). The rate of *Pn*, *gs*, and apparent quantum yield of seedlings under shade-grown tea remained consistently higher as compared to unshaded conditions (Mohotti et al. 2000). It had also been observed that light intensity beyond 1400–1500 μmol m^{-2} s^{-1} resulted photoinhibition that, in turn, declined *Pn* (Mohotti and Lawlor 2002). However, shade reduced photoinhibition by increasing *gs* (De Costa et al. 2007).

Over-shade may reduce yield. It had been emphasized that PS-II, in tea, offered lower efficiency in capturing excitation energy and transport of electrons, apart from having low rates of RuBP carboxylation and RUBISCO activity under the over-shade conditions. Besides, stomatal resistance of CO_2 diffusion towards carboxylation sites made CO_2 unavailable for carboxylation and thus reduced yield (De Costa et al. 2007).

The effect of environmental factors on contents of tea leaves had received meager attentions, and some invaluable breakthroughs had eventually surfaced. Under over-shaded conditions, accumulation of epicatechins, ECG, EGC, and EGCG decreased in newly developing buds because the conversions of glucose, shikimic acid, and phenylalanine were depressed alongside the activity of PAL (Saijo 1980). At the outset, low light intensity inhibited synthesis of flavanols, making decline in gallocatechin and leucoanthocyanin concentrations. The flavogen content of darkened stems, especially in seedlings, was much less decreased than that of the leaves (Forrest 1969). *C. sinensis* var. Yabukita, when grown under shade, appreciably gained phenyl propanoids/benzenoids and phenylalanine whereas upstream metabolites such as shikimic acid, prephenic acid, and phenylpyruvic acid were reduced (Yang et al. 2012).

7.2.5 Chemical Amelioration

Melatonin plays an important role in stress tolerance. In a study, effect of exogenous melatonin on abiotic stress in the tea plant was investigated. Under various abiotic stresses such as cold, salt, and drought stress, increasing malondialdehyde levels and decreasing maximum photochemical efficiency of PSII were observed in tea leaves. Meanwhile, the levels of reactive oxygen species (ROS) increased significantly under abiotic stress. Interestingly, pretreatment with melatonin on leaves alleviated ROS burst, decreased malondialdehyde levels, and maintained high photosynthetic efficiency. Moreover, melatonin-pretreated tea plants showed high levels of glutathione and ascorbic acid and increased the activities of SOD, POD, CAT, and APX under abiotic stress. Notably, melatonin treatments can positively upregulate the genes (*CsSOD, CsPOD, CsCAT,* and *CsAPX*) expression of antioxidant enzyme biosynthesis. Taken together these results confirmed that melatonin protects tea plants against abiotic stress-induced damages through detoxifying ROS and regulating antioxidant systems (Li et al. 2019).

7.2.6 Elemental Stress

Tea encounters typical predicament of element deficiency due to extended life span, unremitting harvest without proper replenishment of nutrients and deterioration of soil fertility due to continuous uptake. Apart from deficiency, heavy metal contaminations due to relentless handling and industrial emission have become a critical alarm to the tea plantation worldwide. Enormous improvements have been made till date in unveiling a handful of elements that wreak havoc resulting in an exponential amount of crop loss. However, fortified with swift accumulation of information along with accelerated breakthrough has enabled scientists to foray into this particular field with full of prospective returns. A summary of different elemental stresses on tea plants is depicted in Fig. 7.1.

7.2.6.1 Nitrogen (N)

Tea plants require huge quantity of N not only for yield (green leaf) but also for amino acid biosynthesis, as well as for synthesis of secondary metabolites (Chamuah

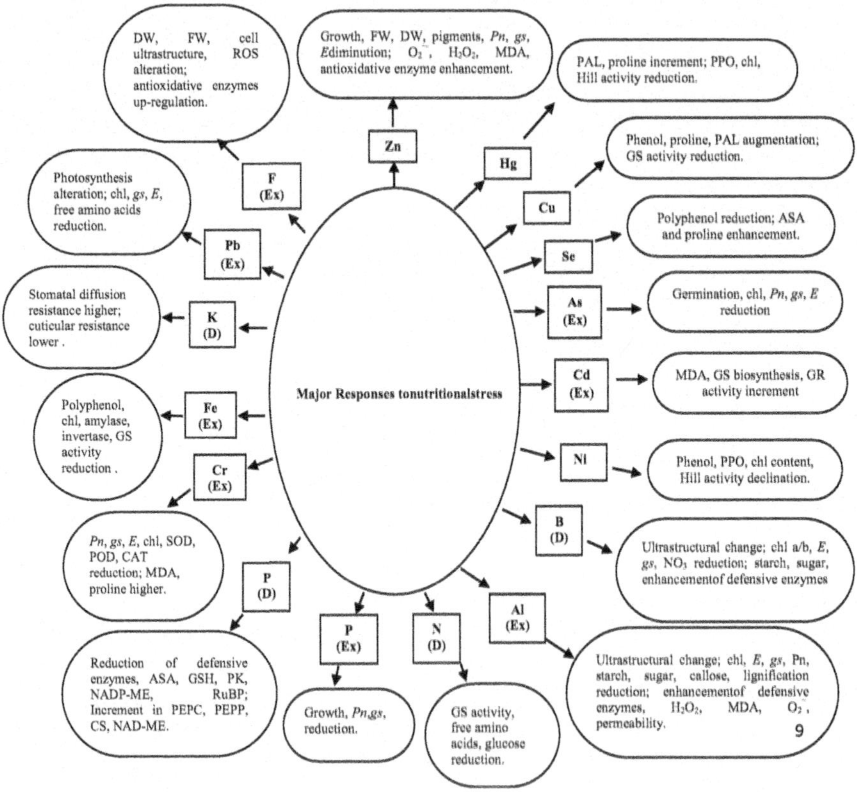

Fig. 7.1 Schematic representation of effects of elemental stresses (Ex = excess and D = deficiency)

1988). Although tea was reckoned low pH tolerant and favored ammonium (NH_4^+) because of quicker absorption, it also took nitrate which noticeably reduced glutamine synthetase (GS) activity, decreased concentrations of total N, free amino acids and glucose, as a consequence, leaves turned yellow reflecting N deficiency (Ruan et al. 2007). Further, N deficiency affected water relations by increasing stomatal resistance and transpiration rate (*E*) but did not alter stomatal opening, leaf water potential, and root resistance probably (Nagarajah 1981). The macro element increased photosynthetic rate (*Pn*) but reduced photoinhibition and hence promoted production (Mohotti and Lawlor 2002).

7.2.6.2 Boron (B)

B deficiency is prevalent in the tea cultivating soils (Mukhopadhyay and Mondal 2015). B deficiency reduced chlorophyll (chl) A/B ratio, net assimilation, *E*, and *gs* of the young leaves whereas increased carotenoids (car) in the older leaves parallel with non-photochemical quenching. Under B deficiency, NO_3^- concentrations diminished, while NO_2^- concentrations increased in the young leaf, and among the antioxidative defensive enzymes, APX and POD increased in older leaf and roots, respectively, in tea (Hajiboland et al. 2011a). B-deficient tea plants mitigated the reduction of photosynthetic energy conversion under intermediate and highly illuminated circumstances. The best possible light intensity for CO_2 assimilation was achieved under intermediate light for the young leaves and high for the old leaves. Activity of APX, SOD, and concentration of proline remained lower under intermediate conditions. In the young leaves, photochemical events were protected additionally under excess light and low B supply. B deficiency-induced activation of antioxidant defense system alleviated the effect of high light stress (Hajiboland et al. 2011b, 2013). B deficiency culminated in an overall inhibition of plant growth as indicated by reduction of *Pn*, *gs*, *E*, pigment concentrations, DW, and FW of shoot and root, but activity of all antioxidant enzymes, total sugar, reducing sugar, starch, ascorbic acid (ASA), and phenolics enhanced, although protein contents decimated under stress. Amount of superoxide anions (O_2^-), malodialdehyde (MDA), electrolyte leakage, and hydrogen peroxide (H_2O_2) augmented under deficiency. Furthermore, B deficiency increased the transcript of antioxidant enzymes, such as POD, SOD, APX, and transcripts of 14-3-3 gene indicating that oxidative metabolism was the prime defense mechanism under B deficiency of young tea plants that brought cascade of biochemical changes and implied the essence of B in upbringing of tea plantlets. In addition, closing of stomata, increasing amount of ABA and proline content in deficient plants indicated that B deficiency induced low moisture stress (Mukhopadhyay et al. 2013a). Transmission electron micrograph (TEM) of cell walls elucidated stress obligation by the presence of thickened middle lamellae of leaf cells (Fig. 7.2), which was a key marker of B deficiency, since the B adequate cells did not possess such structural modifications. Under deficiency, the B-rhamnogalacturonan-II complex formed due to increase in monomers turned out to be influenced that concurrently solidify the middle lamella. Thus, damage of cell wall ultrastructure was a corroboration of B-induced stress to tea plants (Mukhopadhyay et al. 2013a).

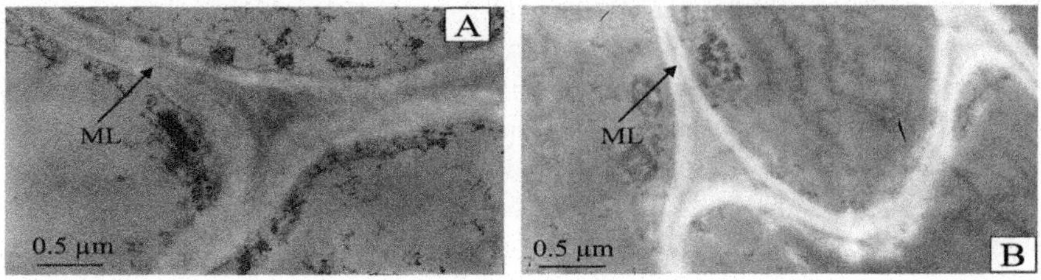

Fig. 7.2 Ultrastructural changes of tea leaf under B deficiency and optimum B concentrations. (**a**) cell wall of B-deficient (0 mg/L) palisade cell with thickened (arrow) middle lamella (ML) (1 cm = 0.5 μm) without excessive thickening (arrow) in palisade cell of tea plantlet treated with B sufficiency (0.15 mg/L). (**b**) Bars indicate the measurements as mentioned. (Adapted from Mukhopadhyay et al. 2013a)

7.2.6.3 Zinc (Zn)

Zn is an essential micronutrient, deficiency of which is quite common in tea cultivation. Apart from other factors, applications of excessive phosphate fertilizer in soil induced Zn deficiency (Nelson 2006; Subba et al. 2014). Deficiency and excess of Zn led to growth inhibition, shoot as well as root FW and DW reduction, and reduction of *Pn*, *E*, and *gs* in conjunction with reduction of both chl a and b, O_2^-, MDA, and H_2O_2. Zn stresses both deficiency and toxic dose increased electrolyte leakage, phenol concentration, and the activities of APX, CAT, SOD, and POD in tea suggesting that antioxidative mechanism was prime defense system under Zn deficiency. Transcripts of the antioxidative enzymes were upregulated under stress corroborating with the biochemical assays. Largely the results implied that Zn stress increased ROS production and activities of the enzymes, but the defensive system en bloc did not afford adequate fortification from oxidative damage (Mukhopadhyay et al. 2013b). Typical toxicity symptom of Zn included prominent browning of leaves although severe toxicity led to the death of the plants. UPASI-9 cultivars, under Zn toxicity of 2000 mg/kg, died within 15 d, while 1000 mg/kg extended the death by 36 d. Elemental interaction had also been studied in tea plants. It had been found that excessive Zn stimulated Mn translocation, whereas antagonized P, Mg, and Fe absorption (Venkatesan et al. 2006). Combination of Zn and Se deficiency increased PPO activity (Qian et al. 2009). The contents of polyphenols, catechins, amino acids, and soluble sugars increased with the increased Zn applications (Huiqun and Yuchen 1987), whereas Zn deficiency enhanced the accumulation of amino acids, glucose, and activity of SOD but decreased MDA and PPO activity (Jinghua et al. 1997) and chl (Owuor 2001). Zn enrichment increased photosynthesis as well as protein synthesis (Huiqun and Yuchen 1987) but reduced the uptake of P and K, besides the reduction of the content of Fe, Mn, Ca, and Mg in tissue (Wu and Fang 1994). Under Zn deficiency, the youngest leaves became narrow, strap-like, erect, and outward appearance became a rosette-like at the stem summit. Leaf blades became yellow, and severely affected leaves became stunted with inward-curling margins; apical growth and branch extension retarded (Nelson 2006) along with dwindled DNA contents in tea seedlings (Wu and Fang 1994).

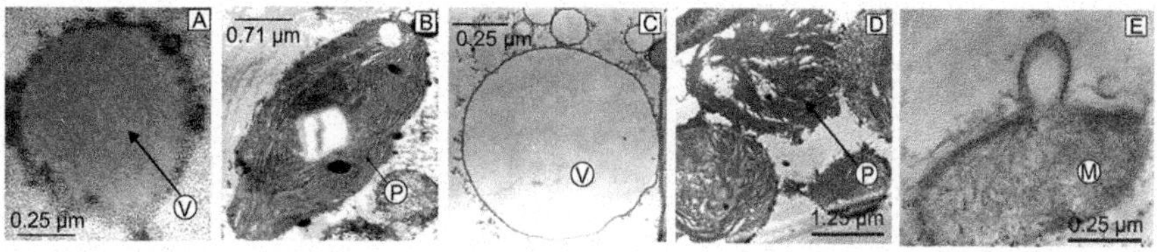

Fig. 7.3 Ultrastructural changes of tea leaf and root under Zn deficiency as well as Zn-excess. (**a**) Vacuole (V) of Zn-deficient (without Zn) palisade cell filled with electron dense phenolic compounds (1 cm = 0.25 μm). (**b**) Normal chloroplast (P) of Zn (0.05 mg/L) treated leaf (1 cm = 0.71 m) and (**c**) normal ultrastructure vacuole (V) (1 cm = 0.25 μm) in leaf Zn-optimum (0.5 mg/L) plants. (**d**) Choloroplast (P) with distorted grana at 5 mg/L Zn treatment (1.25 cm = 1.25 μm) and (**e**) mitochondria (M) (1.25 cm = 0.25 μm) with localized swelling in leaf cells of 5 mg/L Zn treated plant. Bars indicate the measurements as mentioned. (Adapted from Mukhopadyay et al. 2012)

Under Zn deficiency, vacuoles contained electron dense material resembling phenolic compounds that could be reaffirmed by higher content of phenolic compounds. Excess dose of Zn produced deformed chloroplast ultrastructures as indicated by membranes ruptured, thylakoid disintegrated, which adversely influenced photosynthetic activity. Similarly, mitochondria were swelled and cristae were damaged (Fig. 7.3) due to the imposition of stress under redundant supply. These imply that antioxidants did not satisfactorily scavenge superfluous ROS, to safeguard the tissues from free radical injury under Zn stress (Mondal et al. 2014; Mukhopadyay et al. 2012).

Zn modulated drought-mediated biochemical damages in tea plant. It was demonstrated that drought induced the decrease of relative water content (RWC), dry mass of leaf, and antioxidants such as ascorbate and glutathione in the tested tea clones (TV-1, TV-17 and TV-29) by the application of zinc sulfate ($ZnSO_4$) before water withholding for 7 d. Increase of phenolic content with decrease in hydrogen peroxide (H_2O_2) and lipid peroxidation and differential activities of enzymes like SOD, catalase (CAT), POX, PPO, GR, and APX with concomitant increased Zn uptake in leaf were noticed under drought treatment (Upadhyaya et al. 2013).

7.2.6.4 Aluminum (Al)

Tea plants and its wild species are Al hyper-accumulator and can accumulate as high as 30,000 mg/kg Al in mature leaves on DW basis (Matsumoto et al. 1976; Zeng et al. 2011), a concentration that is toxic for other plants. Thus, it attracted the attentions of scientific community to reveal the underlying mechanism. Contemporary investigations highlighted that age and variety determined Al accumulation in tea (Ruan and Wong 2001; Shu et al. 2003), and Al absorption varied on soil conditions and extractable Al (Wong et al. 1998). Uptake of Al^{+++} and NH_4^+ by the roots of tea plants induced H^+ release leading to acidification of surrounding medium. Higher quantity of cations concentrations stimulated more proton release from the roots, because roots took up a huge quantity of cations during the course of growth, and then released them to restore parity of charge (Wan et al. 2012). In the leaves, Al accumulated in the upper epidermal cell walls of mature leaves (Carr et al. 2003). In tea leaves, Al was mainly present in two dominant forms, i.e., as Al-catechin complex (Nagata et al. 1992) and Al-citrate (1:1) complex in the xylem sap (Morita et al. 2004), and accumulated more in Chinary-type plant than Assam types (Fung et al. 2003). The tolerance mechanism rested on the fact that Al retention took place in epidermal leaf apoplast and not in the symplast (Tolrà et al. 2011). Furthermore, N fertilization augmented extractable Al levels but decreased soil pH and exchangeable base cations, but Al quantity in shoots reduced by enormous N fertilizer application (Ruan et al. 2006).

While Al induced the uptake of Ca, Mg, K, and Mn, it reduced the uptake of Fe, Cu, and Zn. Moreover, total phenols and catechins in tea plant tissues increased with increasing availability of Al in soil (Chen et al. 2011a). In hydroponic culture, Al at lower concentration (2 mg/L), improved plant growth, but at higher concentrations, it induced oxidative damage. In accordance with Al application, Al contents increased in the grown-up tissues than their younger complements, and it had been observed that accumulation started after 6 h of exposure of the young root tips to Al solution. Pigment content, *Pn*, *E*, and *gs* were reduced due to Al toxicity. Activated oxygen metabolism was evidenced by increasing MDA, membrane injury, evolution of O_2^-, and H_2O_2 under Al toxicity. Phenol contents initially increased but later reduced at advanced ,levels whereas total sugar and starch concentrations decimated beyond 2 mg/L of Al. Activities of antioxidative enzymes, such as APX, GR, POD, CAT, and SOD, increased with the increase of Al. Expression of citrate synthase (*CS*) transcript seemed to be upregulated in the mature leaves and young and old roots with increased Al content in those parts, perhaps due to Al-citrate complex formation. Simultaneously citrate and malate production were also increased due to Al toxicity. These results considerately indicated that Al at lower levels promoted growth but became a stressor at elevated concentration (Mukhopadyay et al. 2012). Apart from all these defense mechanisms, callose formation was a frequent rejoinder to stresses and lower doses of Al stimulated root elongation and callose reduction but elevated doses stimulated its accumulation (Lian et al. 1998). Similarly, TEM analysis of Al-treated leaves revealed significant membrane damage at 70.67 mg/L of Al because at this level the equilibrium between formation and detoxification of ROS was not much effective (Li et al. 2011a). Interestingly, Al toxicity tolerance depends upon the genotype of tea. For an example, Pingyangtezao and Wuniuzao, two tea cultivars, under Al toxicity performed in a different way as indicated by different levels of chl content with differential activity of SOD, CAT, and POD enzymes among the two cultivars (Yu et al. 2012).

It had also been reported that the content of proline and activities of POD, CAT, and SOD enzymes reduced under high Al:F but increased with the increase of F alone in tea plants (Wang et al. 2009a). Upon application of Al, tea plants accumulated higher levels of ASA, chl, car, phenolics, proline, as well as quality constituents such as gallocatechin, epigallocatechingallate (EGCG), gallocatechingallate, epicatechingallate (ECG), and catechingallate indicating the fact that optimum dose of Al might induced the quality of made tea (Napaporn et al. 2012). Al-induced enhancement of antioxidant enzymes activities, membrane integrity, and delayed lignifications afforded the stimulatory effects of Al on the growth of

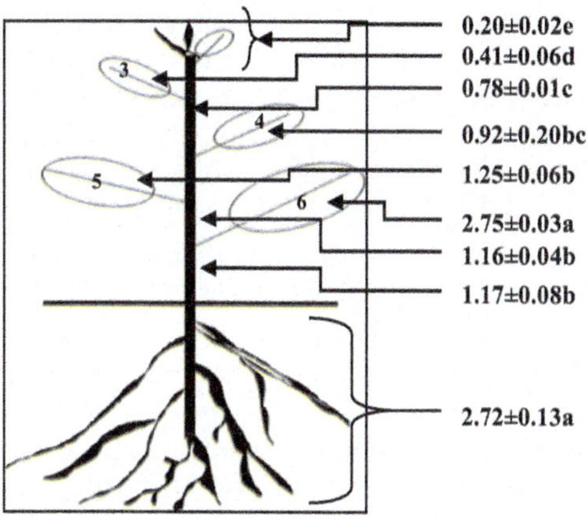

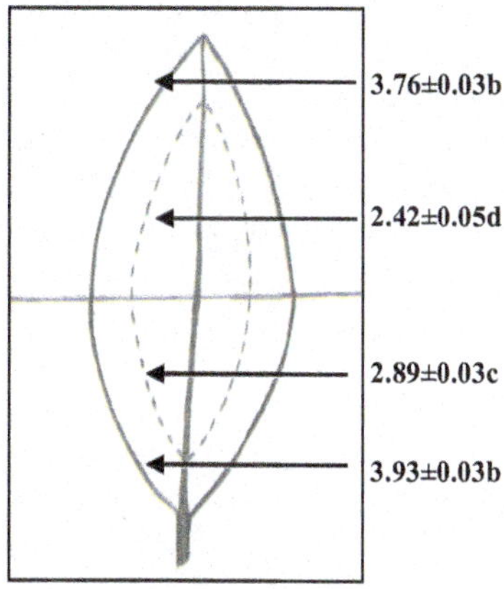

Fig. 7.4 Al content (g kg^{-1} DW) in tea plants treated with 53 g/L of $AlCl_3$ at pH 4.2. (Leaves below the growing shoot apex are indicated as 3, 4, 5, and 6). Distribution Al (g kg^{-1} DW) in the sixth leaf of tea plant treated with 53 mg/L of $AlCl_3$ at pH 4.2. The leaf tissues were cut with scissor along with the line for Al analysis. Each value represents the mean ± SE of three replicates (DW = dry weight). Different letters indicate significant difference at $P < 0.05$. (Adapted from Mukhopadyay et al. 2012)

tea plant (Ghanati et al. 2005). Recent investigations revealed that Al was detoxified by exclusion mechanism forming Al-ligand complexes and thus suppressed Al uptake. It was also proposed that oxalate was a key Al-chelating compound in the tea roots (Morita et al. 2008). Oxalate exudation along with small quantity of malate, citrate, and oxalic acid metabolisms was coupled to Al accumulation in roots (Morita et al. 2001). In tea cultivar TS-78, Al contents increased from young to mature leaf and from immature to full-grown stem suggesting that Al accumulation was specific to developmental stage of tea plant (Fig. 7.4). Proportional variations of Al accumulation in the sixth mature leaf revealed that Al concentration steadily increased from the center towards the margin of the leaf (Fig. 7.4). Maximum quantity of accumulation occurred at margin of the leaf (Mukhopadyay et al. 2012). Interestingly protein played important role to protect the tea plant from Al toxicity as it was found that poly-L-aspartic acid and poly-L-glutamic acid protected growth of tea pollen tube from toxic effects of Al. It was assumed that the polypeptides might prevent Al from either interfering with Ca^{++} transport or binding to calmodulin that was implicated in pollen tube growth (Konishi et al. 1988).

7.2.6.5 Phosphorus (P)

Tea is cultivated in humid and sub-humid tropical, subtropical, and temperate regions of the world, mainly on acidic soils where P deficiency is often experienced (Lin et al. 2009a). Hence, P fertilizers are used regularly in tea plantations to increase productivity (Salehi and Hajiboland 2008). Being an essential element, P played a pivotal role and P deficiency reduces the activities of SOD, APX, GR, CAT, monodehydroascorbate reductase (MDAR), dehydroascorbate reductase (DHAR) enzymes together with the contents of ASA and reduced glutathione (GSH).

Nevertheless, in spite of all biochemical changes, MDA levels remained unchanged indicating the active participation of the antioxidant system to provide substantial safeguard against photo-oxidative injury (Lin et al. 2012). Increased quantity of P inhibited growth due to reduction of *Pn* as well as *gs* that affected the photochemistry of P-deficient leaves (Salehi and Hajiboland 2008). P deficiency-induced malate and citrate discharge accompanied by increasing activity of PEP carboxylase (PEPC), PEP phosphatase (PEPP), CS, and NAD-malic enzyme (NAD-ME) in leaf and decreased activity of pyruvate kinase (PK), NADP-ME, and NADP-isocitrate dehydrogenase activities in roots. Compared to roots, malate accumulation was more in the leaves along with increased activities of NADP-ME, NAD-ME, and PK (Lin et al. 2011). In tea, P deficiency reduced Mg content in tissue but increased C/N ratios (Lin et al. 2009b). P deficiency decreased electron transport due to impairment in the electron transport chain thus ultimately reduced ATP production terminating in a smaller amount RuBP regeneration and CO_2 assimilation as well. It was proposed that energy dissipation was increased to facilitate the protection of P deficiency (Lin et al. 2009b).

7.2.6.6 Lead (Pb)

Pb is a well-known pollutant having the ability to function as a stressor. In Pb-contaminated soil, phytotoxic symptoms appeared in plants together with reduction of growth, caffeine, and amino acids but the enhancements of catechin content. Under such situation, maximum accumulation of Pb occurred in tea roots (Wu et al. 2011). Excess Pb decreased chl content, *Pn*, *gs*, and *E* significantly (Tang et al. 2008a). Tea plant showed the difference in genotypic response for Pb accumulation. For an example, Longjing 43, a genotype showed an enhancement of *Pn* when both Longjing 43 and Zhenong 117 were subjected to Pb stress. POD and SOD activities were increased with higher Pb concentration in Zhenong117 but in Longjing43, POD and SOD activity increased initially and then decreased gradually. It was also observed that application of $CaCO_3$ significantly decreased Pb concentration in tea plants (Han et al. 2007).

In another study, Pb was found to increase the growth of Longjing43 but reduced the growth of Zhenongll7 at the same concentration (430 mg/kg). Conversely at a lower level (160 mg/kg), reduced the net *Pn* of Zhenong117, while increased the *Pn* Longjing43 indicating different genotypes behaved differently under Pb. In Zhenong117, POD and SOD activities increased with increasing Pb availability, whereas in Longjing43, POD activity increased initially and then decreased indicating that the effects of Pb varied on genotypes of tea (Ruan et al. 2012).

Recently an attempt was made to develop database for Pb, Cd, Ni, and Cr in tea soils of South India. This results will be immense helpful on heavy metal contamination of tea soils. They found that there was a positive correlation between externally added heavy metals and their accumulation in different plant parts of tea. In addition, different parts of field-grown tea plants accumulated different concentration of heavy metals during their growth (Seenivasan et al. 2016).

It was reported that heavy metal (Pb, Cd, Cr, As, Mn, Hg, and Cu) pollution has occurred in some tea garden soils in recent years as a result of high soil background values, the application of pesticides and chemical fertilizers containing heavy metals, and industrial activities. In a study, it has been reported that heavy metal such as Pb, Cu, As, Hg, Cd, and Cr accumulated more in mature leaf than the young leaf. They have also reported that order of index of the heavy metal depends on their content in the soil where tea plants are being grown, water that have been used for irrigation (Zhang et al. 2018).

7.2.6.7 Fluorine (F)

Tea accumulates large amounts of F and transports them readily to the leaves. F concentration increased linearly with availability in soil. Interestingly, its uptake and translocation further increased in presence of Al (Ruan et al. 2003). Chinary tea accrued higher F contents compared to Assam variety (Fung et al. 2003). It was suggested that increasing F concentrations in soil decreased FW, DW, chl content, and *Pn*, whereas CAT and guaiacol peroxidase (GPX) activities increased significantly along with the increase of H_2O_2, MDA, and proline (Li et al. 2011b).

Ca applications reduced F uptake, possibly due to the effect of Ca on the properties of cell wall or membrane permeability or may be due to alteration in F speciation (Ruan et al. 2004). Increased F applications often damaged chl membrane, which ultimately reduced the *Pn* (Li et al. 2011b). Conversely, F also mitigated the adverse effect of Al and stimulated pollen tube growth either in combination or in single handedly (Konishi and Miyamoto 1983).

7.2.6.8 Selenium (Se)

Se increases chl, car, phenol, proline, and ASA content. It had been reported that 50 mg/L of Se offered best chemical quality of made tea and highest phenolic contents (Napaporn et al. 2012). Foliar applications of Se augmented yield and quality along with sweetness and aroma, but bitterness decreased significantly. Though polyphenol contents reduced at higher dose of Se, total amino acids and vitamin C content increased extensively by Se application (Hu et al. 2003). In another study, tea enriched with Se, by selenate application, exhibited the maximum inhibition of lipid oxidation, whereas R-tocopherol showed the lowest inhibition. Se-enriched green tea harvested during the early spring had been found to be much superior compared to regular tea indicating that along with other factors, Se might be responsible to improve the quality of green tea (Xu et al. 2003).

7.2.6.9 Arsenic (As) and Cadmium (Cd)

Several studies on the alleviation of heavy metal toxicity by exogenous phytohormones have reported. Enhanced As^{++} in soil decreased germination of tea seeds and subsequently growth of tea seedlings along with instantaneous reduction of chl content, *Pn*, *gs*, and *E* in tea (Tang et al. 2008b). In As and Cd excess soil, feeder roots played the buffering role so that minimum quantity of As get transported to the shoots. In fact, Cd persuaded growth, photosynthesis, and chl synthesis when present in optimum quantity but excess quantity impaired the plants (Tang et al. 2008b). Similarly, Cd toxicity decreased chl as well as protein content but enhanced MDA and upregulated the transcripts of GSH biosynthetic genes such as γ-glutamylcysteine synthetase and glutathione synthetase, and GR, indicating the protective role of these enzymes under excess Cd stress (Mohanpuria et al. 2007). Negative correlations between Cd concentration with *Pn*, *E*, and biomass production were found (Shi et al. 2008). Cd induced higher chl, MDA, and soluble sugar in spring than summer spring tea (Xia and Lan 2008). Depending on the tissue of callus origin, Cd concentration influenced phenol metabolism under in vitro culture indicating that the production of phenol under in vitro conditions was dependent on original tissue of callus origin. In leaf and stem origin calli, Cd decreased the flavans, but uniqueness remain unchanged in calli originated from roots. Simultaneously, lignin content in the calli of root and stem origin increased, although remain unaltered in the leaf derived callus (Zagoskina et al. 2007).

In a recent study, effects of IAA on Cd uptake on tea cultivar were studied. The order of Cd accumulation in tea seedlings was found to be root > stem > matureleaf > tender leaf. Under Cd stress (30 mg/kg), photosynthetic pigment levels, antioxidant enzyme activity, root vigor, root IAA content, and the levels of important metabolites such as caffeine, soluble sugar, and total amino acids were significantly reduced, while levels of malondialdehyde, proline, epicatechin, and some amino acids were increased (Zhang et al. 2020a).

7.2.6.10 Others (Chromium, Iron, Potassium, Mercury, Copper, Magnesium, Manganese, and Nickel)

Despite the fact that nutrients play important role in plant growth, effects of certain elements on tea are less studied which are discussed here.

Chromium (Cr) is a widespread natural contaminant in soil, and its accumulation in soil creates a disparaging effect on tea plants. It not only affects the yield and quality but also jeopardizes human health by either direct contact or food chain. In tea, higher level of Cr diminished chl content, *Pn*, *gs*, and *E* ultimately reduced the number, length as well as weight of the young buds (Tang et al. 2008a). Under Cr toxicity, activity of antioxidant enzymes such as SOD, POD, and CAT was reduced, while proline and MDA content were increased (Tang et al. 2012). Iron (Fe) toxicity reduced chl content and polyphenols along with a decrease of activity of amylase, invertase, aspartate aminotransferase, and glutamate synthase enzyme in tea (Hemalatha and Venkatesan 2011).

Potassium (K)-deficient tea plants demonstrated higher stomatal diffusive resistance along with reduction of night opening. Deficient plants also showed lower cuticular resistance (Nagarajah 1979).

Mercury (Hg) is a heavy metal and can induce oxidative stress in tea. Upon exposure, Hg decreased phenol contents, activity of PPO enzyme, and chl content. On the contrary, PAL activity and proline content were increased (Basak et al. 2001). Tea plants exposed to high Hg showed reduction of *Pn*, root browning, and chlorosis of leaf (Yadav and Mohanpuria 2009).

Copper (Cu) ions can oxidized thiol bonds causing disruptions in protein structure and functions (Yadav and Mohanpuria 2009). While Cu toxicity stress increased phenols, proline content, and PAL activity, yet upon exposure, it reduced the activity of PPO, GS enzyme, and chl content (Basak et al. 2001; Rana et al. 2008). It had been reported that Chinary cultivar was more tolerant than Assam, because Assam type accumulated more ROS and less proline (Yadav and Mohanpuria 2009). Cu-based fungicides are being used for decades to control fungal diseases in tea, which may lead to accumulation of Cu in the soil. Thus, in order to know the biochemical responses to increase concentrations of Cu, leaves of two cultivars of Darjeeling hills were investigated. Exposure to excess Cu resulted in increased lipid peroxidation, reduced chlorophyll content, higher level of phenolic compounds, and an increase in PPO enzyme activity. Two peroxidase isozymes (POD1 and POD2) were detected in plants upon exposure to Cu. In addition, biochemical responses in two tested genotypes, TS-462 and TS-520, differed significantly. TS-520 was found to be more sensitive to Cu. From this study, it appears that Cu exposure led to the production of reactive oxygen species in the leaves resulting in significant lipid peroxidation. Tea plants try to mitigate this oxidative damage through accumulation of phenolic compounds and induction of antioxidant enzymes (Saha et al. 2012). Report on Ni interaction of tea plant is also scanty. Excessive Ni decreased phenols, chl content, and PPO activity but increased PAL activity and proline contents (Basak et al. 2001). Application of Mg in soil improved free amino acids and quality of made tea (Ruan et al. 1999). Nevertheless, Mg inhibited polyphenol content, but up to a certain extent increased chl content. Excess Mg induced coppery color all over the leaf surface; however, it led to defoliations (Venkatesan and Jayaganesh 2010). Mn had significant influence on yield of tea. It had proven that all the chemical constituents of quality such as theaflavin, thearubigin, total color, and caffeine contents increased with the enhancement of Mn level. Mn applications influenced uptake of leaf N, P, and shoot Mn content significantly (Gohain et al. 2000). All these studies indicated that these elements under sufficiency or deficiency conditions become a stressor as indicated by increasing antioxidant enzymes as well as antioxidants.

7.3 Biotic Stress

Tea mosquito bug; mites such as red, pink, as well as purple; thrips; termites; red slug; looper; caterpillar; leafhopper; etc. are major pests of tea (Gurusubramanian and Borthakur 2005), although more than 1000 pests are reported in tea (Das 1965). Significantly, pest infestation also changed important biochemical constituents of plants depending upon some factors including the genotypes, environment, age of the plant parts, and cultivation practices (Chakraborty and Chakraborty 2005; Saikia et al. 2011). Blister blight (*Exobasidium vexans*), a biotrophic fungus, causes a havoc crop loss in tea plantation. Premkumar et al. (2008) reported that upon infestation, leaf area, FW and DW, *Pn*, *gs*, and *E* were reduced together with total sugars, amino acids, total N, polyphenols, and catechins in the susceptible clone than the tolerance genotype. In general, it had been found that under severe blister blight infection, total phenols, catechins, N, amino acids, theaflavins, thearubigins, caffeine, as well as PPO activity along with aroma components, such as hexane-1-ols, phenyl acetaldehyde, linalool, methyl salicylate, geraniol, indoles, β-ionones, and nerolidol, were reduced which led to produce very inferior quality of made tea (Gulati et al. 1999). A simultaneous decrease of 2-phenylethanol content and prephenate dehydratase (PDT) activity was also noticed since it was assumed that loss of PDT activity may be due to palisade tissue injury along with epidermal layers containing the stomata. These ultimately reduced *Pn* thus sinking the carbon flow through the shikimate pathway (Sharma et al. 2011). In sensitive genotype of tea, content of total phenols and ortho-dihydroxy phenols decreased upon infection of Sclerotial blight (*Sclerotium rolfsii*) of tea. Activity of PAL and POD increased specifying their pivotal roles in defense mechanism (Bhagat and Chakraborty 2010). Increased activity of DNase, chitinase, PAL, tyrosine ammonia lyase (TAL), PPO, and POD were reported due to the infection of bird's eye spot disease (*Cercospora theae*) (Gnanamangai et al. 2011).

Similarly, the physiological and biochemical changes due to *Pestalotiopsis* infection were also studied in tea under greenhouse conditions. It had been found that while *Pn*, *E*, water use efficiency, and chl fluorescence were reduced in the infected leaves, yet, stomatal resistance was increased. Biochemical constituents were higher in healthy leaves compared to that of healthy tissue of infected leaves. There was no significant variation in pigments such as chl and car, but in the infected tissue, these constituents reduced significantly. Activity of both PAL and tyrosine ammonia lyase were higher in the healthy zone of infected leaves than that of infected zone of the same leaves. On the other hand, activity of PPO and POD enzymes were higher in the lesion region and healthy portion of infected leaves (Sanjay and Baby 2007).

7.3.1 Relevance of Microbes in Stress Alleviation

Microbial infections alter the biochemical changes as a consequence of stress. Foliar spray of *Bacillus licheniformis* lowered the caffeine content of tea without affecting flavor and aroma (Ramarethinam and Rajalakshmi 2004). Upon application of bioformulations of *Bacillus megaterium*, phenolics and pigment contents were increased noticeably (Chakraborty et al. 2012). *Trichoderma harzianum* and *Azospirillum brasilense* stimulated rooting and acclimatization of in vitro-grown tea shoots and plantlets due to superior nutrient uptake (Thomas et al. 2010). Application of K solubilizing bacteria (*Pseudomonas putida*) in permutation with N, P, and K increased chl, car, N, P, and K contents of the shoots together with total polyphenols, catechins, amino acids, and sugars (Bagyalakshmi et al. 2012). It had been found that *Bacillus megaterium*, isolated from rhizosphere, inhibited brown rot disease (*Fomes lamaoensis*) of tea. Root colonization by *B. megaterium* and subsequent inoculation with *F. lamaoensis* increased the phenols and activity of enzymes such as POD, chitinase, β-1,3-glucanase, and PAL and also produced IAA, siderophore, and antifungal metabolites that promoted tea growth (Chakraborty et al. 2006). Arbuscular mycorrhizal fungi, upon inoculation, increased the content of amino acids and total proteins, respectively, along with total polyphenols, content of caffeine, as well as sugar (Singh et al. 2010). These works collectively suggested that

Table 7.1 Biotic stress-induced alteration of different biochemical parameters of tea plants

Causal organisms	Biochemicals amended	References
H. theivora	ABA ↑ GA_3 ↑ IAA ↓	Saikia et al. (2011)
H. theivora	POD ↑ PPO ↑ APX ↑ PAL ↑ Polyphenols ↓ Catechins ↓	Chakraborty and Chakraborty (2005)
E. vexans	Caffeine ↑ Theaflavins ↑ Thearubigins ↑ Linalool ↑ Geraniol ↑ β-ionone ↑ Methyl salicylate ↑ Total sugar ↓ Aminoacids ↓ PPO ↓ Nitrogen ↓ Catechins ↓	Gulati et al. (1999)
E. vexans	Total sugar ↓ Amino acids ↓ Nitrogen ↓ Polyphenols ↓ Catechins ↓	Premkumar et al. (2008)
E. vexans	2-phenylethanol ↓ Prephenate dehydratase ↓	Sharma et al. (2011)
S. rolfsii	PAL ↑ POD ↑ Total phenol ↓ Ortho-dihydroxy phenols ↓	Bhagat and Chakraborty (2010)
C. theae	DNase ↑ Chitinase ↑ PAL ↑ TAL ↑ POD ↑ PPO ↑	Gnanamangai et al. (2011)

beneficial microbes upon infection promoted growth of tea plants and simultaneously reduced the adverse effects of stress (Table 7.1).

7.3.2 Autotoxins

Autotoxicity, prevailing both in natural and agricultural ecosystem, is a process by which a species or its decomposing residues discharge phytotoxins to inhibit growth of the other plants (Eden 1940). Analysis of the effects of autotoxins designate that the chemicals possess both inhibitory and stimulatory effects on antioxidant activities in which stimulatory actions can be seen at lower concentrations, but an increment over time unveils obvious inhibitory effects. It had also been observed that adverse effects of autotoxins were counteracted by the enzymes of antioxidant defense system resulted in higher MDA formation in the leaf of tea (Cao et al. 2011a).

7.3.3 Effect of Plant Growth Regulators (PGRs)

PGRs such as 2,4-D and NAA determined the growth of in vitro culture of tea plants (Mondal et al. 1998). Under stress, external applications of salicylic acid induced expansion of leaf area suggesting that it acted as phytohormone. Total soluble solid of leaf and invertase activity increased appreciably, thus making the way for potential plant resistance against environmental stresses (Kaveh et al. 2004). Brassinolides stimulated rapid healthy shoot growth by modifications in stomatal response during the period of low moisture stress (Ng'etich and Bore 2001). Ploidy levels of tea also influenced the photosynthesis. It had been found that net CO_2 assimilation and stomatal conductance were lower in the tetraploids than in either diploids or triploids and were also lower in triploids than diploids (Ng'-Etich and Wachira 2003).

Effects of exogenous application of ABA on physiological and metabolic changes in tea were investigated. The highest improvement of yield was evident in response to regular application (a day after every harvest) of ASA at 600 mg/L followed by 400 mg/L. Foliar application of ASA (600 mg/L) at regular intervals enhanced chl *a* and *b* contents besides a significant increase of total polyphenols and catechins when compared to the untreated control. Exogenous application of ASA at alternate harvesting rounds increased the activity of PPO, whereas APX remained unchanged. An increase of PAL activity was noticed with foliar application of ASA irrespective of its concentrations. TFs and TRs of made tea were increased when ABA was applied with 400 mg/L and 600 mg/L at alternate dose. The scores of brewed tea liquor characteristics, viz., infusion, total color, strength, and briskness, were higher even at the lower concentration of ASA treatment. Results suggested that foliar application of ASA (400 mg/L) proved to be a useful measure to improve the yield, physiological attributes, and antioxidant properties of tea (Murugan et al. 2012).

7.4 Effects of Stress on Quality of Made Tea

Aroma is an indispensable aspect affecting the quality of tea and pivotal for its marketing as well as price realization. Many intrinsic and extrinsic factors interplay to achieve the most desirable feature of made tea. Quality of made tea depends upon the age of leaf, i.e., younger leaf produced better quality. In immature leaves, a relatively large proportion was used to produce catechins, a compound responsible for quality of black tea (Sanderson and Sivapalan 1966).

Adequate supply of Mg nutrient promoted the synthesis of theanine in roots and its accumulation in the young leaves of tea plants. Mg status in soil was an important factor influencing the mobility of amino acids and sugars via xylem and phloem especially when N and C reserves remobilized to support spring growth of young shoots (Ruan et al. 2012). Application of $MgSO_4$ followed by kieserite induced higher content of chl, car, quantity of TFs, TRs and polymerized substances (Jayaganesh et al. 2011). Activity of theanine synthetase, an enzyme involved in theanine biosynthesis, could be influenced by salt treatments. Salt treatments positively affect the levels of theanine and total free amino acids, and the level of theanine product was corroborated with that of theanine synthatase transcript expression (Deng et al. 2012). However, although salinity is not a major problem for tea cultivation worldwide.

7.5 Emerging Physiological Stresses

Several physiological factors are related to the quality as well as yield of tea and any alteration of them become stressor. A brief account of them is enumerated here.

The CO_2 assimilation capacity of tea plants decreased with the increase of temperature due to high irradiance while maximum *Pn* found to be around 30–35 °C yet, no net photosynthesis can be seen beyond 42 °C (Mohotti and Lawlor 2002). Respiration rate also influenced growth rate due to removal of the photo-assimilates as substrate. Mature tea used almost 67% of the photo-assimilates for respiration (Tanton 1979). In tea, root respiration rate increased with increasing N supply, owing to increase respiratory cost for NO_3^- uptake (Anandacoomaraswamy et al. 2000). Tea plants produced lower biomass ascribed to continuous elimination of young shoots. It had been found that unplucked tea produced 36% more total biomass and 64% more woody tissue than plucked tea. It had also been established that biomass production rate and tea yield maintained inverse relationships (Magambo and Cannell 1981). Plucking of young shoots before maturity reduced the sink capacity of the tea bush and thus reduced biomass production (Tanton 1979). Photoperiod affected shoot growth of tea, and the growth of tea shoots was repressed with 11 h photoperiod combined with night temperature around 20 °C. Additionally it had been proposed that tea shoots become dormant below a photoperiod of 11.16 h (De Costa et al. 2007).

Rising atmospheric carbon dioxide, a recent consequence of climate change, has multifarious effects on crop yields and quality. Tea being a perennial non-deciduous plant operates massive physiologic, metabolic, and transcriptional reprogramming to adapt to increasing CO_2. Rate of photosynthesis increased when grown at CO_2-enriched environment which is attributed to increased maximum carboxylation rate of RuBisCO and maximum rates of RuBP regeneration. Elevated CO_2-induced photosynthesis enhances the energy demand which triggers respiration. Stimulation of photosynthesis and respiration by elevated CO_2 promotes biomass production. Moreover, elevated CO_2 increases total carbon content, but it decreases total nitrogen content, leading to an increased ratio of carbon to nitrogen in tea leaves. Increased CO_2 alters the tea quality by differentially influencing the concentrations and biosynthetic gene expression of tea polyphenols, free amino acids, catechins, theanine, and caffeine. Despite enhanced synthesis of defense compounds, infection of some insects and pathogens is compromised under elevated CO_2 (Ahammed et al. 2020).

The role of caffeine in tea plant defense against pathogenic fungus has been reported under elevated CO_2. A reduction in caffeine content in tea leaves under elevated CO_2 conditions increases the susceptibility of the tea plants to *Colletotrichum gloeosporioides*, which causes anthracnose, brown blight, and die-back disease of tea in different geographical locations depending on the weather conditions (Li et al. 2016).

Due to climate change, acid rain become quite frequent. Tea grown across the world now are often exposed to acid rain though it is detrimental for the growth of the plant. In a study, influence of simulated acid rain (SAR) on various morphological, biochemical, and physiological parameters were studied. Results showed that SAR at pH 4.5 did not hinder plant development because growth characteristics, photosynthesis, and APX and CAT activities did not decrease at this pH compared to those at the other investigated pH values. However, at pH 3.5 and pH 2.5, the activities of antioxidase and concentrations of malondialdehyde and proline increased significantly in response to the decrease of photosynthetic pigments and Fv/Fm. Thus it indicated that SAR at pH 3.5 and pH 2.5 could restrict photosynthesis and the antioxidant defense system, causing metabolic disorders and ultimately affecting plant development and growth, but SAR at pH 4.5 had no toxic effects on tea seedlings when no other stress factors are involved (Zhang et al. 2020b) (Table 7.2).

7.6 Impact of Various Stresses on Wild Species of Tea

Like any other plants, different *Camellia* species too have different growth habits under different climatic conditions (He et al. 2012). A brief account of effect of stresses on growth of some *Camellia* species is discussed here.

The relationship between the low temperature and the cell injury in eight species of *Camellia* showed that *C. oleifera* was found to be best suitable species for high-temperature zones, whereas *C. vietnamensis* was more suitable for temperate regions. *C. oleifera* required optimum concentrations of Al and P for its growth (Tan et al. 2011) but under Al toxicity and P inadequacy reduced growth and photosynthesis. Lime in combination with P led to increase in *Pn* and *gs* and decrease in intercellular CO_2 concentration. Simultaneously content of pigments such as car and chl was reduced significantly, indicating its crucial role of car in photo-protection (He et al. 2010). Another study revealed that P deficiency led to higher activity of acid phosphatase in some tea varieties (Chen et al. 2011c). In fact, leaf age and leaf position were found to be two important influencing factors that determined the *Pn* in *C. oleifera* (Wang et al. 2009b). Density of planting also influenced *Pn* because lower planting yielded higher *Pn* than compact planting in *C. oleifera,* a fact which is well-established (He and Lu 2008). Under high illumination, BAP reduced the accumulation of chl as well as protein content in *C. oleifera* (Hu 1998). Low Al content increased the content of chl a and b, *gs*, and *Pn* but decreased when Al concentration increased. Simultaneously *Pn* also affected biomass accumulation (He et al. 2010). It had been found that under γ radiation (up to the dose of 1000 rad), SOD activity was increased but decreased beyond 1000 rad. Similarly, POD activity was also found to be influenced by intensity of the radiation (Huang et al. 2009). Leaf blight of *C. oleifera* is one of the major leaf diseases, which causes premature defoliation, hence reduction of growth of the plants. It had been detected that *Bacillus subtilis* offered antagonism against this disease (Li et al. 2011c). High temperature stress increased the contents of soluble protein, sugar, and free proline in *C. oleifera* leaves (Wang et al. 2012). Low moisture stress also affected antioxidants system of *C. oleifera* plant due to reduction of PPO, POD, and CAT activity, while MDA and relative conductivity were increased gradually (Cao

Table 7.2 Changes of physiochemical parameters of tea plants under various abiotic stresses

Stress	Physiological changes	Biochemical changes	References
Low moisture stress	Water potential ↑ Growth ↓	–	Yang et al. (1987)
	Diffusion resistance RWC ↑	–	Sandanam et al. (1981)
	Diffusion resistance ↑ RWC, *Pn* ↓	–	Wijeratne et al. (1998)
	F_0, thermal dissipation Fm, Fv, Fv/Fm, Fv/F_0, ↓photosynthesis ↑	–	Guo et al. (2009)
	–	Peroxidase, ↑ POD ↑	Jeyaramraja et al. (2002b)
	E, *gs*, root: shoot, CO_2 assimilation ↓	chl ↑	Bore (2008)
	–	Proline, H_2O_2, MDA, O^{2-}, ↓ CAT, SOD chl, car ABA, GPX, GR ↑	Upadhyaya and Panda (2004)
	–	MDA, O_2^-,POD, GR, CAT, phenol, H_2O_2 ↑ ASA, glutathione ↓	Upadhyaya et al. (2008)
	–	Electrical conductivity ↑ POD, PPO, *Pn* ↓	Yang et al. (1987)
	–	Proline ↑	Wu and Pan (1995)
	–	H_2O_2, APX, SOD ↑	Chen et al. (2011b)
Low temperature	–	SOD, APX, GR ↑	Vyas et al. (2007)
	–	MDA, Proline, car ↑	Netto et al. (2005)
	–	Number of chloroplasts ↓	Du et al. (2008)
	–	–	Mamati et al. (2006)
	–	EGCG ↑	Zheng et al. (2008)
	Callus-forming cell size ↓	Soluble phenols ↑	Zagoskina et al. (2003)
	PS-II ↑	Phenolic compounds ↑	Zagoskina et al. (2005)

(continued)

Table 7.2 (continued)

Stress	Physiological changes	Biochemical changes	References
Low light intensity	–	EGCG, Epicatechin, ECG, ⬇ PAL, EGC	Saijo (1980)
	–	Flavanols ⬇	Forrest (1969)
	–	Phenylpropanoids ⬆, phenylalanine, shikimic acid ⬇, prephenic acid, phenylpyruvic acid	Yang et al. (2012)
2,4-D	Growth ⬆	–	Nikolaeva et al. (2009)
NAA	Growth, biomass ⬇	Phenol, flavan ⬆	Nikolaeva et al. (2009)
ABA	–	TSS, starch, protein ⬆	Sharma et al. (2004)
SA	Leaf area, FW, DW ⬆	TSS, invertase ⬆	Kaveh et al. (2004)

et al. 2011b). In a preliminary study, effect of low water stress on *C. oleifera* was studied in details, and it had been found that leaf was more vulnerable to dehydration than root and stem (Cao et al. 2011c). Simulated acid rains inhibited seed germination, increased MDA, and reduced POD, SOD activity, chl a, chl b content, as well as their ratio in young seedlings. Spraying paclobutrazol on the leaves of *C. oleifera* seedling increased shoot as well as root growth with increase of chl, *Pn*, and protein content (Hu et al. 1993). The growth of *C. sasanqua* pollen grains was found to be slow, but when soaked in acetone or diethyl ether the same grew faster (Iwanami 1973). Under light stress, *C. nitidissima* changed both qualitative and quantitative characters. Low light intensity induced development of short narrow leaves, while high light intensity stimulated longer and wider leaves of *C. nitidissima* (Hu et al. 2010).

Effect of cold on growth and development was compared between *C. japonica* and *C. rusticana*. Under prolonged snow cover, the leaves of *C. rusticana* showed no visible changes in comparison with *C. japonica*, while *C. japonica showed a slow growth* after 200 d. The decline in the rate of soluble carbohydrate content in *C. rusticana* remained about one-sixth of that in *C. japonica*. The *Pn* of *C. japonica* decreased to about half of its initial value after 140 d, while that of the *C. rusticana* remained unchanged even after almost 1 year. Under dark treatment, the stomata of *C. rusticana* were closed but in *japonica*, even after 90 d also, remained open (Kume et al. 1998). Collectively together, these physiological studies in different species are sporadic although wild species harbor many important characters of ABstreses, yet systematic study is lacking.

7.7 Conclusion

Tea plants, being woody perennial, experience several biotic and abiotic stresses (ABstreses). Although conventional breeding is successful for varietal improvement of tea, they are focused onto limited traits such as improvement of yield, quality, drought, and few diseases (Mukhopadhyay et al. 2016). While significant amount of achievements have been made in improving yield and quality parameters of tea, other attempts have marginally improved the target. This is primarily due to the fact that most of the ABstreses are complex in nature, controlled by several genetic elements except few cases of biotic stress. Thus, prior to any attempt to progress further, it is essential to understand the physiological as well as biochemical basis of such stresses, which later formulate the basis for studying the molecular mechanism. In tea, several studies have been accomplished which identified major physiochemical parameters. Detection of the key biochemical molecules associated with an assortment of traits and identification of the physiological parameters would aid to nursery selection for agronomically important traits and will assist molecular breeding ultimately. Although in several aspects of tea, physiochemical study needs to be investigated; few specific areas are (1) tea is a hyper-accumulator of Al and accumulates Al in such an extent that is toxic to other plants. Thus identification of the mechanism of tea plants will facilitate to understand the mechanism which in turn will assist to develop the Al-tolerant plant in other species; (2) mild infestation of green leafhopper (*Empoasca vitis* Göthe) in tea leaf increases the aroma of made tea. Until now, the physiochemical basis is not documented. However, most importantly, in tea identification of extreme genotypes tolerant to ABstreses or at least development of trait specific "core" panel through the approach of phenomics should be done in priority. This will be immensely useful for several applications including identification of major QTLs/genes. Besides, association mapping can only be successful if morphological, physiological, as well as biochemical scorings are successful.

References

Ahammed GJ, Li X, Liu A, Chen S (2020) Physiological and defense responses of tea plants to elevated CO_2: a review. Front Plant Sci 11:305

Anandacoomaraswamy A, De Costa WAJM, Shyamalie HW, Campbell GS (2000) Factors controlling transpiration of mature field-grown tea and its relationship with yield. Agric Forest Meteor 103:375–386

Apel K, Hirt H (2004) Reactive oxygen species: metabolism, oxidative stress, and signal transduction. Annu Rev Plant Biol 55:373–399

Bagyalakshmi B, Ponmurugan P, Marimuthu S (2012) Influence of potassium solubilizing bacteria on crop productivity and quality of tea (*Camellia sinensis*). Afr J Agric Res 7:4250–4259

Barman TS, Baruah U, Saikia JK (2008) Irradiation influences tea leaf (*Camellia sinensis* L.) photosynthesis and transpiration. Photosynthetica 46:618–621

Barua DN (1969) Light as a factor in metabolism of tea plant. In: Luckwill LC, Cuttings CV (eds) Physiology of tree Crops. Academic Press, New York, pp 306–322

Barua DN (1989) Science and practice in tea culture. Tea Research Association, Kolkata, p 509

Basak M, Sharma M, Chakraborty U (2001) Biochemical responses of *Camellia sinensis* (L.) O. Kuntze to heavy metal stress. J Environ Biol 22:37–41

Bhagat I, Chakraborty BN (2010) Defense response triggered by *Sclerotium rolfsii* in tea plants. Ecoprint 17:69–76

Bhat KV, Mondal TK, Gaikwad AB, Kole PR, Chandel G, Mohapatra T (2020) Genome-wide identification of drought-responsive miRNAs in grass pea (*Lathyrus sativus* L). Plant Gene 21:100210

Bore JK (2008) Physiological responses of grafted tea (*Camellia sinensis* L.) to water stress. JKUAT abstracts of post graduate thesis. http://elearning.jkuat.ac.ke/journals/ojs/index.php/pgthesis_abs/article/view/323

Burgess PJ, Carr MVK (1996) Responses of young tea (*Camellia sinensis*) clones to drought and temperature. II. Dry matter production and partitioning. Exp Agric 32:377–394

Cao P, Liu C, Li D (2011a) Effects of different autotoxins on antioxidant enzymes and chemical compounds in tea (*Camellia sinensis* L.) Kuntze. Afr J Biotechnol 10:7480–7486

Cao Z, Zhang S, Liu C, Hu J, Zhang X, Shu Q (2011b) Effects of dehydration degree on the physiological and chemical characteristics of *Camellia oleifera* seedlings. J Anhui Agric Univ 3:2011–2003

Cao Z, Hu J, Shu Q, Wen J, Zhan W, Xu G, Bu J (2011c) Effects of water stress on physical characteristics and survival rate of container seedling in *Camellia oleifera*. Nonwood Forest Res:2011–2004

Carr HP, Lombi E, Küpper H, McGrath SP, Wong MH (2003) Accumulation and distribution of aluminium and other elements in tea (*Camellia sinensis*) leaves. Agronomie 23:705–710

Chakraborty U, Chakraborty BN (2005) Impact of environmental factors on infestation of tea leaves by *Helopeltis theivora*, and associated changes in flavonoid flavour components and enzyme activities. Phytoparasitica 33:88–96

Chakraborty U, Dutta S, Chakraborty BN (2002) Response of tea plants to water stress. Biol Plant 45:557–562

Chakraborty U, Chakraborty BN, Basnet M (2006) Plant growth promotion and induction of resistance in *Camellia sinensis* by *Bacillus megaterium*. J Basic Microbiol 46:186–195

Chakraborty U, Chakraborty BN, Chakraborty AP (2012) Induction of plant growth promotion in *Camellia sinensis* by *Bacillus megaterium* and its bioformulations. World J Agric Sci 8:104–112

Chamuah GS (1988) The effect of nitrogen in root growth and nutrient uptake of young tea plant (*Camellia sinensis*) grown in sand culture. Fert Res 16:59–65

Chen YM, Tsao TM, Liu CC, Lin KC, Wang MK (2011a) Aluminium and nutrients induce changes in the profiles of phenolic substances in tea plants {*Camellia sinensis* CV TTES, no. 12 (TTE)}. J Sci Food Agric 91:1111–1117

Chen Q, Yang L, Ahmad P, Wan X, Hu X (2011b) Proteomic profiling and redox status alteration of recalcitrant tea (*Camellia sinensis*) seed in response to desiccation. Planta 233:583–592

Chen L, Chen Y, Wang R, Ma L, Peng S, Wang X, Tang W (2011c) The effects of phosphorus deficiency stress on activities of acid phosphatase in different clones of *Camellia oleifera* Abel. Chin Agric Sci Bull 6:2011–2031

Chen M, Zhu X, Zhang Y et al (2020) Drought stress modify cuticle of tender tea leaf and mature leaf for transpiration barrier enhancement through common and distinct modes. Sci Rep 10:6696

Cheruiyot EK, Mumera LM, Ngetich WK, Hassanali A, Wachira FN (2007) Polyphenol as potention indicators for drought tolerance in tea (*Camellia sinensis L.*). Biosci Biotechnol Biochem 71:2190–2197

Cheruiyot EK, Mumera LM, Ng'etich WK, Hassanali A, Wachira FN, Wanyoko JK (2008) Shootepicatechin and epigallocatechin contents respond to water stress in tea {*Camellia sinensis (L.) O.* Kuntze}. Biosci Biotechnol Biochem 72:1–8

Das GM (1965) Pests of tea in north East India and their control. Memorandom no. 27. Tocklai Experimental Station, Tea Research Association, Jorhat, Assam, India, pp 169–173

Das A, Mukhopadhyay M, Sarkar B, Saha D, Mondal TK (2015) Influence of drought stress on cellular ultrastructure and antioxidant system in tea cultivars with different drought sensitivities. J Environ Biol 36:875–882

De Costa WAJM, Mohotti AJ, Wijeratne MA (2007) Ecophysiology of tea. Braz J Plant Physiol 19:299–332

Deng WW, Wang S, Chen Q, Zhang ZZ, Hu XY (2012) Effect of salt treatment on theanine biosynthesis in *Camellia sinensis* seedlings. Plant Physiol Biochem 56:35–40

Du YY, Chen H, Zhong WL, Wu LY, Ye JH, Lin C, Zheng XQ, Lu JL, Liang YR (2008) Effect of temperature on accumulation of chlorophylls and leaf ultrastructure of low temperature induced albino tea plant. Afr J Biotechnol 7:1881–1885

Eden T (1940) Studies in the yield of tea IV-the effect of cultivation and weeds on crop. Empire J Exp Agric 7:269–279

Forrest GI (1969) Effects of light and darkness on polyphenol distribution in the tea plant (*Camellia sinensis* L.). Biochem J 113:773–781

Frederick JE, Snell HE, Haywood EK (1989) Solar ultraviolet radiation at the Earth's surface. Photochem Photobiol 50:443–450

Fung KF, Zhang ZQ, Wong JWC, Wong MH (2003) Aluminum and fluoride concentrations of three tea varieties growing at Lantau island, Hong Kong. Environ Geochem Health 25:219–232

Ghanati F, Morita A, Yokota H (2005) Effects of aluminum on the growth of tea plant and activation of antioxidant system. Plant and Soil 276:133–141

Gnanamangai BM, Ponmurugan P, Yazhini R, Pragadeesh SK (2011) PR enzyme activities of *Cercospora theae* bird's eye spot disease in tea plants (*Camellia sinensis* (L.) O. kuntze). Plant Pathol J 10:13–21

Gohain T, Barbora AC, Deka A (2000) Effect of manganese on growth, yield and quality of tea [*Camellia sinensis L.* (O) Kuntez]. Res Crops 1:91–97

Gulati A, Gulati A, Rabindranat SD, Gupta AK (1999) Variation in chemical composition and quality of tea (*Camellia sinensis*) with increasing blister blight (*Exobasidium vexans*) severity. Mycol Res 103:1380–1384

Guo CF, Sun Y, Tang YH, Zhang MQ (2009) Effect of water stresson chlorophyll fluorescence in leaves of tea plant. Chin J EcoAgric 17:560–564

Gupta D, Bhardwaj R, Nagar PK, Kaur S (2004) Isolation and characterization of brassinosteroides from leaves of *Camellia sinensis (L.)* O. Kuntze. Plant Growth Regul 42:97–100

Gurusubramanian G, Borthakur M (2005) Integrated management of tea pests. In: Dutta AK, Baruah SK, Ahmed N, Sarma AK, Burugohain D (eds) Field management in tea. Tocklai Experimental Station, TRA, Assam Printing Works, Jorhat, Assam, pp 159–172

Hajiboland R, Bastani S (2012) Tolerance to water stress in boron deficient tea (*Camellia sinensis*) plants. Folia Horti 24:41–45

Hajiboland R, Bastani S, Rad SB (2011a) Photosynthesis, nitrogen metabolism and antioxidant defense system in B-deficient tea (*Camellia sinensis* (L.) O. Kuntze) plants. J Sci 22:311–320

Hajiboland R, Bastani S, Rad SB (2011b) Effect of light intensity on photosynthesis and antioxidant defense in boron deficient tea plants. Acta Biol Szeged 55:265–272

Hajiboland R, Bahrami-Rad S, Bastani S (2013) Phenolics metabolism in boron-deficient tea [*Camellia sinensis (L.) O. Kuntze*] plants. Acta Biol Hung 64:196–206

Hajra NG, Kumar R (2002) Responses of young tea clones to subtropical climate: effect on photosynthesis and biochemical characteristic. J Plant Biol 29:257–264

Han WY, Shi YZ, Ma LF, Ruan JY, Zhao FJ (2007) Effect of liming and seasonal variation on lead concentration of tea plant (*Camellia sinensis* (L.) O. Kuntze). Chemosphere 66:84–90

Handique AC, Manivel L (1990) Effect of certain anti-transpirants in tea. Two Bud 37:20–23

He Y, Lu F (2008) Study on the photosynthesis of the different density of *Camellia Oleifera*. Mod Agric Sci 5:2003–2008

He G, Liu Q, Peng S (2010) Effect of aluminum toxicity on photosynthetic characters of wild *Camellia oleifera* under acidic conditions. Hubei Agric Sci:2010–2017

He XY, Ye H, Ma JL, Zhang RQ, Chen GC, Xia YY (2012) Semi-lethal high temperature and heat tolerance of eight *Camellia* species. Int J Exp Bot 81:177–180

Hemalatha K, Venkatesan S (2011) Impact of iron toxicity on certain enzymes and biochemical parameters of tea. Asian J Biotechnol 6:384–394

Hernandez I, Alegre L, Munné-Bosch S (2006) Enhanced oxidation of flavan-3-ols and proanthocyanidinaccumulation in water-stressed tea plants. Phytochemistry 67:1120–1126

Hu Z (1998) Delayed action of 6-BA on detached-leaves senescence in *Camellia oleifera*. J Fujinan Coll For 3:1998–1901

Hu Z, Shen W, Zhang Y (1993) Physiological effect of paclobutrazol on the growth of *Camellia oleifera* seedling. J Fuji Coll For:1993–1902

Hu Q, Xu J, Pang G (2003) Effect of selenium on the yield and quality of green tea leaves harvested in early spring. J Agric Food Chem 51:3379–3381

Hu XH, Li JW, Jiang QS, Zhao RF (2010) Responses of leaf characters of *Camellia nitidissima* to different light environments. Guihaia 5:201–203

Huang Y, Yuan Z, Zhong W, Zhang H, Dong B (2009) Effects of ~ (60)co-γ radiation on isozyme activity of *Camellia Oleifera* seedling. Nonwood For Res 4:200–202

Huiqun M, Yuchen R (1987) Relationship between zinc and the metabolism of tea plant. J Tea Sci (China) 7:35–40

Iwanami Y (1973) Acceleration of the growth of *Camellia sasanqua* pollen by soaking in organic solvent. Plant Physiol 52:508–509

Jayaganesh S, Venkatesan S, Senthurpandian VK (2011) Impact of different sources and doses of magnesium fertilizer on biochemical constituents and quality parameters of black tea. Asian J Biochem 6:273–281

Jeyaramraja PR, Jayakumar D, Pius PK, Kumar RR (2002a) Screening of certain tea cultivars for productivity and drought hardiness using biochemical markers. J Plant Crop 30:23–26

Jeyaramraja PR, Jayakumar D, Pius PK, Kumar RR (2002b) Application of rubisco, peroxidase and polyphenol oxidase as markers for productivity and drought tolerance in tea. J Plant Biol 29:315–320

Jeyaramraja PR, Pius PK, Kumar RR, Jayakumar D (2003a) Soil moisture stress-induced alterations in bioconstituents determining tea quality. J Sci Food Agric 83:1187–1191

Jeyaramraja PR, Raj Kumar R, Pius PK, Thomas J (2003b) Photoassimilatory and photorespiratory behavior of certain drought tolerant and susceptible tea clones. Photosynthetica 41:579–582

Jinghua Y, Yongping C, Hesong L (1997) Effect of Zn^{++} on quality and lipid peroxidation in leaves of tea. J Anhui Agric Sci 25:30–32

Kakkar RK, Nagar PK (1997) Distribution and changes in endogenous polyamines during winter dormancy in tea (*Camellia sinensis* (L.) O. Kuntze). J Plant Physiol 151:63–67

Kaveh SH, Bernard F, Samiee K (2004) Growth stimulation and enhanced invertase activity induced by salicylic acid in tea cuttings (*Camellia sinensis* L.). In: Proceedings of the fourth international Iran and Russia conference, Shahr-e-Kord, Iran, pp 113–116

Kefei Z, Ma Q, Zang H (1997) Effect of water deficit on physiological activities of paddy rice and upland rice seedlings. J Shandong Agr Univ 28:53–57

Konishi S, Miyamoto S (1983) Alleviation of aluminum stress and stimulation of tea pollen tube growth by fluorine. Plant Cell Physiol 24:857–862

Konishi S, Ferguson IB, Putterill J (1988) Effect of acidic polypeptides on aluminium toxicity in tube growth of pollen from tea (*Camellia sinensis* L.). J Plant Sci 56:55–59

Kumar V, Yadav SK (2009) Proline and betaine provide protection to antioxidant and methylglyoxal detoxification systems during cold stress in *Camellia sinensis* (L.) O. Kuntze. Acta Physiol Plant 31:261–269

Kume A, Tanaka C, Matsumoto S, Ino Y (1998) Physiological tolerance of *Camellia rusticana* leaves to heavy snowfall environments: the effects of prolonged snow cover on evergreen leaves. Ecol Res 13:117–124

Li C, Xu H, Xu J, Chun X, Ni D (2011a) Effects of aluminium on ultrastructure and antioxidant activity in leaves of tea plant. Acta Physiol Plant 33:973–978

Li C, Zheng Y, Zhou J, Xu J, Ni D (2011b) Changes of leaf antioxidant system, photosynthesis and ultrastructure in tea plant under the stress of fluorine. Biol Plant 55:563–566

Li H, Zhou G, Zhang H, Song G, Liu J (2011c) Study on isolated pathogen of leaf blight and screening antagonistic bacteria from healthy leaves of *Camellia oleifera*. Afr J Agric Res 6:4560–4566

Li X, Ahammed GJ, Li Z, Tang M, Yan P, Han W (2016) Decreased biosynthesis of jasmonic acid via lipoxygenase pathway compromised caffeine induced resistance to colletotrichum gloeosporioides under elevated CO_2 in tea seedlings. Phytopathology 106:1270–1277

Li J, Yang Y, Sun K, Chen Y, Chen X, Li X (2019) Exogenous melatonin enhances cold, salt and drought stress tolerance by improving antioxidant defense in tea plant (*Camellia sinensis (L.)* O. Kuntze). Molecules 24:1826

Lian CU, Wake YO, Yokota H, Wang G, Konishi S (1998) Effect of aluminum on callose synthesis in root tips of tea (*Camellia sinensis L.*) plants. Soil Sci Plant Nutr 44:695–700

Lin Z-H, Chen L-S, Chen R-B, Peng A (2009a) Effects of phosphorus deficiency on nutrient absorption of young tea bushes. J Tea Sci 29:295–200

Lin Z-H, Chen L-S, Chen R-B, Zhang F-Z, Jiang H-X, Tang N (2009b) CO_2 assimilation, ribulose-1,5-bisphosphate carboxylase/oxygenase, carbohydrates and photosynthetic electron transport probed by the JIP-test, of tea leaves in response to phosphorus supply. BMC Plant Biol 9:43–55

Lin Z-H, Chen L-S, Chen R-B, Zhang F-Z, Jiang H-X, Tang N, Smith BR (2011) Root release and metabolism of organic acids in tea plants in response to phosphorus supply. J Plant Physiol 168:644–652

Lin Z-H, Chen L-S, Chen R-B, Zhang F-Z (2012) Antioxidant system of tea (*Camellia sinensis*) leaves in response to phosphorus supply. Acta Physiol Plant 34:2443–2448

Mackerness AHS, Jordan BR, Thomas B (1997) UV-B effects on the expression of genes encoding proteins involved in photosynthesis. In: Lumsden PJ (ed) Plant and UV-B: responses to environmental change. Cambridge University Press, Cambridge, pp 113–114

Magambo MJS, Cannell MGR (1981) Dry matter production and partition in relation to yield of tea. Exp Agric 17:33–38

Mamati GE, Liang YR, Lu JL (2006) Expression of basic genes involved in tea polyphenol synthesis in relation to accumulation of catechins and total tea polyphenols. J Sci Food Agric 86:459–464

Matsumoto H, Hirasawa E, Morimura S, Takahashi E (1976) Localization of aluminum in tea leaves. Plant Cell Physiol 17:627–631

Mohanpuria P, Rana NK, Yadav SK (2007) Cadmium induced oxidative stress influence on glutathione metabolic genes of *Camellia sinensis* (L.) O. Kuntze. Environ Toxicol 22:368–374

Mohotti AJ, Lawlor DW (2002) Diurnal variation of photosynthesis and photoinhibition in tea: effects of irradiance and nitrogen supply during growth in the field. J Exp Bot 53:313–322

Mohotti AJ, Dennett MD, Lawlor DW (2000) Electron transport as a limitation to photosynthesis of tea {*Camellia sinensis* (L.) O. Kuntze}: a comparison with sunflower (*Helianthus annus* L.) with special reference to irradiance. Trop Agric Res 12:1–10

Mokgalaka NS, McCrindle RI, Botha BM (2004) Multi-element analysis of tea leaves by inductively coupled plasma optical emission spectrometry using slurry nebulization. J Anal Atom Spectr 19:1375–1378

Mondal TK (2003) Frost management of tea. Assam Rev Tea New 92:8–12

Mondal TK (2009) Tea. In: Prydarsini M, Jain SM (eds) Breeding plantation tree crops tropical species, pp 545–587

Mondal TK, Saha D (2013) Molecular mechanism involved in abiotic stress tolerance in plant. In: Pullaiah T (ed) Abiotic stress and biotechnology. Regency Publication, New Delhi. (ISBN 978-81-89233-87-7), pp 1–28

Mondal TK, Bhattacharya A, Sood A, Ahuja PS (1998) Micropropagation of tea using thidiazuran. Plant Growth Regul 26:57–61

Mondal TK, Bhattacharya A, Laxmikumaran M, Ahuja PS (2004) Recent advance in tea biotechnology. Plant Cell Tiss Org Cult 75:795–856

Mondal TK, Ganie SA, Neraj RR, Rana MK (2014) Cloning and in silico analysis of a gene encoding a putative β-carbonic anhydrase from cowpea (Vigna unguiculata L. Walp). J Plant Inter 9(1):504–513

Morita A, Fujii Y, Yokota H (2001) Effect of aluminium on exudation of organic acid anions in tea plants. In: Horst WJ et al (eds) Plant nutrition-food security and sustainability of agro-ecosystems, pp 508–509

Morita A, Horie H, Fujii Y, Takatsu S, Watanabe N, Yagi A, Yokota H (2004) Chemical forms of aluminum in xylem sap of tea plants (*Camellia sinensis* L.). Phytochemistry 65:2775–2780

Morita A, Yanagisawa O, Takatsu S, Maeda S, Hiradate S (2008) Mechanism for the detoxification of aluminum in roots of tea plant (*Camellia sinensis* (L.) Kuntze). Phytochemistry 69:147–153

Mukhopadhyay M, Mondal TK (2015) Effect of zinc and boron on growth and water relations of *Camellia sinensis* (L.) O. Kuntze cv. T-78. Natl Acad Sci Lett 38(3):283–286

Mukhopadhyay M, Ghosh PD, Mondal TK (2013a) Effect of boron deficiency on photosynthesis and antioxidant responses of young tea (*Camellia sinensis* (L.)O. Kuntze) plantlets. Russ J Plant Physiol 60:633–639

Mukhopadhyay M, Das A, Subba P, Bantawa P, Sarkar B, Ghosh PD, Mondal TK (2013b) Structural, physiological and biochemical profiling of tea plantlets (*Camellia sinensis* (L.) O. Kuntze) under zinc stress. Biol Plant 57:474–480

Mukhopadhyay M, Mondal TK, Chand PK (2016) Biotechnological advances in tea (*Camellia sinensis* [L.] O. Kuntze): a review. Plant Cell Rep 35(2):255–287

Mukhopadyay M, Bantawa P, Das A, Sarkar B, Bera B, Ghosh PD, Mondal TK (2012) Changes of growth, photosynthesis and alteration of leaf antioxidative defence system of tea (*Camellia sinensis* (L.) O. Kuntze) seedling under aluminum stress. Biometals 25:1141–1154

Murugan AC, Thomas J, Rajagopal RK, Mandal AKA (2012) Metabolic responses of tea (*Camellia* sp.) to exogenous application of ascorbic acid. J Crop Sci Biotech 15:53–57

Nagar PK (1996) Changes in en dog e nous abscisic acid and phenols during winter dormancy in tea (*Camellia sinensis* (L.) O. Kuntze). Acta Physiol Plant 18:33–38

Nagar PK, Kumar A (2000) Changes in endogenous gibberellin activity during winter dormancy in tea (*Camellia sinensis* (L.) O. Kuntze). Acta Physiol Plant 22:439–443

Nagar PK, Sood S (2006) Changes in endogenous auxins during winter dormancy in tea (*Camellia sinensis* L.) O. Kuntze. Acta Physiol Plant 28:165–169

Nagarajah S (1979) The effect of potassium deficiency on stomatal and cuticular resistance in tea (*Camellia sinensis*). Physiol Plant 47:91–94

Nagarajah S (1981) The effect of nitrogen on plant water relations in tea (*Camellia sinensis*). Physiol Plant 51:304–308

Nagarajah S, Ratnasooriya GB (1977) Studies with antitranspirants on tea (*Camellia sinensis* L.). Plant and Soil 48:185–197

Nagata T, Hayatsu M, Kosuge N (1992) Identification of aluminum forms in tea leaves by ^{27}Al NMR. Phytochemistry 31:1215–1218

Nandi SK, Palni LMS (1993) Shoot growth and winter dormancy in tea. J Plant Crops 21:328–333

Napaporn SL, Orapin K, Natta L (2012) Chemical qualities and phenolic compounds of Assam tea after soil drench application of selenium and aluminium. Plant and Soil 356:381–393

Nelson S (2006) Zinc deficiency in tea (*Camellia sinensis*). Plant Dis:PD 34

Netto LA, Jayaram KM, Haridas P, Puthur JT (2005) Characterization of photosynthetic events and associated changes in various clones of tea (*Camellia sinensis* L.) under low temperature conditions. J Plant Biol 48:326–331

Ng'etich WK, Bore JK (2001) The effect of brassinolides (plant growth regulators) on yield and yield attributes of clonal tea (*Camellia sinensis*). Tea 22:8–12

Ng'-Etich WK, Wachira FN (2003) Variations in leaf anatomy and gas exchange in tea clones with different ploidy. J Hortic Sci Biotech 78:173–176

Nikolaeva TN, Zagoskina NV, Zaprometov MN (2009) Production of phenolic compounds in callus cultures of tea plant under the effect of 2,4-D and NAA. Russ J Plant Physiol 56:45–49

Owuor PO (2001) Effects of fertilizers on tea yields and quality: a review with special reference to Africa and Sri Lanka. Int J Tea Sci 1:1–11

Premkumar R, Ponmurugan P, Manian S (2008) Growth and photosynthetic and biochemical responses of tea cultivars to blister blight infection. Photosynth 46:135–138

Qian D, Chang-quan W, Bing L, Huan-xiu L, Yang L (2009) Effects of se, Zn and their interaction on polyphenol oxidase activity of tea leaves in summer season. Acta Metall Sin 15:930–935

Rahimi M, Kordrostami M, Mortezavi M (2018) Evaluation of tea (*Camellia sinensis L.*) biochemical traits in normal and drought stress conditions to identify drought tolerant clones. Physiol Mol Biol Plants 25(1):59–69

Ramarethinam S, Rajalakshmi N (2004) Caffeine in tea plants [*Camellia sinensis* (L) O. Kuntze]: in situ lowering by *Bacillus licheniformis* (Weigmann) Chester. Indian J Exp Biol 42:575–580

Rana NK, Mohanpuria P, Yadav SK (2008) Expression of tea cytosolic glutamine synthetase is tissue specific and induced by cadmium and salt stress. Biol Plant 52:361–364

Ruan J, Wong MH (2001) Accumulation of fluoride and aluminum related to different varieties of tea plant. Environ Geochem Health 23:53–63

Ruan JY, Wu X, Hardter R (1999) Effect of potassium and magnesium nutrition on the quality components of different types of tea. J Sci Food Agric 79:47–52

Ruan J, Ma L, Shi Y, Han W (2003) Uptake of fluoride by tea plant (*Camellia sinensis* L) and the impact of aluminium. J Sci Food Agric 83:1342–1348

Ruan J, Ma L, Shi Y, Han W (2004) The impact of pH and calcium on the uptake of fluoride by tea plants (*Camellia sinensis* L.). Ann Bot 93:97–105

Ruan J, Ma L, Shi Y (2006) Aluminium in tea plantations: mobility in soils and plants, and the influence of nitrogen fertilization. Environ Geochem Health 28:519–528

Ruan J, Gerendás J, Härdter R, Sattlemacher B (2007) Effect of nitrogen form and root-zone pH on growth and nitrogen uptake of tea (*Camellia sinensis*) plants. Ann Bot 99:301–310

Ruan J, Ma L, Yang Y (2012) Magnesium nutrition on accumulation and transport of amino acids in tea plants. J Sci Food Agric 92:1375–1383

Saha D, Mandal S, Saha A (2012) Copper induced oxidative stress in tea (*Camellia sinensis*) leaves. J Environ Biol 33:861–866

Saijo R (1980) Effect of shade treatment on biosynthesis of catechins in tea plants. Plant Cell Physiol 21:989–998

Saikia JK, Baruah U, Barman TS, Saikia H, Bandyopadhyay T (2011) Impact of tea mosquito infestation on endogenous hormones of tea (*Camellia sinensis* L.). Sci Cult 77:412–415

Salehi SY, Hajiboland RA (2008) A high internal phosphorus use efficiency in tea (*Camellia sinensis* L.) plants. Asian J Plant Sci 7:30–36

Sandanam S, Gee GW, Mapa RB (1981) Leaf water diffusion resistance in clonal tea (*Camellia sinensis* L.): effects of water stress, leaf age and clones. Ann Bot 47:339–349

Sanderson GW, Sivapalan K (1966) Translocation of photosynthetically assimilated carbon in tea plants. Tea Q 37:140–153

Sanjay R, Baby UI (2007) Physiological and biochemical changes in tea leaves due to *Pestalotiopsis* infection. J Plant Crop 35:15–18

Seenivasan S, Anderson TA, Muraleedharan N (2016) Heavy metal content in tea soils and their distributionin different parts of tea plants, *Camellia sinensis (L).* O Kuntze. Environ Monit Assess 188:428

Sharma P, Pandey S, Bhattacharya A, Nagar PK, Ahuja PS (2004) ABA associated biochemical changes during somatic embryo development in *Camellia sinensis* (L.) O. Kuntze. J Plant Physiol 161:1269–1276

Sharma V, Joshi R, Gulati A (2011) Seasonal clonal variations and effects of stresses on quality chemicals and prephenate dehydratase enzyme activity in tea (*Camellia sinensis*). Eur Food Res Technol 232:307–317

Shi Y, Ruan J, Ma L, Han W, Wang F (2008) Accumulation and distribution of arsenic and cadmium by tea plants. J Zhejiang Univ Sci B 9:265–270

Shu W, Zhang ZQ, Lan CY, Wong MH (2003) Fluoride and aluminium concentrations of tea plants and tea products from Sichuan Province, PR China. Chemosphere 52:1475–1482

Singh S, Pandey A, Kumar B, Palni LMS (2010) Enhancement in growth and quality parameters of tea [*Camellia sinensis* (L.) O. Kuntze] through inoculation with arbuscular mycorrhizal fungi in an acid soil. Biol Fertil Soils 46:427–433

Smirnoff N (1993) The role of active oxygen in the response of plants to water deficit and desiccation. New Phytol 125:27–58

Smith BG, Stephens W, Burgess PJ, Carr MKV (1993) Effects of light, temperature, irrigation and fertilizer on photosynthetic rate in tea (*Camellia sinensis*). Exp Agric 29:291–306

Smith BG, Burgess PJ, Carr MKV (1994) Effects of clone and irrigation on the stomatal conductance and photosynthetic rate of tea (*Camellia sinensis*). Exp Agric 30:1–16

Stapleton AE (1992) Ultraviolet radiation and plants: burning questions. Plant Cell 4:1352–1358

Subba P, Mukhopadhyay M, Mahato SK, Bhutia KD, Mondal TK, Ghosh SK (2014) Zinc stress induces physiological, ultra-structural and biochemical changes in mandarin orange (*Citrus reticulata Blanco*) seedlings. Physiol Mol Biol Plants 20(4):461–473

Tan X, Yuan J, Li Z, Ye S, Jiang Z (2011) Effects of aluminum and phosphate on material distribution and roots characteristics and activity of *Camellia oleifera* seedlings. J Central South Univ Forest Tech 20:11–12

Tang Q, Li X, Zhu X, Huang Y, Yang H (2008a) Effects of plumbum and chromium stress on the growth of tea plants. Southwest J Agric Sci 21:156–162

Tang Q, Zhu X, Li X, Tan H (2008b) Effects of As and Cd stress on the growth of tea plants. J Henan Agric Sci:208–211

Tang J, Xu J, Wu Y, Li Y, Tang Q (2012) Effects of high concentration of chromium stress on physiological and biochemical characters and accumulation of chromium in tea plant (*Camellia sinensis* L.). Afr J Biotechnol 11:2248–2255

Tanton TW (1979) Some factors limiting yields of tea (*Camellia sinensis*). Exp Agric 15:187–192

Thomas J, Ajay D, Raj Kumar R, Mandal AKA (2010) Influence of beneficial microorganisms during in vivo acclimatization of *in vitro*-derived tea (*Camellia sinensis*) plants. Plant Cell Tiss Org Cult 101:365–370

Tolrà R, Vogel-Mikuš HR, Kump P, Pongrac P, Kaulich B, Gianoncelli A, Babin V, Barceló J, Regvar M, Poschenrieder C (2011) Localization of aluminium in tea (*Camellia sinensis*) leaves using low energy X-ray fluorescence spectro-microscopy. J Plant Res 124:165–172

Upadhyaya H, Panda SK (2004) Responses of *Camellia sinensis* to drought and rehydration. Biol Plant 48:597–600

Upadhyaya R, Panda SK (2013) Abiotic stress responses in tea [*Camelliasinensis* L (O) Kuntze]: an overview. Rev Agric Sci 1:1–10

Upadhyaya H, Panda SK, Dutta BK (2008) Variation of physiological and antioxidative responses in tea cultivars subjected to elevated water stress followed by rehydration recovery. Acta Physiol Plant 30:457–468

Upadhyaya H, Panda SK, Dutta BK (2011) $CaCl_2$ improves post-drought recovery potential in *Camellia sinensis* (L) O. Kuntze. Plant Cell Rep 30:495–503

Upadhyaya H, Dutta BK, Sahoo L, Panda SK (2012) Comparative effect of Ca, K, Mn and B on post-drought stress recovery in tea [*Camellia sinensis* (L.) O Kuntze]. Am J Plant Sci 3:443–460

Upadhyaya H, Dutta BK, Panda SK (2013) Zinc modulates drought induced biochemical damages in tea [*Camellia sinensis (L) O Kuntze*]. J Agric Food Chem 61:6660–6670

Venkatesan S, Jayaganesh S (2010) Characterization of magnesium toxicity, its influence on amino acid synthesis pathway and biochemical parameters of tea. Res J Phytochem 4:66–67

Venkatesan S, Hemalatha KV, Jayaganesh S (2006) Zn toxicity and its influence on nutrient uptake in tea. Am J Plant Physiol 1:185–192

Vyas D, Kumar S (2005a) Purification and partial characterization of a low temperature responsive Mn-SOD from tea (*Camellia sinensis (L.)*O. Kuntze). Biochem Biophys Res Commun 329:831–838

Vyas D, Kumar S (2005b) Tea (*Camellia sinensis* (L.) O. Kuntze) clone with lower period of winter dormancy exhibits lesser cellular damage in response to low temperature. Plant Physiol Biochem 43:383–388

Vyas D, Kumar S, Ahuja PS (2007) Tea (*Camellia sinensis*) clones with shorter periods of winter dormancy exhibit lower accumulation of reactive oxygen species. Tree Physiol 27:1253–1259

Waheed A, Hamid FS, Shah AH, Ahmad H, Khalid A, Abbasi FM, Ahmad N, Aslam S, Sarwar S (2012) Response of different tea (*Camellia sinensis* L.) clones against drought stress. J Master Environ Sci 3:395–410

Wan Q, Xu RK, Li XH (2012) Proton release by tea plant (*Camellia sinensis* L.) rootsas affected by nutrient solution concentration and pH. Plant Soil Environ 58:429–434

Wang X, Liu P, Luo H, Xie Z, Xu G, Yao J, Chen K (2009a) Effect of Al and F interaction on physiological characteristics of tea plants. Acta Hort Sin 7:207–209

Wang R, Chen Y, Wang X, Peng S, Yang X, Wang Y, Yang Y (2009b) Influencing factors on photosynthetic characteristic of superior clones of *Camellia oleifera*-leaf age and leaf position. Chin Agric Sci Bull 2:209–217

Wang LY, Wei K, Jiang YW, Cheng H, Zhou J, He W, Zhang CC (2011) Seasonal climate effects on flavanols and purine alkaloids of tea (*Camellia sinensis* L.). Eur Food Res Technol 233:1049–1055

Wang G, Chen L, Kou L, Yang Y, Feng F, Cao F, Fang Y (2012) Effects of high temperature stress on osmotic adjustment substances of 25 varieties of *Camellia oleifera*. J Henan Agric Sci:2012–2004

Wijeratne MA, Formham R, Anandacumaraswamy A (1998) Water relations of clonal tea (*Camellia sinensis* L.) with reference to drought resistance: II. Effect of water stress. Trop Agric Res Ext 1:74–80

Wong MH, Zhang ZQ, Wong JWC, Lan CY (1998) Trace metal contents (Al, cu and Zn) of tea: tea and soil from two tea plantations, and tea products from different provinces of China. Environ Geochem Health 20:87–94

Wu C, Fang X (1994) Effect of zinc on carbon and nitrogen metabolism in tea plant (*Camellia sinensis* L.). Sci Agric Sin 27:72–77

Wu B, Pan G (1995) Studies on physiological and biochemical response to water stress in tea plant. J Zhejiang Agric Univ 4:1995–2005

Wu Y, Liang Q, Tang Q (2011) Effect of Pb on growth, accumulation and quality component of tea plant. Proc Eng 18:214–219

Xia J, Lan H (2008) Effects of cadmium stress on growth of tea plant and physiological index in leaves of mengshan tea. J Tea Sci 3:205–207

Xu J, Yang F, Chen L, Hu Y, Hu Q (2003) Effect of selenium on increasing the antioxidant activity of tea leaves harvested during the early spring tea producing season. J Agric Food Chem 51:1081–1094

Yadav SK, Mohanpuria P (2009) Responses of *Camellia sinensis* cultivars to cu and Al stress. Biol Plant 53:737–740

Yang Y, Zhuang X, Hu H (1987) Effect of the soil water content on the physiological process of tea plant. J Tea Sci 7:23–28

Yang Z, Kobayashi E, Katsuno T, Asanuma T, Fujimori T, Ishikawa T, Tomomura M, Mochizuki K, Watase T, Nakamura Y, Watanabe N (2012) Characterisation of volatile and non-volatile metabolites in etiolated leaves of tea (*Camellia sinensis*) plants in the dark. Food Chem 135:2268–2276

Yu C, Pan Z, Chen J, Fan D, Wang X (2012) Effects of Al^{3+} on growth and physiological characteristics of tea plant (*Camellia sinensis*). Plant Nutr Fert Sci:2012–2021

Zagoskina NV, Dubravina GA, Alyavina AK, Goncharuk EA (2003) Effect of ultraviolet (UV-B) radiation on the formation and localization of phenolic compounds in tea plants callus cultures. Russ J Plant Physiol 50:270–275

Zagoskina NV, Alyavina AK, Gladyshko TO, Lapshin PV, Egorova EA, Bukhov NG (2005) Ultraviolet rays promote development of photosystem II photochemical activity and accumulation of phenolic compounds in the tea callus culture (*Camellia sinensis*). Russ J Plant Physiol 52:731–739

Zagoskina NV, Goncharuk EA, Alyavina AK (2007) Effect of cadmium on the phenolic compounds formation in the callus cultures derived from various organs of the tea plant. Russ J Plant Physiol 54:237–243

Zeng QL, Chen RF, Zhao XQ, Wang HY, Shen RF (2011) Aluminium uptake and accumulation in the hyperaccumulator *Camellia Oleifera* Abel. Pedosphere 21:358–364

Zhang J, Yang R, Chen R, Peng Y, Wen X, Gao L (2018) Accumulation of heavy metals in tea leaves and potential health risk assessment: acase study from Puan county, Guizhou Province, China. Int J Environ Res Pub Health 15:133

Zhang C, He Q, Wang M, Gao X, Chen J, Shen C (2020a) Exogenous indole acetic acid alleviates Cd toxicity in tea (*Camellia sinensis*). Ecotoxicol Environ Saf 190:1–12

Zhang C, Yi X, Gao X, Wang M, Shao C, Lv Z, Chen J, Liu Z, Shen C (2020b) Physiological and biochemical responses of tea seedlings (*Camellia sinensis*) to simulated acid rain conditions. Ecotoxicol Environ Saf 192:110315

Zheng XQ, Jin J, Chen H, Du YY, Ye JH, Lu JL, Lin C, Dong JJ, Sun QL, Wu LY, Liang YR (2008) Effect of ultraviolet B irradiation on accumulation of catechins in tea (*Camellia sinensis* (L) O. Kuntze). Afr J Biotechnol 7:3283–3287

Functional Genomics

8

8.1 Introduction

Functional genomics plays an important role to understand the fundamental question of plant improvement (Mondal 2013; Mukhopadhyay et al. 2013). In tea, functional genomics work was initiated with the isolation of chalcone synthase gene from a Japanese green tea cultivar 'Yabukita' (Takeuchi et al. 1994a). Since then a significant amount of genomic work has been done in tea and its wild relatives targeting isolation and characterization of trait or tissue-specific genes and their expression under varying conditions (Table 8.1). It has been evident that two initial types of efforts have been made: (1) cloning of individual gene associated with particular trait and (2) differential gene expression which leads to identification of genes that are associated to a trait. Later with the advancement of next-generation sequencing technology, several workers made attempts to identify the differentially expressed genes (Mondal et al. 2004), the associated pathways using RNAseq technology. Further with the availability of tea genome sequence recently, multiple gene family has been characterized in tea which are discussed in this chapter.

8.2 Cloning and Characterization of Individual Genes

Cloning and characterization are the initial step of functional genomics [particularly important when the genome sequence is not available (Mondal et al. 2014)]. Interestingly, majority of the genes cloned so far belongs to quality followed by yield as these two traits have major demand in the commercial tea cultivations.

8.2.1 Quality-Related Genes

Tea is valued for its cup quality. Therefore, research efforts of tea are biased towards the improvement of quality. Similarly, in the area of genomics, several individual

T. K. Mondal, *Tea: Genome and Genetics*,
https://doi.org/10.1007/978-981-15-8868-6_8

Table 8.1 Genes cloned from *Camellia* species

Stress	Gene	Encoding product	Expressed tissue	Reference
Growth and development	*CsTUA 1*	Tubulin	Pollen tube	Fang et al. (2006)
	CsLOX1	Lipoxygenase	Flower	Liu and Han (2010)
	Cs14-3-3	14-3-3 protein	Late stage flower bud	Liu et al. (2010)
	CsELIP	Early light-induced proteins	Leaf	Li et al. (2013a)
	CsTUA	α-tubulin	Leaf	Paul et al. (2012a)
	Cs26srRNA	26S rRNA gene	Different plant parts	Singh et al. (2004)
	CsH3	Histone H3.1	Leaf	Singh et al. (2009d)
	CsQM	Ribosomal protein L10	Leaf	Singh et al. (2009e)
	CsTUB	β-tubulin	Leaf	Takeuchi et al. (1994b)
	CsCYS	Cystatin	Leaf	Wang et al. (2005)
	CsBDP	BURP domain-containing protein	Leaf	Wang et al. (2011a)
	CsARP1	Auxin-repressed protein	Dormant bud	Wang et al. (2011b)
	CsCYC1	Cyclin	Sprouting bud	Wang et al. (2011c)
	CsCDK1	Cyclin-dependent kinase	Bud	Wang et al. (2012c)
	CsACT	Actin	Leaf	Yang et al. (2012b)
	CsRbcS	Ribulose-1,5-bisphosphate carboxylase/oxygenase	Leaf	Ye et al. (2009)
	CsCYP	Cyclophilin	Meristem tissue	Zhang et al. (2007)
	CsS-RNase	Self-incompatibility RNase	Flower	Zhang et al. (2016a)
	Putative S-RNase gene	Self-incompatibility RNase	Style in flower	Zhang et al. (2016b)
	CsSCLP	Serine carboxypeptidase-like protein	Flower	Chen et al. (2016)
	CrGA20ox1	Gibberellin 20-oxidase	Shoot tip	Wang et al. (2018f)

(continued)

Table 8.1 (continued)

Stress	Gene	Encoding product	Expressed tissue	Reference
	CsQM	QM-like protein	Apical leaf bud	Singh et al. (2009e)
	CnFLS1	Flavonol synthase-like gene	Different floral organs	Zhou et al. (2013)
	CsARGOS	*ARGOS* gene induced by auxin	Root, stem, and leaf	Wang et al. (2017a)
	PRL-1	Pathogenesis-related protein	Pistil in flower	Tomimoto et al. (1999)
Abiotic stress	*CsCBF*	C-repeat-binding factor	Leaf	Chen et al. (2009a)
	CsCBF1	C-repeat/dehydration-responsive element binding factors	Shoot tip	Wang et al. (2012d)
	CsICE1	Inducer of CBF expression 1	Shoot tip	Wang et al. (2012d)
	CsCSH1	H1-histone	Leaf	Fang et al. (2009), Fang et al. (2013)
	CsCOR1	CsCOR1	Leaf	Li et al. (2010)
	CsDHY	Dehydrin1, dehydrin2	Leaf	Paul and Kumar (2013)
	CsNAM	Nuclear-localized protein	Stem, flower, bud	Paul et al. (2012b)
	CsGS	Glutamine synthetase	Leaf, bud	Rana et al. (2008a)
	CsVDE	Violaxanthin de-epoxidase	Leaf	Wei et al. (2003)
	CsSMT	Selenocysteine methyltransferase Stearoyl-acyl carrier protein desaturase	NA	Zhu et al. (2008b)
	CsSAD	Glutathione peroxidase	Leaf	Ding et al. (2016)
	CsGPX2	Transcription factor	Root, leaf, stem, flower petal	Fu (2014)
	CsWRKY2	Transcription factor	Different plant parts	Wang et al. (2015c)
	CsERF-B3	AP2-ERF transcription factor	Root	Wu et al. (2014b)
	CsbZIP1	Basic leucine zipper proteins	NA	Li et al. (2014)

(continued)

Table 8.1 (continued)

Stress	Gene	Encoding product	Expressed tissue	Reference
	CsGolS1	Galactinol synthase	Different plant parts	Zhou et al. (2017b)
	CsHSP17.2	Heat shock protein	Different plant parts	Mingle et al. (2015)
	CsMTP11	Metal tolerance protein 11	Old leaf and root	Yu et al. (2017)
	CsBADH1	Betaine aldehyde dehydrogenase	NA	Li et al. (2013b)
	Co-sHSP	Small heat shock protein	Seed	Chen et al. (2009b)
	CsNAC	*N*-acetyl cysteine	NA	Tang et al. (2013)
	CsPCS1	Phytochelatin synthase	Third leaves from the top buds	Wang et al. (2017b)
	CsDREB	Drought response element-binding protein	Fourth leaves from the top buds	Wang et al. (2017c)
	CsbZIP6	Basic region/leucine zipper transcription factor	Apical mature leaves	Wang et al. (2017d)
Biotic stress	*CsCHIi*	Chitinase	Shoot	Roy and Chakraborty (2009)
	CsPR	*PR-1 like protein*	Petals, anthers, style and young leaf	Tomimoto et al. (1999)
	CsCAD	Cinnamoyl alcohol dehydrogenase	Leaf	Deng et al. (2013a)
	CsGolS2 *CsGolS3*	Galactinol synthase	Leaf	Zhou et al. (2017b)
	CsHPL	Fatty acid hydroperoxide lyase	Leaf	Deng et al. (2016)
	CbTps1	*Terpene synthase*	Flower	Hattan et al. (2016)
Quality	*CsLCYE*	Lycopene-epsilon-cyclase	Leaf	Borchetia et al. (2011)
	CsPSY	Phytoene synthase	leaf	Borthakur et al. (2008)
	CsTS	Theanine synthetase	Root and shoot	Deng et al. (2012); Deng et al. (2013b)
	CsSAM	S-adenosylmethionine synthetase	Leaf	Feng and Liang (2001)

(continued)

Table 8.1 (continued)

Stress	Gene	Encoding product	Expressed tissue	Reference
	CsTPS1	Sesquiterpene synthase	Root, leaf	Fu (2012)
	CsTBS	Theobromine synthase	Leaf	Ishida et al. (2009)
	CsGUS	β-glucosidase	Cytoplasm	Jiang et al. (2005)
	CsTCS1	Caffeine synthase	Leaf	Kato et al. (2000)
	CsO-MTF	O-methyltransferase	Shoot	Kirita et al. (2010)
	CsCS	Caffeine synthase	Leaf	Li et al. (2004b, 2007)
	CsFLS	Flavonol synthase	Leaf	Lin et al. (2007), Su Xia et al. (2009)
	CsLAR	Leucoanthocyanidin reductase	Leaf	Ma et al. (2010)
	CsPAL	Phenylalanine ammonia-lyase	Leaf	Matsumoto et al. (1994b)
	CsPMO	β-primeverosidase	Leaf	Mizutani et al. (2002)
	CsLCR	Leucoanthocyanidin reductase	Leaf	Park et al. (2004)
	Cs4CL	p-Coumarate:CoA ligase	Shoot	Rani et al. (2009)
	CsCHI	Chalcone isomerase	Leaf	Rani et al. (2012)
	CsF3′-5′H	Flavanone-3-hydroxylase	Leaf	Rani et al. (2012)
	CsANS	Anthocyanidin synthase	Leaf	Rani et al. (2012)
	CsF3′H	Flavanone-3-hydroxylase	Leaf	Singh et al. (2008)
	CsDFR	Dihydroflavonol 4-reductase	Leaf	Singh et al. (2009a), Punyasiri et al. (2004)
	CsANR	Anthocyanidin reductase	Leaf	Singh et al. (2009b), Zhang et al. (2012); Pang et al. (2013)
	CsC4H	Cinnamate-4-hydroxylase	Leaf	Singh et al. (2009c)
	CsPAL	Phenylalanine ammonia-lyase	Leaf	Singh et al. (2009c)
	CsCHS1, *CsCHS2*, *CsCHS3*	Chalcone synthase	Leaf	Takeuchi et al. (1994a)
	CsCHS	Chalcone synthase	Leaf	Takeuchi et al. (1994a), Rani et al. (2012)

(continued)

Table 8.1 (continued)

Stress	Gene	Encoding product	Expressed tissue	Reference
	CsPPO	Polyphenol oxidase	Leaf	Wu et al. (2010)
	CsACCSy	ACC synthase	Leaf	Zhang et al. (2008a)
	CsACCO	ACC oxidase	Leaf	Zhang et al. (2008b)
	CsPPO	Polyphenol oxidase	Leaf	Zhao et al. (2001)
Ion homeostasis	*CsNDPK1*	Nucleotide diphosphate kinase	Leaf	Prabu et al. (2012)
	CsGS1	Cytosolic glutamine synthetase	Root	Taniguchi and Tanaka (2004)
	CsAMT	Ammonium transporter protein	Root	Taniguchi and Tanaka (2004)
	CsGDH2	Glutamate dehydrogenase α-subunit	Leaf	Wang et al. (2012e)
	CsAPS1, *CsAPS2*	Adenosine phosphosulfate	Leaf and root	Zhu et al. (2008a)
Development/ stress	*CsPALa*, *CsPALb*, *CsPALc*, *CsPALd*, *CsPALe*, *CsPALf*	Phenylalanine ammonia-lyase	Different plant organs	Wu et al. (2017c)
Metabolism/ biotic stress	AF537127	β-glucosidase gene	Leaf	Jun et al. (2005)
	CsAOC	Allene oxide cyclase	Leaf	Wang et al. (2016g)
	CsOPR3	Oxophytodienoate reductase	Different plant organs	Xin et al. (2017)
Stress signaling metabolism	*CsMYB1*	Transcription factor	Shoot tip explants	Yang et al. (2012a, b)
	CsMYB4a	R2R3-MYB transcription factor	Leaf	Li et al. (2017g)
	CsC4Ha *CsC4Hb* *CsC4Hc*	Cinnamate 4-hydroxylase	Leaf and tender stem	Xia et al. (2017b)
	CsANR	Anthocyanidin reductase	Apical buds and leaves	Singh et al. (2009b)
	CsF3Ha *CsF3Hb*	Flavonone-3-hydroxylase	Tender leaf	Han et al. (2017)
	CsLAR	Leucoanthocyanidin reductase	Leaf	Zhang et al. (2016c)
	CsSAMT	Salicylic acid carboxyl methyltransferase	Different tea organs	Deng et al. (2017)

(continued)

Table 8.1 (continued)

Stress	Gene	Encoding product	Expressed tissue	Reference
	CsANR1 *CsANR2*	Anthocyanidin reductase	Leaf	Pang et al. (2013)
	CsGS	Glutamine synthetase	Bud (youngest leaf)	Rana et al. (2008b)
	UGT73A17	Glucosyltransferase	Different plant parts	Ohgami et al. (2014)
	CsDFR	Dihydroflavonol 4-reductase	Leaf from apical bud	Singh et al. (2009a)
	CsLARa *CsLARb* *CsLARc*	Leucocyanidin reductase	Leaf	Wang et al. (2018g)
	CHS1 *CHS2* *CHS3*	Chalcone synthase	Leaf	Zhang et al. (2016c)
	CsANS	Anthocyanidin synthase	Leaf	Zhang et al. (2016c)
	CsANR1 *CsANR2*	Anthocyanidin reductase	Leaf	Zhang et al. (2016c)
	Co-accC *Co-accD*	Acetyl co-A carboxylase	Different tissues	Wang et al. (2018h)
	Csp450	Cytochrome P450 monooxygenase	Young shoots and first two leaves	Eminoglu et al. (2018)
	CsMYB6A	R2R3-MYB transcription factor	Leaf and buds	He et al. (2018)
	CsUGT72AM1	Novel UGT gene	Leaf and buds	He et al. (2018)
	CjF3Ha *CjF3Hb*	Flavonone 3-hydroxylase	Flower petal	Tateishi et al. (2010)
	CjDFR	Dihydroflavonol 4-reductase	Flower petal	Tateishi et al. (2010)
	CjANS	Anthocyanidin synthase	Flower petal	Tateishi et al. (2010)
	CnFLS1	Flavonol synthase	Flower	Zhou et al. (2013)
	CsLIS/NES	Terpene synthase	Leaf	Liu et al. (2018c)
	ClGA2ox1 *ClGA2ox2* *ClGA2ox3*	Gibberellin 2-oxidase	Stem, mature leaf and root	Xiao et al. (2015)
	CoFBA1 *CoFBA2* *CoFBA3* *CoFBA4*	Fructose-1,6-bisphosphate aldolase	Seed	Zeng et al. (2014)
	CoACP	Acyl carrier protein	Seed	Zeng et al. (2014)

(continued)

Table 8.1 (continued)

Stress	Gene	Encoding product	Expressed tissue	Reference
	CoSAD	Stearoyl-ACP desaturase		Zeng et al. (2014)
	CoALDH	Aldehyde dehydrogenase	Seed	Ze et al. (2017)
	CsFOMT	Flavonoid O-methyltransferase	NA	Ying et al. (2013)
	CoFAD2-1	FAD2	NA	Tan et al. (2008)
	CsSDR	Short-chain dehydrogenase/ reductase	All tissue	Ying et al. (2015)
	Partial PPO gene	Polyphenol oxidase	Flower	Samynathan et al. (2015)
	CsF3′H	Flavonoid 3′-hydroxylase	Apical bud and terminal leaf	Zhou et al. (2016)
	CsAPS1 *CsAPS2*	ATP sulfurylase	One leaf and one bud	Zhu et al. (2008a)
	CbTPS1 *CsTPS1*	Hedycaryol synthase (terpene synthase)	Leaf	Hattan et al. (2016)
	CsCuAO	Copper-containing amine oxidase	Flower, bud and leaf	Bharalee et al. (2012)
	CsHPL	Hydroperoxide lyase	Leaf	Deng et al. (2016)

genes of various biochemical pathways are targeted, cloned, and characterized which are discussed below (Fig. 8.1).

8.2.1.1 Theanine Biosynthesis-Related Genes

Theanine, a glutamate derivative, is the most dominant amino acid in tea plant, which although is found in all the organs and tissues of tea plant, its maximum amount is found in tender leaves (Sugiyama and Sadzuka 2004). Theanine is one of the constituents which provide the briskness of tea, a parameter which is used to determine the quality of black tea. It had been documented that theanine makes up approximately 50% of the total amino acids of tea. It is synthesized in the roots and is transported to the tender parts of the plant for storage. In the presence of sunlight, it is converted to polyphenols (Anonymous 2005). Apart from its contribution in the taste of tea, it has some medicinal values as well (Sugiyama and Sadzuka 2003). Glutamine synthase, an important enzyme of theanine biosynthetic pathway, is involved in assimilation of ammonia generated by various biochemical processes

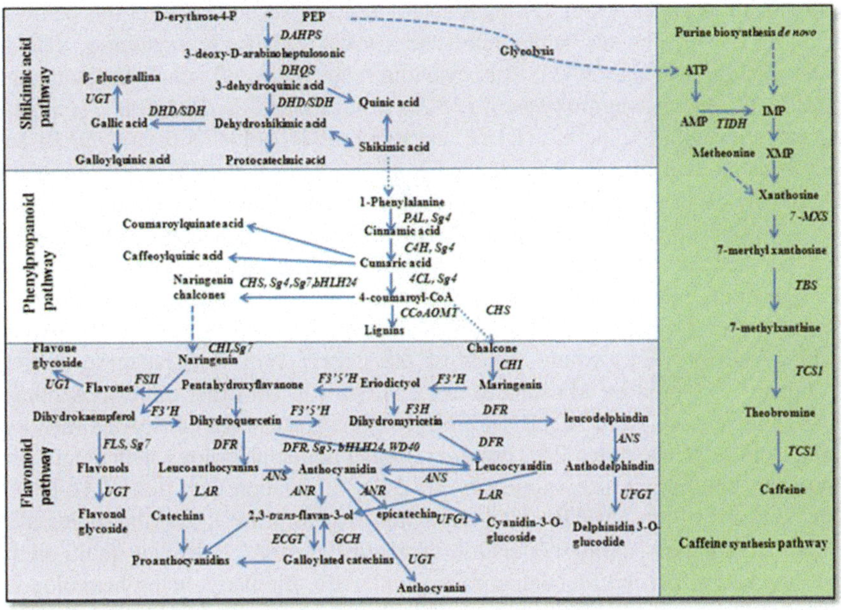

Fig. 8.1 Summary of major biochemical pathways involved in the quality of made tea and genes of enzyme that are cloned (partially or fully) from tea, denoted by italics. ANR, anthocyanidin reductase; ANS, anthocyanidin synthase; C4H, cinnamate-4-hydroxylase; 4CL, 4-coumaroyl-CoA ligase; CHS, chalcone synthase; CHI, chalcone isomerase; CCoAOMT, caffeoyl-CoA 3-O-methyl transferase; DAHPS, 3-deoxy-D-arabinoheptulosonate 7-phosphate synthase; DFR, dihydroflavonol 4-reductase; DHQS, 3-dehydroquinate synthase; DHD, 3-dehydroquinate dehydratase; ECGT, epicatechins 1-O-galloyl-β-D-glucose O-galloyltransferase; F3H, flavanone 3-hydroxylase; F3′H, flavonoid 39-hydroxylase; F3′5′H, flavonoid 39,59-hydroxylase; FLS, flavonol synthase, FSII, flavone synthase II; GCH, galloylated catechins hydrolase; IMP, inosine-5-monophosphate; LAR, leucoanthocyanidin reductase; 7-MXS, 7 methyl xanthosine synthase; PAL, phenylalanine ammonia-lyase; SDH, shikimate dehydrogenase; TBS, theobromine synthase; TCS1, caffeine synthase, TIDH, inosine-5_-monophosphate dehydrogenase, UFGT, UDP-glucose: flavonoid 3-O-glucosyltransferase, UGGT, UDP glucose galloyl-1-O-β-D-glucosyltransferase; UGT, uridine diphosphate glycosyltransferase; XMP, xanthosine-5′-monophosphate; Sg4, Sg7, WD40, and bHLH24 are the transcription factors as reported by Zhao et al. (2013)

and had been cloned from tea (Rana et al. 2008a, b). Based on their cellular locations of isozymes, they were classified into two groups, *GS3A* (cytosolic GS) and *GS2* (chloroplast GS). The distinct, cell-specific patterns of expression conferred by the promoters for *GS2* and *GS3A* suggest the different roles in metabolism (Edwards et al. 1990).

8.2.1.2 Flavonoid Biosynthesis-Related Genes

Polyphenol content in tea leaf accounts for almost 15–35% of the total dry weight depending on the genotypes, environment on which they grow, etc. Polyphenols are chemical compounds such as flavonoids and tannins found naturally in tea. Among the polyphenols, catechins are the most abundant and account for 80% of the total polyphenols (Mondal 2001). Owing to its importance, several genes of this pathway such as phenylalanine ammonia-lyase (*PAL*), chalcone synthase (*CHS*), anthocyanidin synthase (*ANS*), anthocyanidin reductase (*ANR*), chalcone isomerase (*CHI*), flavanone-3-hydroxylase (*F3′H*), flavanone-3-hydroxylase (*F3′5′H*), leucoanthocyanidin reductase (*LAR*), and dihydroflavonol 4-reductase (*DFR*) had been isolated, cloned, and characterized from tea plants (Table 8.1). Among them, PAL catalyzes the initial step for catechin biosynthesis, via phenylpropanoid pathway, by deamination of phenylalanine-evolving ammonium ion and trans-cinnamic acid. On the other hand, CHS and CHI (chalcone isomerase) catalyze the formation of flavonoids from flavanone (2S)-naringenin by sequential reactions of 4-coumaroyl-CoA and three molecules of malonyl-CoA (Matsumoto et al. 1994a).

In a pioneer work, young leaves of the widely cultivated Japanese cultivar 'Yabukita' was used by Matsumoto et al. (1994b) to construct the cDNA library and hybridized with rice *PAL* cDNA probe to clone the tea PAL gene. Alignments of amino acid sequence of tea PAL protein revealed higher homology with dicots than monocots, confirming its evolutionary proximity in relation with dicot *PAL* genes. Later, a 280 bp from 3′ UTR of this gene was used as probe to test the segregation pattern among the mapping population. It showed that *PAL* in tea was multi-allelic and inherited as per the Mendelian monogenic ratio. Similarly, using heterologous probe, *CHS* gene (*CsCHS*) of tea was cloned by screening the cDNA library which showed 85% similarity between parsley *CHS* and *CsCHS* (Takeuchi et al. 1994a). Later several genes related to biosynthesis of flavonoids including LAR (Ma et al. 2010) had been cloned in tea (Table 8.1).

Further DFR from tea (*CsDFR*) was also cloned. In vivo expression analysis through Q-PCR indicated younger leaves had higher level of expression than older leaves under controlled condition. However, it was downregulated under dehydration stress and ABA treatment, but upregulated upon wounding. Interestingly, all the regulation of this gene was modulated with catechin content in the respective tissue or stress (Singh et al. 2009a). Anthocyanidin reductase, a key enzyme of epicatechins pathway of tea (*CsANR*), had been cloned (Singh et al. 2009b). The functional protein, upon its expression in *E. coli*, catalyzed the conversion of cyanidin to epicatechins in the presence of NADPH. PAL and C4H are two important enzymes in catechin biosynthesis. Fragment of *C4H* gene was isolated from tea by differential display, and later full-length gene was cloned using Rapid Amplification of cDNA End (RACE) technology. It had been found that catechin contents decreased in response to drought and ABA and GA_3 treatments but increased in response to wounding. Interestingly, the expression of *CsPAL* and *CsC4H* transcripts was in similar trend along with catechin contents under the above treatments. Therefore, a positive correlation between catechin contents and gene expression was established during catechin biosynthesis (Singh et al. 2009c). Flavanone 3-hydroxylase (F3′H), an important enzyme of catechin biosynthetic pathway, catalyzes the stereospecific hydroxylation of (2S)-naringenin to form (2R, 3R)-dihydrokaempferol. The *CsF3′H* was cloned from tea and expressed in *E. coli* which yielded a functional protein. Furthermore, a positive correlation was detected between the concentration of catechins and *CsF3′H* expression at different developmental stages of tea plants (Singh et al. 2008).

Polyphenol oxidase is an important enzyme which plays vital role during fermentation of black tea processing. Using degenerate primers, Zhao et al. (2001) were first to clone this gene from young tea leaf. Phylogenetic analysis indicated a close proximity of *CsPPO* to other woody plants. Later, in order to understand the allelic variation, *CsPPO* was cloned from five different genotypes of tea, and potential genic-SNPs were detected. Wu et al. (2010) also cloned PPO from tea. It had a signal peptide targeting the chloroplast and three Cu-binding domains. The mature *PPO* showed high expression and enzyme activity in *E. coli* strain BL21 under pET30c expression vector. 4-Coumaroyl-CoA (4CL) is another important gene of catechin biosynthetic pathway. The full-length cDNA cloning of *Cs4CL* and its association with catechin accumulation in tissue was established. Altering catechin content through drought stress, ABA and GA_3 treatments, and wounding established a strong positive correlation coefficient between catechin content and the expression of *Cs4CL* (Rani et al. 2009). Pang et al. (2013) characterized the gene of key enzymes involved in polyphenolic proanthocyanidins pathway of tea. Recombinant protein of tea leucoanthocyanidin reductase (*CsLAR*) is a key enzyme of proanthocyanidins pathway of tea which was expressed in *E. coli*, found to be functional when leucocyanidin taken as substrate. This could produce the product 2R, 3S-*trans*-flavan-ol (+)-catechin under in vitro conditions. Two genes encoding anthocyanidin reductase, *CsANR1* and *CsANR2*, were also expressed in *E. coli*, to get recombinant proteins that exhibited similar kinetic properties. Both the enzymes converted cyanidin to a mixture of (+)-epicatechin and (-)-catechin, although in different proportions, indicating that both enzymes possess epimerase activity and hence functional in nature (Pang et al. 2013).

8.2.1.3 Purine Biosynthesis-Related Genes

Caffeine, theobromine, and theophylline are considered as purine alkaloids and flavor-forming chemicals present in some species of *Camellia*. Caffeine and theobromine are purine alkaloids that are present in high concentrations in tea. However, most species of the genus *Camellia* do not contain any purine alkaloids. The major steps of caffeine biosynthetic pathway are xanthosine → 7-methylxanthosine → 7-methylxanthine → theobromine → caffeine (Suzuki 1972; Ashihara and Kubota 1987; Ashihara et al. 1996, 1997; Kato et al. 1996; Deng et al. 2008). Among the pivotal enzymes, S-adenosylmethionine synthase, converts methionine to S-adenosylmethionine. It donates methyl group in transmethylation reactions and also acts as the precursor in ethylene and polyamine biosynthesis. So, isolation and cloning of *SAM* might pave the way for further study and manipulation of caffeine biosynthesis, stress, and senescence physiology. Accordingly a *CsSAM* gene was cloned, and ORF was found to be 1303 bp in length with 349 amino acids. This is an essential step to manipulate the caffeine biosynthesis in tea plant (Feng and Liang 2001).

In tea, among the three genes, namely, 7-N-methyltransferase (*7-NMT*), 3-N-methyltransferase (*3-NMT*), and theobromine N-methyltransferase (*1-NMT*), two

(*3-NMT* and *1-NMT*) are highly homologous due to similarity of many properties and are considered as single enzyme, caffeine synthase (*CS*). Initially using degenerate primer, partially *CsCS* gene and later full length was cloned from young leaf of tea using RACE-PCR (Kato et al. 1999). The substrate specificity of the recombinant enzyme, after expression of cDNA in *E. coli*, remained similar to that of the original tea enzyme. The cloning of *CS* gene led to the development of transgenic caffeine-deficient tea plants by antisense technology (Kato et al. 2000).

Tracer experiments using [8-^{14}C] adenine and [8-^{14}C] theobromine showed that the purine alkaloid pathway is not fully functional in leaves of purine alkaloid-free species. Interestingly, in purine alkaloid-free *Camellia* species, sufficient evidence was obtained to show the occurrence of genes that are homologous to caffeine synthase. Recombinant enzymes derived from purine alkaloid-free species showed only theobromine synthase activity. Unlike the caffeine synthase gene, these genes were expressed more strongly in mature tissue than in young tissue (Ishida et al. 2009).

Mohanpuria et al. (2009) suggested that caffeine metabolism in tea appears to be dependent on developmental and seasonal factors. Expression of caffeine synthase as well as caffeine content was found to be higher in harvestable plant parts such as apical bud, first leaf, second leaf, and young stem and was lower in older leaf during non-dormant growth periods compared to dormant growth phase.

8.2.1.4 Floral Aroma Formation-Related Genes

Aroma is an important yardstick to determine quality of black tea, and primarily the aroma-forming compounds are produced by endogenous enzymes such as β-glucosidases and β-primeverosides. In order to increase the quality of made tea, it is obvious to pursue the ways that ensure an increment in tea aroma-forming compounds. The β-glucosidase, a hydrolase, was found to be involved in the formation of floral aroma (Takeo 1981). Later on many aroma-forming compounds had been isolated and identified from green as well as black tea leaves. Of late, the concept of hydrolyzation of polyglycoside by β-glucosidase vis-a-vis β-primeverosides and subsequent removal of floral aroma was established. It is a disaccharide-specific glycosidase which hydrolyzes aroma precursors of β-primeverosides to liberate aroma compounds. Using degenerate primer, full-length cDNA sequence of β-primeverosidase genes was cloned and characterized (Mizutani et al. 2002). It had 50–60% identical protein sequence with β-glucosidases of other plants and classified this as family 1-glycosyl hydrolase. It was confirmed that the β-primeverosidase selectively hydrolyzed the β-glycosidic bond of β-primeverosides between the disaccharide and the aglycons (Mizutani et al. 2002). Later, the β-glucosidase gene of tea was also cloned (Li et al. 2004a). It shared 40–60% similarity in nucleotide sequence with other plants. Upon expressing in heterologous system in *E. coli*, it possessed normal bioactivity by breaking down the glycosidic bonds. Subsequently, the spatiotemporal expression pattern of the aforesaid genes in different leaves of young shoot was determined in tea (Zhao et al. 2006a). This information will help in the genetic manipulation of tea plant with the respective genes.

8.2.2 Abiotic Stress-Related Genes

Abiotic stress is very important for tea which not only affects the yield but also affects its quality. Differential expression of a cytosolic glutamine synthetase (*CsGS*) of tea under developmental stages and light or dark conditions on the utilization of nitrate as well as ammonia had been reported (Rana et al. 2010). The *CsGS* transcript expression level was highest in apical bud, gradually reduced towards the older leaf, and was found to be lowest in the fourth leaf. Exposure to both nitrate and ammonia N-sources showed enhancing effect on *CsGS* enzyme activity during light, but under dark, ammonium increased *CsGS* activity and nitrate had inhibitory effect (Rana et al. 2010).

Ammonium transporters (*AMT*) are involved in nitrogen absorption of plant roots, therefore important for utilization in nitrogenous fertilizer. This gene had also been cloned from tea root (Taniguchi and Tanaka 2004). Cold is an important abiotic stress for tea cultivation as it hampers the productivity of tea significantly. A well-known cold-responsive gene, *CsCOR1*, had been cloned from tea leaves recently (Li et al. 2010). The deduced amino acid sequence contained hydrophobic N-terminus as a signal peptide and a characteristic glycine, arginine, and proline-rich C-terminal domain. The gene was found to be localized in the cell wall of transgenic tobacco by *CsCOR1::GPF* fusion approach. Its (*CsCOR1*) expression was upregulated under cold and dehydration stress.

Sulfate assimilation is an important metabolic pathway for plants, and ATP sulfurylase, the first enzyme of this pathway, converts ATP and sulfate to adenosine phosphosulfate. Two cDNAs encoding ATP sulfurylase (*APS1* and *APS2*) were cloned from tea (Zhu et al. 2008a). While ORF of *APS1* was found to be 1415 bp in length with 360 amino acids, *APS2* ORF was found to be 1706 bp in length which encoded 465 amino acids. They shared 59.6% identity with each other, indicating that these two paralogues are quite diverse.

A higher level of selenium is toxic to plants. Enzyme selenocysteine methyltransferase (*SMT*) methylates selenocysteine to se-methylselenocysteine and thus plays a vital role for removing the toxicity. Full-length cDNA of *CsSMT* had been cloned from tea (Zhu et al. 2008b). This sequence of amino acids shared 74% and 69% identity with *Astragalus bisulcatus* SMT and *Brassica oleracea* SMT, respectively. Expression of *CsSMT* correlated with the presence of SMT enzyme activity in cell extracts, and bacterial recombinant *CsSMT* had higher tolerance to selenate and selenite.

8.2.3 Biotic Stress-Related Genes

Pests and pathogens are harmful agents that act upon the bush health and productivity. Although both pest and diseases are important for tea cultivation, not much works have been done, primarily due to lack of tolerant genotype of tea for various pests and diseases. However, Wang et al. (2005) cloned the cystatin gene of tea with the help of degenerate primers. The deduced amino acid sequence contained the motif QXVXG conserved among the most members of cystatin superfamily. Eight wound-/pathogen-inducible cDNAs were cloned and their structures analyzed in silico (Yoshida and Homma 2005). It had been found that they had high sequence homology with other wound-inducible genes of herbaceous plants.

8.2.4 Energy Metabolism-Related Genes

Green plants, by virtue of chlorophyll, absorb solar energy to produce organic substance and release oxygen through the process of photosynthesis. Though solar energy is the basis for photosynthesis, excessive energy may result in photooxidative damage to photosynthetic apparatus and other cell components. There exist some mechanisms to protect the plants from such adverse effects. From tea plants, two photosynthesis-related genes had been cloned, which were photosystem II protein D1 and violaxanthin de-epoxidase genes (*VDE*). Among them, *VDE* had been studied extensively. Under excess light intensity, violaxanthin converted to antheraxanthin and ultimately to zeaxanthin by *VDE*, whereas under low light intensity zeaxanthin epoxidase (ZE) catalyzes the reverse reaction. Zeaxanthin can protect photosynthetic apparatus from excessive irradiation. Thus VDE can be considered as a vital enzyme in the xanthophyll cycle-dependent photoprotective mechanism. Using degenerate primers, Wei et al. (2003) cloned full-length cDNA of *VDE* gene. It was demonstrated that the gene contained lipocalin signature which is considered to be the most active domains in VDE (Wei et al. 2004).

RUBISCO (ribulose-1,5-bisphosphate carboxylase) is a key enzyme in energy metabolism of the Calvin cycle pathway of photosynthesis. The full-length RUBISCO small subunit (*rbcS*) was cloned from tea. The amino acid sequence had a high identity compared to those of other plant *rbcS* genes. It had three conserved domains as well as a protein kinase C phosphorylation site, one tyrosine kinase phosphorylation site, and two N-myristoylation sites. The Q-PCR analysis showed that the expression of *rbcS* was the highest in youngest tea leaf (Ye et al. 2009). Although reason is not known, this could be due to the fact that chlorophyll biogenesis mostly occurs in young leaf.

8.2.5 Developmentally Regulated Genes

Microtubules composed of α-tubulin and β-tubulin dimers are very important constituents of eukaryotic cytoskeleton. It is an important structural element involved in mitosis, cytokinesis, and vesicular transport. Sediment directions of microfilaments on cell walls, which control the growth of cells, are also regulated by microtubules. The β-tubulin gene had been cloned from tea (Takeuchi et al. 1994b). The putative amino acid sequence had a high similarity to β-tubulin genes from other plants.

The eukaryotic nucleosome octamer core consists of the four histones, H3, H2A, H2B, and H4. Using differential display and RACE-PCR, a full-length histone H3.1

cDNA (*CsH3*) was cloned from tea leaves. During active growth, higher expression was observed in apical buds and decreased gradually with maturity of the leaf. During dormancy, drought stress, and ABA treatments, the expression of *CsH3* was severely downregulated but upregulated by GA_3 treatment. Positive co-relations between *CsH3* and active cellular growth suggested its role in plant growth and development (Singh et al. 2009d).

On the other hand, lipoxygenase (*LOXs*) is a large gene family which catalyzes the hydroperoxidation of free polyunsaturated fatty acids into different biologically active compounds and had been isolated from tea plant (*CsLOX1*). Heterologous expression in yeast showed that *CsLOX1* protein conferred a dual positional specificity, and hence it was named 9/13-*CsLOX1*. Analysis of the isolation and expression of the *LOX* gene in tea plant indicated that the acidic *CsLOX1* together with its primary and final products played an important role in regulating cell death related to flower senescence and the JA-related defensive reaction of the plant to phloem feeders (Liu and Han 2010). Pollen tube growth in tea was promoted by *Tua 1* gene. The cDNA-AFLP technique was used by Fang et al. (2006) to isolate transcript-derived fragments corresponding to these genes. The complete cDNA sequence of this gene had been cloned which had two N-glycosylation sites and four protein kinase C phosphorylation sites.

8.2.6 Other Genes

Ribosomes play an important role in plant evolution with 5.8S, 16S, and 26S rRNA being the prime components of ribosome. The QM like protein gene encodes for ribosomal protein which has a function of ribosome stability. Full-length QM cDNA from tea (*CsQM*) had been cloned (Singh et al. 2009e) which shared 71–87% and 85–91% nucleotide identity and amino acid sequences, respectively, with QM genes isolated from other woody plants.

8.3 Differentially Expressed Transcripts

Several studies to identify the differentially expressed genes in tea have been conducted, but collectively they show that earlier attempts were made by low-throughput techniques such as suppression subtractive hybridization (SSH), DDRT, etc., which later was conducted by RNAseq technology, discussed below.

8.3.1 Low-Throughput Transcriptome Analysis

Understanding the changes of global transcripts under a particular stress, cell types, tissue, or developmental stage is a robust approach. Although several techniques are available (Mondal and Sutoh 2013; Das et al. 2013), SSH and cDNA-AFLP have been widely used to study the differential gene expression in tea. For example, SSH had been used to understand the molecular regulation of secondary metabolism pathways in the young tea leaves (Park et al. 2004; Chen et al. 2005), to identify the cold-responsive genes (Wang et al. 2009), to identify dormancy-associated genes (Wang et al. 2010; Krishnaraj et al. 2011; Paul and Kumar 2011; Yang et al. 2012a; Phukon et al. 2012; Thirugnanasambantham et al. 2013), to identify light-regulated phenylpropanoid metabolism in tea calli (Wang et al. 2012a), to identify gray blight disease infection genes (Senthilkumar et al. 2012), and to identify the drought-responsive genes (Sharma and Kumar 2005; Das et al. 2012; Gupta et al. 2012, 2013; Muoki et al. 2012). In Taiwan, "oriental beauty" is a very popular high-flavored tea which is made from leafhopper-infested leaf. In order to know the genes that are responsible for this unique flavor, gene expression and biochemical profile were done by DNA microbead array, a fluorescence labeled-based technique where different fluorescence DNAs are sorted out by cell sorter machine. Interestingly several biotic stress-responsive genes were found to be upregulated in infested tea leaves indicating that they may involve to activate the aroma formation pathway (Choi et al. 2007). Eungwanichayapant and Popluechai (2009) reported that the expression of phenylalanine ammonia-lyase 1, chalcone synthase, dihydroflavonol 4-reductase, leucoanthocyanidin reductase, and flavanone 3-hydroxylase genes were higher in the young shoots than in the mature leaves which were obvious as most of the quality compounds were produced by these genes. These genes are normally found to be more expressive in young leaves. To understand the molecular mechanism of enhanced aroma under mild infestation of green leafhopper (*Empoasca vitis* Göthe) in tea leaf, a SSH library was constructed, and several genes were identified (Yang et al. 2011), and interestingly it was found that insect infection triggered several genes that were responsible for quality. Further, a forward SSH library under drought stress was constructed from TV-23 (a drought-tolerant cultivar) roots, to ascertain 123 drought-responsive genes (Das et al. 2012). Recently seven cDNA libraries from various organs of tea plants were used to generate 17,458 ESTs of tea (Taniguchi et al. 2012). Interestingly, to investigate the molecular mechanisms by which light regulates phenylpropanoid metabolism, a light-induced SSH library of tea calli was constructed. Several genes in lignin biosynthetic pathways were identified in the light-induced library (Wang et al. 2012a). Furthermore, through cDNA-AFLP approach, they identified 108 transcript-derived fragments that were expressed differentially in the leaf of drought-tolerant tea cultivars (Gupta et al. 2013). Tea geometrid moth is an important insect of tea. Nineteen genes were found to be upregulated with *Ectropis obliqua* infection as detected by SSH, differential screening, and Q-PCR analysis. These genes were involved in JA synthesis, cell wall modification, metabolism of secondary metabolites as well as carbohydrate, pathogen-defensive response, and oxidative stress protection. The results showed that *E. obliqua* infections activated defense-related genes and induced tea defense responses (Qiao et al. 2011). Darjeeling teas of India are high altitude tea which is popular for its flavor, aroma, and quality. Apart from the genetic makeup of the plant, reports suggested that insect infestation, particularly jassids and thrips, triggered the aroma and flavor formation in Darjeeling tea. Several genes and transcription factors were identified through the SSH library of leaf that were highly infested by these insects (Gohain et al. 2012).

Expression of genes encoding enzymes involved in flavan-3-ol biosynthetic pathway such as *CHS, CHI, F3H, F3′-5′H, DFR, ANS, ANR*, and *LAR* was investigated. Transcripts of all genes, except *LAR*, were found to be more abundant in leaves and stems than in roots and cotyledons. In tea, flavan-3-ols are produced by a naringenin-chalcone to naringenin to dihydrokaempferol pathway. Dihydrokaempferol is a branch point in the synthesis of (-)-epigallocatechin-3-O-gallate and other flavan-3-ols which can be formed by routes beginning with either a flavonoid 30-hydroxylase-mediated conversion of the flavonol to dihydroquercetin or a flavonoid 30,50-hydroxylase-catalyzed conversion to dihydromyricetin with subsequent steps involving sequential reactions catalyzed by dihydroflavonol-4-reductase, anthocyanidin synthase, anthocyanidin reductase, and flavan-3-ol gallate synthase (Ashihara et al. 2010). *PAL, CHS*, and *DFR* were differentially expressed under red and blue light; however, their expressions were lower in mature leaf than in the younger leaf (Takechi and Matsumoto 2003).

R2R3-MYB, bHLH, and *WD40* transcription factors are well-known to control multiple enzymatic steps in the biosynthetic pathway responsible for the production of flavonoids, important secondary metabolites of tea. The presence of 73 *R2R3-MYB*, 49 *bHLH*, and 134 *WD40* transcripts was identified through in silico analysis of 127,094 unigenes of tea. In silico analysis of protein sequences of CsMYB4-1, CsMYB4-2, CsMYB4-3, CsMYB4-4, CsMYB5-1, and CsMYB5-2 indicated the presence of [DE]Lx2[RK]x3Lx6Lx3R motif, potentially contributing to the specificity of the bHLH partner in the stable MYB-bHLH complex. Q-PCR analysis validated selected genes and their expression profiles under various developmental stages and treatment conditions, including hormones (Zhao et al. 2012).

The plant hormone auxin plays a key role in adventitious rooting. To increase the understanding of genes involved in adventitious root formation, Wei et al. (2013) identified differentially expressed transcripts in single nodal cuttings of tea treated with or without IBA. A total of 77 differentially expressed transcripts were identified that were expressed under IBA treatment. This could be involved in adventitious rooting of tea.

Tea plantation at lower altitude uses shade for better cultivation. Thus, influences of shade on flavonoid biosynthesis in relation to expression of the flavonoid pathway genes in tea leaves were analyzed. It was found that shade had notable effects on both flavonoids (including catechins, O-glycosylated flavonols) and PAs and lignin biosynthesis, but had no significant effect on anthocyanin accumulation. Among all the compounds, the concentration of PAs and O-glycosylated flavonols under shaded leaves reduced more than other compounds, compared to the sunlight-exposed leaves. This observation was correlated with gene expression. Expressions of *PAL*, flavanone 3-hydroxylase, flavonoid 3-hydroxylase, *DFR*, and *ANR1* were notably correlated with the concentration of PAs in leaves, and expressions of *CHS* and flavonoid 3,5-hydroxylase (*F3′-5′H*) were correlated with the concentration of O-glycosylated flavonols. It was suggested that polymerization of catechins and glycosylation of flavonols might be key pathways of flavonoid metabolism in tea leaves affected by shade treatment. On the other hand, phenolic acids were increased under shade, and a negative correlation with lignin accumulation suggested that phenolic acids might compete for the same substrate with lignins and flavonoids in tea leaves under different illumination conditions (Wang et al. 2012b).

Callus grown in tissue culture has different morphology from white loose fibrous to compact hard green. In order to identify the differentially expressed genes related to morphological variations of callus, Yang et al. (2012a) analyzed two contrasting callus lines (Yunjing 63Y and Yunjing 63X) that showed different morphological characteristics and catechin contents. Yunjing 63Y callus was yellow and compact, while Yunjing 63X callus was white and loose. Using cDNA AFLP, they identified 68 genes that were differentially expressed between the two lines. Gray blight disease (*Pestalotiopsis theae*) is an economically important disease of tea which accounts for an enormous loss to the tea industry.

Three tea clones with varying degree of polyphenol content were selected to identify genes that were related to polyphenol biosynthetic pathway. Totally, 1680 gene tags were obtained from the cDNA library to develop the first cDNA microarray of tea plant. Finally, two genes that were closely related to tea aroma were selected for Q-PCR analysis to validate the cDNA microarray. The newly developed cDNA microarray could be applied in various tea research fields to make the high-throughput detection in the gene expression profiling (Zhao et al. 2006b). Cross-species hybridization was also attempted to understand the gene expression of tea. Using *Arabidopsis* cDNA array, Venkatesh et al. (2006) found that highest level of induction of CHS gene in the tea leaf is observed/recorded during the second flush (a period where flavonoid content of tea was found to be the highest in tea plantation of Assam, India). On the other hand, expressions of cinnamyl alcohol dehydrogenase (CAD), cinnamyl CoA reductase (CCR), and POD involved in the synthesis of lignin were reduced during the second flush. These observations suggested that CHS may be a key enzyme involved in catechin biogenesis in tea leaf. Similarly, to identify the aluminum (Al) uptake-responsive genes in tea, Murayama et al. (2007) used cross-species microarray analysis. For the target gene of microarray analysis, approximately 8500 cDNA fragments were identified from *Arabidopsis thaliana*. According to the signal intensity of cDNA detected by microarray analysis, putative genes including auxin-induced protein, zinc finger family protein, and blue copper binding protein were found to be expressed differentially under Al treatment in tea. Therefore, it was suggested that antioxidant enzyme genes and synthesis of auxin as well as cell wall genes might be induced by Al treatment in tea root tips.

There are albino cultivars in tea. Scientists are always keen to know albinism. To understand the molecular mechanisms of albinism, tea cultivar 'Anji Baicha' was selected. A total of 671 differentially expressed genes in leaf at different albescent stages were identified using cDNA microarray analysis. The corresponding genes were involved in energy metabolism, carbon fixation, cell expansion, secondary metabolism, plant growth and defense, and other physiological processes including protein, nucleotide synthesis, etc. Particularly, some differentially expressed genes encoding important catalyzing enzymes or regulatory proteins which took part in chlorophyll biosynthesis or chloroplast development were identified. The present study gave some useful clues for genes worthy of further understanding the albino phenotype of 'Anji Baicha' and also provided a model for utilization of the microarray technology in the tea plant (Ma et al. 2012). L-theanine is an important component of tea, content of which is tissue and genotype specific. Expression analysis of 17 genes was tried to co-relate with L-theanine content in different tissues of tea and found that three genes such as *CsTS2*, *CsGS1*, and *CsGDH2* showed positive correlation with the L-theanine contents in tea (Liu et al. 2017a).

8.3.2 High-Throughput Transcriptome Analysis

Low-throughput techniques such as SSH, DDRT, and MPSS have major limitations for generating limited number of transcripts signature. On the other hand, due to technological superiority, recently, the next-generation sequencing technology, specially RNAseq, is widely used for identification of differentially expressed genes (Mondal and Rana 2013; Mukhopadhyay et al. 2013). The work on identifying the differentially expressed genes of RNAseq started with Jiang et al. (2011) who used next-generation platform such as 454 GS FLX for transcriptome sequencing of *C. oleifera*, *C. chekiangoleosa*, and *C. brevistyla* to generate 182,766, 190,545, and 132,147 reads, respectively. These reads were assembled into 49,909 contigs. They could estimate common genes among the three species along some transcript-based SNPs. Since then, several RNAseq experiments had been conducted with tea for various applications such as differential gene expression, SSR marker development, detection of alternative splicings, etc. which are discussed here under category of quality aspects, biotic stress and abiotics, growth and development, heterosis breeding, manufacturing, tea seed oils, etc.

8.3.2.1 Quality of Made Tea

In tea, quality depends upon several factors such as climatic conditions, quality of harvest, genotype, and age of the tissue that influence the accumulation of various chemical constituents (Mondal 2004). Similarly, the quality of made tea also depends upon the manufacturing process. In tea, quality depends upon the secondary metabolites, including flavonoids, theanine, and caffeine, which are the primary sources of the rich flavors, fresh taste, and health benefits of tea. Thus not only it is polygenic, but also a complex network exists which determines the quality of made tea (Zhang et al. 2018a). Thus several workers time and again tried to study the quality aspects of tea using RNAseq technology. In the beginning, the focus was to standardize RNAseq protocol from young tea leaf and discover genes related to major metabolic pathways. This transcriptome dataset was very important at that time as an important public information platform for gene expression, genomics, and functional genomics studies of tea (Shi et al. 2011). Later, to know the genes involved secondary metabolism, RNAseq of tea leaf tissue was done by several workers. All these studies basically tried to identify differentially expressed genes between the tissues, the seasons, or from different genotypes having varying degree of catechins. Wu et al. (2013) found that some important genes are related to developmentally regulated in plants, including ubiquitin/26S proteasome, lipid transfer protein, PPR containing protein, small GTPase, expansin, transport inhibitor response 1 and thioredoxin for catechins synthesis. Zhang et al. (2018b) reported that three TFs homologous to *ANL2*, *WRKY44*, and *AtMYB113* might play key roles in catechin regulation. Guo et al. (2017a) found that a variety of processes such as the regulation of the cell cycle, starch and sucrose metabolism, photosynthesis, phenylpropanoid biosynthesis, phenylalanine metabolism, and flavonoid biosynthesis were involved in catechin formation. Some of the transcription factors such as *MYB*, *bHLH*, and *MADS* are involved in the regulation of catechin biosynthesis and were identified through co-expression analysis of transcription factors and structural genes.

Wu et al. (2014a) reported that catechin content has high positive correlation with *ANS*, *ANR*, and *LAR* genes of tea. Caffeine is a crucial secondary metabolic product in tea plants. For the elucidation of the caffeine biosynthesis and catabolism in Camellia plants, fresh leaves from four *Camellia* plants with low to high caffeine concentrations were used. Transcriptome study discovered a degradation pathway of caffeine to theobromine, thus assisting researchers in understanding the caffeine-related mechanisms in *Camellia* plants containing low, normal, and high caffeine content (Zhu et al. 2019a). Externally supplied sucrose induced the synthesis of terpene, aromatic derivatives, lipid, and other pathway genes. Some of the specific genes were glutathione S-transferase (GST), ATP-binding cassette transporters (ABC transporters), and multi-drug and toxic compound extrusion transporters (MATE transporters) (Qian et al. 2018). The ratio of dihydroxylated to trihydroxylated catechins (RDTC) is an important indicator of tea quality and biochemical marker for the study of genetic diversity. Transcriptome analysis reveals key flavonoid 3′-hydroxylase and flavonoid 3′,5′-hydroxylase genes in affecting the ratio of dihydroxylated to trihydroxylated catechins in tea (Wei et al. 2015). The correlation analysis showed that *CsCHIc*, *CsF3′H*, and *CsANRb* expression levels are associated significantly with the concentration of soluble PA as well as the expression levels of CsPALc and CsPALf with the concentration of insoluble PA (Wang et al. 2016a). In order to study the gene expression related to quality parameters, RNAseq was done from different tissues of three different genotypes, i.e., Him Sphurti (China type), TV23 (Assam type), and UPASI-9 (Cambod type). Seasonal variation in these cultivars was studied during active, mid-dormant, dormant, and mid-active stages in two developmental tissues, viz., young and old leaf. Developmental stages appear to affect gene expression more than the seasonal variation and cultivar types. Further, detailed transcript and metabolite profiling has identified key genes of catechin biosynthesis, caffeine biosynthesis/catabolism. Moreover, differential expression of genes involved in histone and DNA modification further suggested the role of epigenetic mechanisms in coordinating global gene expression during developmental and seasonal variation in tea (Kumar et al. 2016).

Major secondary metabolites, including flavonoids, caffeine, and theanine, are important components of tea products and are closely related to the taste, flavor, and health benefits of tea. The gene network responsible for the regulation of the secondary metabolic pathways was analyzed to elucidate the possible crosstalk in gene regulation between the secondary metabolite biosynthetic pathways in *C. sinensis* (Li et al. 2015a). Around 7650 differential expression transcripts (DETs) along with 28,980 alternative splicing events were predicted during leaf development genes that were involved in the flavonoid, theanine, and caffeine biosynthetic pathways. Additionally, 9052 long non-coding RNAs (lncRNAs) were also predicted which plays a crucial role in the regulation of the secondary metabolism of the tea plant (Qiao et al. 2019).

In order to identify the differences in biochemical pathways in oolong tea, transcriptomic, HPLC, and GC-MS analyses indicated that the expression levels of genes related to secondary metabolism and high contents of catechins, anthocyanin, caffeine, and limonene may contribute to the formation of the quality oolong tea (Guo et al. 2019).

Epigallocatechin gallate (EGCG) accumulation in pruned tea trees is found to be significantly higher than that of unpruned tea trees. *SCPL1A* expression (encoding a class of serine carboxypeptidase), which has been reported to have a catalytic ability during EGCG biosynthesis, together with *LAR*, was upregulated in the pruned tea trees. Moreover, metabolic flux enhancement and transcriptome analysis revealed low EGCG accumulation in the leaves of unpruned tea trees. Because of the bitter and astringent taste of EGCG, these results provide a certain understanding of the lower bitterness and astringency in teas from "ancient tea trees," growing in the wild with no trimming, than teas produced from pruned plantation trees (Sun et al. 2018).

Methyl jasmonate (MeJA) could improve the quality of tea aroma by promoting flavor volatiles in tea leaves, although the gene involved was not known. The α-linolenic acid degradation pathway was firstly responded, resulting in the activation of the JA-pathway inner tea leaves and the MEP/DOXP pathway significantly exaggerated. Notably, the expression level of jasmonate O-methyltransferase, which is a major enzyme in JA biosynthetic pathway, was increased by 7.52-fold under MeJA treatment. Several other genes such as genes related to the terpenoid backbone biosynthetic pathway, 1-deoxy-D-xylulosephosphate synthase (DXS), all-trans-nonaprenyl-diphosphate synthase, geranylgeranyl reductase, geranylgeranyl diphosphate synthase (type II), hydroxymethylglutaryl-CoA reductase, and 4-hydroxy-3-methylbut-2-enyl diphosphate reductase also increased by two- to fourfold under MeJA treatment, indicating that MeJA can trigger aroma formation in tea (Shi et al. 2015).

Previous studies on the biosynthesis of major secondary metabolites using next-generation sequencing technologies limited the accurately prediction of full-length (FL) splice isoforms. This was resolved with single-molecule sequencing to pooled tea plant tissues which identified 94 FL transcripts and four AS events for enzyme-coding genes involved in the biosynthesis of flavonoids, theanine, and caffeine. According to the comparison between long-read isoforms and assemble transcripts, it was found that resulting FL transcripts, together with the improved assembled transcripts and identified AS events, enhance understanding of genes involved in the biosynthesis of characteristic secondary metabolites of tea (Xu et al. 2017a).

Apart from made tea quality, RNAseq was also employed in wild species of *Camellia*. For example, *C. chekiangoleosa* is an important species of genus

Camellia. It provides high-quality edible oil and has great ornamental value. The flowers are big and red which bloom between February and March. RNAseq study revealed that 15 genes are involved in anthocyanin biosynthesis, among which nine anthocyanin biosynthetic pathway genes (*PAL*, *CHS1*, *CHS2*, *CHS3*, *CHI*, *F3H*, *DFR*, *ANS*, and *UFGT*) were identified and cloned. The spatiotemporal expression patterns of these genes were also analyzed. The study results not only enrich the gene resource but also provide valuable information for further studies concerning anthocyanin biosynthesis (Wang et al. 2014a). To better understand the key pathways determining tea flavor and enhance tea tree breeding programs, RNAseq for *C. taliensis* was conducted as this species is being used for making tea in some area of China. To gain insights into the evolution of these genes, comparative analysis was done with previously cloned orthologous genes of tea and found that considerable nucleotide variation within several genes involved in important secondary metabolic biosynthetic pathways, of which flavone synthase II gene (FNSII) is the most variable between these two species. Moreover, comparative analyses revealed that *C. taliensis* showed a remarkable expansion of *LEA* genes, compared to tea, which might contribute to the observed stronger stress resistance of *C. taliensis* (Zhang et al. 2015). Transcriptome study was conducted in young leaf and bud of *C. sinensis* and *C. oleifera* and found that genes encoding enzymes involved in flavonoid, theanine, and caffeine pathways exhibited considerably different expression levels in tea compared to oil tea. It could be the reason why *C. sinensis* produces tea through low-level expression and the same is also noticed in oil tea, indicating that there is a special regulatory mechanism in oil tea which is why they are expressed in low level despite the presence of these genes (Tai et al. 2015). Wang et al. (2018a) analyzed 36 genes from 12 gene families of tea. During this process, some intron retention events of the *CsCHI* and *CsDFR* genes were found. Furthermore, the transcriptome sequencing of various tea tissues and subcellular location assays revealed co-expression and colocalization patterns. The correlation analysis showed that *CsCHIc*, *CsF3′H*, and *CsANRb* expression levels are associated significantly with the concentration of soluble PA as well as the expression levels of *CsPALc* and *CsPALf* with the concentration of insoluble PA. This work provides insights into catechin metabolism in tea.

8.3.2.2 Biotic

Tea being monoculture is exposed to several pests and diseases. Additionally the shade tree also serves as an alternative host of insects and pathogens due to which biotic stress is pathologically as well as commercially important.

Blister blight (BB), being a fungal leaf diseases, is very important as it directly causes economic loss of tea cultivation. This disease is caused by an obligate biotrophic fungus, *Exobasidium vexans*. Transcriptomic data of a BB-tolerant genotype SA6 indicated that salicylic acid and jasmonic acid possibly induce synthesis of antimicrobial compounds, required to overcome the virulence of *E. vexans* (Jayaswall et al. 2016).

Anthracnose caused by *Colletotrichum camelliae* is another important fungal disease of tea. Transcriptome sequencing in a tolerant tea genotype was challenged with *C. camelliae* and revealed that secondary metabolism-related pathways, plant hormone biosynthesis and signaling, as well as plant-pathogen interaction pathways genes were involved in the resistance of this variety (Wang et al. 2016b). On the contrary, when the transcriptomic study was conducted to an anthracnose susceptible genotype, it was observed that endogenous salicylic acid pathway becomes active which was corroborated by the fact of higher accumulation of salicylic acid in infected leaf (Shi et al. 2019). Further in a comparative analysis of transcriptomic sequences, between a susceptible and tolerant cultivars of anthracnose, it is indicated that hydrogen peroxide (H_2O_2) metabolism, cell death, secondary metabolism, and carbohydrate metabolism are involved in the defense of tolerant cultivar. Histochemical analysis subsequently validated the strong hypersensitive response and H_2O_2 accumulation that occurred around the hyphal infection sites in tolerant genotype, i.e., Zhongcha 108 (Wang et al. 2018b).

Ectropis obliqua is an important insect of tea which damages tea leaf. Through RNAseq analysis differentially expressed genes were identified from the infected tissue and were found to be signal transduction genes; anti-insect responsive transcription factors, jasmonate/ethylene synthesis, and signaling transduction appeared to be particularly active for insect tolerance in geometrid-damaged tea plants (Wang et al. 2015a, b, c). Interestingly several secondary metabolite-related pathways were also found to be active in the insect-infected leaf indicating that these genes are induced by this insect infection (Wang et al. 2016c) which is why tea made out of this leaf produced better quality. Further, systemic response that triggered in undamaged roots upon infection is also investigated through transcriptomic analysis. It has been found that carbohydrate dynamics that leaf herbivory activated did systemic carbon reallocation to enhance resource investment for secondary metabolism. The crucial role of jasmonic acid and the involvement of other potential hormone signals for local and systemic signaling networks were supported by phytohormone accumulation and dynamic expression analysis of phytohormone-related genes (Yang et al. 2019).

8.3.2.3 Abiotic Stress

With the changes of climatic factors, abiotic stress becomes very important for agricultural crop including tea. Several abiotic stresses are important for tea as they cause huge economic loss, and thus attempts have been made to understand the genes that are underlying for the tolerance of this trait in tea and are discussed below.

8.3.2.3.1 Cold

Low temperature stress is an important limitation of tea cultivation as they grow well in high altitude hilly terrain which has cold weather. Changes that occur at the molecular level in response to low temperature are poorly understood in tea plants. To elucidate the molecular mechanisms of cold acclimation, Wang et al. (2013) identified differentially expressed transcripts from the leaf of tea plant treated with cold. These genes belong to the group of cold sensor or signal transduction genes, cold-responsive transcription factor genes, plasma membrane stabilization-related genes, osmosensing-responsive genes, and detoxification enzyme genes. Pathway analysis indicated that the "carbohydrate metabolism pathway" and the "calcium signaling pathway" play a vital role in tea plants' responses to cold stress. In a comparative study between cold-tolerant and cold-sensitive tea plant, it was found that several DEGs were enriched in pathways of photosynthesis, hormone signal transduction, and transcriptional regulation of plant-pathogen interactions in tolerant cultivar. Notably while expression of *Lhca2* was decreased yet expression of *SnRK2.8* was increased in cold-tolerant tea genotype under cold treatment although detail function was not mentioned (Li et al. 2019a). The mitogen-activated protein kinase (MAPK)-dependent ethylene and calcium signaling pathways were two major early cold-responsive mechanisms involved in sprouting shoots and were followed by the induction of the inducer of CBF expressions (ICE)-C-repeat binding factors (CBF)-cold-responsive (COR) signaling pathway to augment cold tolerance. During the cold shock, growth, photosynthesis, and secondary metabolism—mainly involving flavonoid biosynthesis—were remarkably affected. Notably, the increased starch metabolism, which might be dependent on the high expression of β-amylase3 (BAM3) induced by CBF, played crucial roles in protecting young shoots against freezing cold (Hao et al. 2018).

Transcript profiles of young and mature leaves exposed to cold temperatures revealed that the genes were predominantly related to the cellular component terms of chloroplasts and cell membranes, oxidation-reduction process, carbohydrate metabolism pathway and the calcium signaling pathway, as well as glutathione metabolism and photosynthesis, indicating that these components and pathways may contribute to the cold hardiness of mature leaves. Conversely, the inhibited expression of genes related to cell membranes, carotenoid metabolism, photosynthesis, and ROS detoxification in young leaves under cold conditions might lead to the disintegration of cell membranes and oxidative damage to the photosynthetic apparatus (Li et al. 2018a). RNAseq study further indicates that zinc finger genes and *LEA* genes were induced under cold stress in tea when tea cultivar 'Suchazao' was treated with moderately low temperature stress (ML), severely low temperature stress (SL), moderately high temperature stress (MH), and severely high temperature stress (SH). In addition, energy metabolisms were inhibited by SH, SL, and ML. Furthermore, the expression of anthocyanin synthesis genes was different under the different low temperature stresses indicating that anthocyanin accumulation depends on a particular temperature (Shen et al. 2019a). Apart from this, unsaturated fatty acid biosynthesis and jasmonic acid biosynthetic pathways were deduced to be involved in the low temperature responses in *C. japonica* (Li et al. 2016a). Wild oil-tea camellia is an essential genetic resource for the breeding of cultivated oil-tea camellia, one of the four major woody oil crops in the world. Thus, to identify the variations in transcriptomes of wild oil-tea camellia from different latitudes and elevations, which has different temperature, RNAseq study was undertaken. Large amounts of SSRs, SNPs, and InDels were identified. The phylogenetic analysis based on SNPs showed genetic differentiation between latitudes. Such a result suggests that the sequence variations identified can be used to develop molecular markers for analyzing genetic differentiations along latitude and elevation gradients in wild oil-tea camellia. Our study clearly showed that candidate genes for cold acclimation may be predominantly involved in transmembrane transporter activities (Chen et al. 2017).

In field conditions, tea plants are often simultaneously exposed to various abiotic stresses such as cold and drought, which have profound effects on plant growth and tea quality. However, most studies of gene expression in stress responses focus on a single inciting agent, and the confounding effect of multiple stresses on crop quality and leaf senescence remains unearthed. Thus global transcriptome profiles of tea leaves under separately cold and drought stress were compared with their combination using RNAseq technology. This revealed that tea plants shared a large overlap in unigenes, displayed "similar" (26%) expression pattern, and avoid antagonistic responses (lowest level of "prioritized" mode: 0%) to exhibit very congruent responses to co-occurring cold and drought stress; 31.5% differential expressed genes and 38% of the transcriptome changes in response to combined stresses were unpredictable from cold or drought single-case studies. Finally 319 candidate genes were identified for enhancing plant resistance to combined stress. These results showed that drought-induced leaf senescence were severely delayed by (1) modulation of a number of senescence-associated genes and cold-responsive genes, (2) enhancement of antioxidant capacity, (3) attenuation of lipid degradation, (4) maintenance of cell wall and photosynthetic system, (5) alteration of senescence-induced sugar effect/sensitivity, as well as (6) regulation of secondary metabolism pathways that significantly influence the quality of tea during combined stress (Zheng et al. 2016).

Nitric oxide (NO) functions as a critical signaling molecule in cold stress in plants. It polarized the growth of pollen tube in tea. Despite this, the potential mechanisms underlying the participation of NO in pollen tube responses to low temperature remain unclear. Thus gene expression of pollen tubes exposed to low temperature stress and NO reveals differentially expressed genes related to the inhibition of tea pollen tube growth under low temperature and NO were identified (Pan et al. 2016).

8.3.2.3.2 Drought

In general *Camellia* species is canonically water-loving plants; thus, low moisture stress is an important constrain for their growth. In order to understand the genes regulating the drought tolerance in *Camellia*, several workers did transcriptomic experiment to find out the differentially expressed genes in *C. oleifera* (Dong et al. 2017) and tea (Liu et al. 2016a; Jin et al. 2018). These experiments generated information regarding drought-responsive genes along with various related pathways. Collectively they found genes involved in ABA-dependent and ABA-independent pathways, ethylene and jasmonic acid biosynthesis, potassium as well as ABC membrane transporters, and antioxidant defense system and signaling pathways which were generally upregulated under drought stress. Besides, drought stress also helps to accumulate major bioactive ingredients, especially catechins, caffeine, and theanine. RNAseq analysis of four diverse tea genotypes with inherent contrasting genetic response to drought generated differentially expressed transcripts that were related to ROS scavenging genes in tolerant genotypes, while it seems to be either absent or expressed very low in sensitive genotypes (Wang et al. 2016d; Parmar et al. 2019). However, drought is complex trait and co-occurs with many other stress such as cold, salinity, etc. which also effects the quality of made tea. Thus RNAseq analysis was conducted separately from cold and drought stress leaf separately which revealed that tea plants shared 319 candidate genes for enhancing plant resistance to combined stress. Further, investigation on combined effect of cold and drought on tea quality and leaf senescence showed that drought-induced leaf senescence was severely delayed by (1) modulation of a number of senescence-associated genes and cold-responsive genes, (2) enhancement of antioxidant capacity, (3) attenuation of lipid degradation, (4) maintenance of cell wall and photosynthetic system, (5) alteration of senescence-induced sugar effect/sensitivity, as well as (6) regulation of secondary metabolism pathways that significantly influence the quality of tea during combined stress. Nevertheless more in-depth study is required for studying the drought stress in tea (Zheng et al. 2016).

8.3.2.3.3 Aluminum

Tea is a hyperaccumulator of Al, known for a long time, though molecular mechanism is not known till now. Recently, RNAseq analysis revealed the DEGs that are related to cell wall-modifying enzymes, actin, cytoskeleton, cyclin and H^+-ATPase, transporters, transcription factors, cytochrome P450, ubiquitin ligase, organic acid biosynthesis, heat shock proteins suggesting that these pathways were involved in better root growth under different Al concentration (Li et al. 2017a; Fan et al. 2019a).

8.3.2.3.4 Nitrogen

To understand the genes involved for the nitrogen (N) uptake and utilization, RNAseq analysis of root and shoot tissue was done from a tea plant which was fed on different N source. Among the differentially expressed genes, *AMT*, *NRT*, and *AQP* for N uptake and *GOGAT* and *GS* for N assimilation were found to be the key genes. Importantly, only *AQP* and three novel DEGs associated with stress, manganese binding, and gibberellin-regulated transcription factor were common in N responses across all tissues and varieties (Li et al. 2017b). Nitrogen remobilization during flower development in tea has been studied for gene expression which is particularly important during late autumn. It has been found that the N content in the attached leaf decreased during flower development, which can be increased upon removal of the flower. RNAseq analysis indicated that the expression of genes associated with autophagy, sucrose transporters, amino acids permease, glutamine synthetase, and asparagine synthetase was significantly upregulated in leaves during the flowering process and was strongly modulated by the removal of flower buds (Fan et al. 2019b). Further, nitrogen as nutrition significantly affected quality of tea though little is known about the physiological and molecular mechanisms underlying the effects of short-term repression of N metabolism on tea roots and leaves. The results showed that GS activities in tea roots and leaves that were exposed to a specific inhibitor of glutamine synthetase (GS), methionine sulfoximine (MSX), for a short time (30 min), were significantly inhibited upon MSX treatment, and both tissue types showed a sensitive metabolic response to GS inhibition. In tea leaves, the hydrolysis of theanine results the increase in theanine and free ammonium content. The biosynthesis of all other amino acids was repressed, and the content of N-containing lipids declined, suggesting that short-term inhibition of GS reduces the level of N reutilization in tea leaves. Metabolites related to glycolysis and the tricarboxylic acid (TCA) cycle accumulated after GS repression, whereas the content of amino acids such as glycine, serine, isoleucine, threonine, leucine, and valine declined in the MXS-treated group. Thus they speculate that the biosynthesis of amino acids is affected by glycolysis and the TCA cycle in a feedback loop. Thus it seems GS repression in tea plant leads to the reprogramming of amino acid and lipid metabolic pathways (Liu et al. 2019a).

8.3.2.3.5 Fluorine

Tea plant is also a typical fluoride (F) hyperaccumulator enriching most F in old leaves. Thus there is association between the risk of fluorosis and excessive consumption of teas prepared using the old leaves. A comprehensive RNAseq analysis of tea plants grown at various F levels for different durations by hydroponics provides major pathways that are involved in the mechanism of F metabolism in tea plant (Li et al. 2018b). Interestingly, F accumulation in tea leaves is also gradually increased under exogenous F application. Interestingly, flavonoid and caffeine content also increased in the F treatment. However, in contrast, the polyphenol content, free amino acids, and the total catechin content decreased significantly. Special amino acids, such as sulfur-containing amino acids and proline, had the opposite trend of free amino acids. These results suggest that the F accumulation and content of bioactive compounds were dramatically affected by F treatment. Furthermore, DEGs related to the metabolism of catechins, caffeine, and theanine biosynthetic pathways and amino acids were identified and found to be upregulated upon F application indicating that they play important role for F tolerant to tea plant (Zhu et al. 2019b). F treatment induced the expression of defense genes such as receptor-like kinases (RLKs) and U-box domain-containing protein. Based on the present study, F uptake is considered to be related to calcium transporting ATPase, especially autoinhibited Ca^{2+} ATPase (ACAs) which was activated by the RLKs and worked as a carrier in uptake of F by tea plant (Li et al. 2017c).

8.3.2.3.6 Selenium

Selenite is the main form of Se absorbed and utilized by tea plant. Tea plant has strong enrichment ability for selenium (Se), yet mechanism of selenite absorption and accumulation in tea plant is still unknown. Thus RNAseq was used to understand the molecular mechanism of selenite absorption and accumulation in tea plant. There were 60,582 and 1362 differentially expressed genes (DEGs) in roots and leaves, respectively. Based on GO terms, it was found that the unigenes were mainly involved in cell binding and metabolic process. KEGG pathway enrichment analysis showed that predominant pathways included ribosome and protein processing in endoplasmic reticulum. Further analysis revealed that sulfur metabolism, glutathione metabolism, selenocompound metabolism, and plant hormone signal transduction responded to selenite in tea plant. Additionally, a large number of genes of higher expressions associated with phosphate transporters, sulfur assimilation, antioxidant enzymes, antioxidant substances, and responses to ethylene and jasmonic acid were identified. Stress-related plant hormones might play a signaling role in promoting sulfate/selenite uptake and assimilation in tea plant. This study provides a possibility for controlling Se accumulation in tea plant (Cao et al. 2018).

8.3.2.3.7 Heat

Tea is a shade-loving plant, and hence high temperature is harmful for the growth. Till today there is no high heat-tolerant tea cultivar available. Thus, scientists have made attempts (1) to ameliorate the heat stress by external application of chemicals and (2) identification of genes that are responsive to heat tolerance in tea. Thus several chemicals have been identified which modulate the heat tolerance upon external application. For an example, exogenous calcium application on the physiological characteristics of heat-stressed tea was studied. It has been found that a calcium pretreatment increased the proline, soluble sugar, Ca^{2+}, and chlorophyll contents but decreased the malondialdehyde content and relative electrical conductivity in tea leaves under heat stress. Several DEGs were identified which are primarily related to signal transduction, transcriptional regulation, and post-translational modification (Wang et al. 2018c). Further, it was reported that heat shock protein genes *Hsp90* and *Hsp70* played more critical roles in tea plants in adapting to thermal stress than cold. However, despite the fact that it is a very important trait for tea industry, no further study has been conducted perhaps due to nonavailability of tolerant genotype of tea.

8.3.2.3.8 Growth and Development

Vegetative cutting is predominant method of propagation in tea across the world. Auxin such as IBA is routinely used to induce the adventitious root. However, to understand potential mechanisms involved in adventitious root formation, transcriptome analysis was conducted from IBA-treated cuttings. Functional annotation analysis showed that many genes were involved in plant hormone signal transduction, secondary metabolism, cell wall organization, and glutathione metabolism, indicating potential contributions to adventitious rooting (Wei et al. 2014).

There are some albino tea germplasm which are cultivated in China because of its bright color and high amino acid content. For an example, 'ZH1' is an albino cultivar of tea. RNAseq study revealed that "phenylpropanoid biosynthesis," "glutathione metabolism," "phenylalanine metabolism," "photosynthesis," and "flavonoid biosynthesis" were active in their leaf. Altered gene and protein levels in these pathways may account for the increased amino acid content and reduced chlorophyll and flavonoid content of ZH1 (Wang et al. 2015b). 'Huangjinya' is another albino cultivar of tea which produces tea with high amino acids. Thus RNAseq analysis revealed that the increased levels of amino acids in the chlorotic leaves are likely due to increase protein catabolism and/or decrease glycolysis and diminished biosynthesis of nitrogen-containing compounds other than amino acids, including chlorophyll, purines, nucleotides, and alkaloids (Zhang et al. 2017a; Li et al. 2017d). Further, it has been observed that in Huangjinya, the flavonoid and carotenoid levels increased after moderate shading treatment. RNAseq analysis of 'Huangjinya' plants exposed to sunlight and shade were analyzed. Shading 'Huangjinya' made its leaf color turn green. Differentially expressed genes were identified which were mainly involved in flavonoid and carotenoid biosynthesis. Shade-grown tissue showed lower expression of flavonoid biosynthesis-associated genes and induced carotenoid biosynthesis-related genes. Thus, change of leaf color under shading in 'Huangjinya' is largely determined by the combined effects of flavonoid and carotenoid biosynthesis (Song et al. 2017). RNAseq analysis of 'Anji Baicha' leaves, another albino tea at the yellow-green, albescent, and re-greening stages, identified differentially expressed genes which were mainly related to metabolic pathways, biosynthesis of secondary metabolites, and phenylpropanoid biosynthesis. Chemical analyses revealed higher β-carotene and theanine levels but lower chlorophyll a levels, in the albescent stage than in the green stage. Furthermore, genes were involved in carotenoid, chlorophyll, and theanine biosynthetic pathways, and the expression patterns of the differentially expressed genes in these biosynthetic pathways were characterized. Through co-expression analyses, key genes of theses pathways were identified. These genes may be responsible for the metabolite biosynthesis differences among the different leaf color and developmental stages of 'Anji Baicha' tea plants (Li et al. 2016b). Albino cultivar 'Yu-Jin-Xiang' ('YJX') has the capability to shift the phenotype from a young pale/yellow leaf phenotype to a green leaf phenotype under shade. For a better understanding of mechanisms underlying the phenotype shift and the altered catechin and theanine production, RNAseq analysis was done which indicated that shaded leaf greening resulted due to upregulation of genes that were related to leaf chlorophyll, carotenoid abundance, and chloroplast development. Metabolic flux redirection and transcriptomic reprogramming were found in flavonoid and carotenoid pathways of the 'YJX' pale leaves and shaded green leaves to different extents compared to the non-albino green tea cultivar 'Shu-Cha-Zao.' Enhanced production of the antioxidant quercetin rather than catechin biosynthesis was correlated positively with the enhanced transcription of *flavonol synthase* and *flavanone/flavonol hydroxylases* leading to quercetin accumulation and negatively correlated to suppressed *leucoanthocyanidin reductase*, *anthocyanidin reductase*, and *synthase* leading to more catechin biosynthesis. The altered levels of quercetin and catechins in 'YJX' will impact on its tea flavor and health benefits (Liu et al. 2017b).

RNAseq-based transcriptomic comparisons of the bud and two youngest leaves in Zijuan (ZJ), a purple-colored tea, and YunKang (YK), a green-colored tea, identified 93%, 9%, and 5% expressed genes that were shared in YK- and ZJ-specific cultivars, respectively. A comparison of both transcript abundance and particular metabolites revealed that the high expression of gene UFGT for anthocyanin biosynthesis is responsible for purple coloration, which competes with the intermediates for catechin-like flavanol biosynthesis. Genes with differential expression are enriched in response to stress, heat, and defense and are casually correlated with the environmental stress of ZJ plant origin in the Himalayas. In addition, the highly expressed

C4H and LDOX genes for synthesizing flavanol precursors, ZJ-specific CLH1 for degrading chlorophyll, alternatively spliced C4H and FDR, and low photosynthesis also contributed to the altered color and flavor of ZJ (Wang et al. 2019a).

Tea flowers are white in color, but a natural mutant of tea with pink color has been identified recently. Rothenberg et al. (2019) investigated the molecular pathways that lead to pink coloration. They found that some compounds such as cyanidin O-syringic acid, petunidin-3-O-glucoside, pelargonidin-3-O-beta-D-glucoside, along with 21 other genes, are responsible for pink coloration which have direct impact on biosynthesis and accumulation of three flavonoid compounds, namely, cyanidin 3-O-glucoside, petunidin-3-O-glucoside, and epicatechin gallate. Similarly, purple shoot tea attributing to the high anthocyanin accumulation is of great interest for its wide health benefits. To better understand the molecular mechanisms of purple color of tea, differentially expressed genes were identified through RNAseq and found that anthocyanin biosynthesis genes and genes involved in anthocyanin transportation were largely expressed in those pink color leaf but the early biosynthetic genes were less or non-affected (Wei et al. 2016).

Regeneration system of tea plant has been explored for long time though success depends on several factors (Mukhopadhyay et al. 2013), but often it was found that some of the culture are highly regenerative and others are not. Thus to understand the molecular switch, transcriptomic analysis on initial explants of tea plant and their dedifferentiated and redifferentiated tissues was done which revealed that genes encoding the auxin- and cytokinin-responsive proteins, transcription factor MYB15, and ethylene-responsive transcription factor ERF RAP2-12 were responsible for such switch (Gao et al. 2019).

Tea is highly self-incompatible. In this study, the transcriptomes of self- and cross-pollinated pistils of two tea cultivars 'Fudingdabai' and 'Yulv' were compared to elucidate the SI mechanism of tea plants. Differentially expressed genes were unidentified from pollinated and unpollinated pistil of self- and cross-pollinated flower. Functional annotation indicated that three genes encoding UDP-glycosyltransferase 74B1 (UGT74B1), mitochondrial calcium uniporter protein 2 (MCU2), and G-type lectin S-receptor-like serine/threonine-protein kinase (G-type RLK) might play important roles during SI process in tea plants (Ma et al. 2018).

The effects of blue (BL) and green light (GL) treatment during the dark period were examined in tea as a first step to understanding the spectral effects of artificial BL and GL on plant secondary metabolism and light signaling interactions. BL could induce the expression of CRY2/3, SPAs, HY5, and R2R3-MYBs to promote the accumulation of anthocyanins and catechins in tea plants. GL, on the other hand, could stimulate the accumulation of several functional substances (e.g., procyanidin B2/B3 and L-ascorbate) and these BL responses via downregulation of CRY2/3 and PHOT2. Furthermore, the molecular events triggered by BL and GL signals were partly overlapped with abiotic/biotic stress responses. This result shows the possibility of a targeted use of BL and GL to regulate the amount of functional metabolites to enhance tea quality and taste and to potentially trigger defense mechanisms of tea plants (Zheng et al. 2019a).

To capture the global gene profiles during floral bud development in *C. azalea*, RNAseq analysis at three developmental stages, i.e., floral bud initiation, floral organ differentiation, and bud outgrowth, identified nine co-expression clusters with distinctive patterns. It has been found that transition from floral bud initiation to floral organ differentiation required changes of genes in flowering timing regulation, while transition to floral bud outgrowth was regulated by various pathways such as cold and light signaling, phytohormone pathways, and plant metabolisms. Further analyses of dormancy-associated MADS-box genes revealed that SVP- and AGL24-like genes displayed distinct expression patterns suggesting divergent roles during floral bud development (Fan et al. 2015a). RNAseq analysis assembles gene-related information involved in reproductive growth of tea. Gene Ontology (GO) analysis of the annotated unigenes revealed that the majority of sequenced genes were associated with metabolic and cellular processes, cell and cell parts, catalytic activity, and binding. Kyoto Encyclopedia of Genes and Genomes (KEGG) pathway, biosynthesis of secondary metabolites, and plant hormone signal transduction were enriched among the DEGs. Furthermore, 207 flowering-associated unigenes were identified. Some transcription factors, such as *WRKY*, *ERF*, *bHLH*, *MYB*, and MADS-box, had shown upregulation in floral transition, which might play the role of progression of flowering (Liu et al. 2017c). Tea being a highly cross-pollinated crop, self-pollen tubes grow slower than cross. Thus to identify SI-related genes, RNAseq of self- and cross-pollinated styles was employed to explore the molecular mechanisms of SI at different time points after pollination. Consequently, several ubiquitin-mediated proteolysis, Ca^{2+} signaling, apoptosis, and defense-associated genes were obtained. The temporal expression pattern of genes following CP and SP was analyzed; six peroxidase, one polyphenol oxidase, and seven salicylic acid biosynthetic process-related genes were identified (Zhang et al. 2016a).

Tea plant leaves are a dietary source of ascorbic acid (AsA) for humans. However, biosynthesis, recycling pathway, and distribution of AsA during leaf development in tea plants are unclear. To gain insight into the mechanism and distribution of AsA in the tea plant leaf, Li et al. (2016c) identified 18 related genes and recycling pathway based on the transcriptome database of tea plants. AsA contents in tea plant leaves at three developmental stages were measured and found a positive correlation between expression levels of these genes and AsA contents during the development of tea plant leaves. Further, results indicated that the L-galactose pathway might be the primary pathway of AsA biosynthesis in tea plant leaves. Two genes, namely, *CsMDHAR* and *CsGGP* might play a regulatory role in AsA accumulation in the leaves of three cultivars of tea plants. These findings may provide a further glimpse to improve the AsA accumulation in tea plants and the commercial quality of tea (Li et al. 2016c).

The yellow-leaf tea cultivar Zhonghuang 2 (ZH2) produces quality tea due to higher contents of theanine and free amino acids, whereas the contents of carotenoids, catechins, and anthocyanin were lower in it. Thus through RNAseq, it has been found that differentially expressed genes belonged to 24 pathways as being significantly regulated, including cysteine and methionine metabolism; glycine, serine, and threonine metabolism; flavonoid biosynthesis; porphyrin and chlorophyll metabolism; and carotenoid biosynthesis (Wang et al. 2014b).

The signaling networks dominated by nitric oxide (NO) during cold stress that inhibited tea pollen tube growth are investigated in vitro. Cytological analysis shows that cold-induced NO is involved in the inhibition of pollen tube growth along with disruption of the cytoplasmic Ca^{2+} gradient, increase in ROS content, acidification of cytoplasmic pH, and abnormalities in organelle ultrastructure and cell wall component distribution in the pollen tube tip. Furthermore, DEGs related to signaling pathway, such as NO synthesis, cGMP, Ca^{2+}, ROS, pH, actin, cell wall, and MAPK cascade signal pathways, are identified and quantified, which indicated a potential molecular mechanism for the above cytological results. Thus these findings suggest that a complex signaling network dominated by NO, including Ca^{2+}, ROS, pH, and RACs signaling and the crosstalk among them, is stimulated in the *C. sinensis* pollen tube in response to cold stress, which further causes secondary and tertiary alterations, such as ultrastructural abnormalities in organelles and cell wall construction, ultimately resulting in perturbed pollen tube extension (Wang et al. 2016e).

Molecular insights commencing self-incompatibility (SI) and cross-compatibility/fertilization (CC) in self-pollinated (SP) and cross-pollinated (CP) pistils of tea were studied. The fluorescence microscopy analysis revealed ceased/deviated pollen tubes in SP, while successful fertilization occurred in CP at 48 h after pollination. RNAseq analysis of SP and CP pistils generated 109.7 million reads with overall 77.9% mapping rate to draft tea genome. Furthermore, concatenated de novo assembly resulted into 48,163 transcripts. Functional annotations and enrichment analysis (KEGG and GO) resulted into 3793 DEGs of which 195 DEGs involved in pollen-pistil interaction Interestingly, the presence of 182 genes in a major hub of the protein-protein interactome network suggests a complex signaling cascade commencing SI/CC. Furthermore, it was suggested that LSI initiated in style and was sustained up to the ovary with the active involvement of *csRNS*, *SRKs*, and *SKIPs* during *SP*. Nonetheless, *COBL10*, *RALF*, *FERONIA-rlk*, *LLG*, and *MAPKs* were possibly facilitating fertilization. The current study comprehensively unravels molecular insights of phase-specific pollen-pistil interaction during SI and fertilization, which can be utilized to enhance breeding efficiency and genetic improvement in tea (Seth et al. 2019).

C. reticulata, which is native to Southwest China, is famous for its ornamental flowers and high-quality seed oil. Several works did transcriptomic study to find out the genes related to oil production. They are genes involved in its oil production, pigment biosynthesis, and photoperiod flowering. Among many of those genes, *MYBF1*, a transcription regulator gene of the FlaBS pathway, was found with great sequence variation and alteration of expression patterns, probably resulting in functional evolutionary differentiation in *C. reticulata*. *MYBA1_a* and some anthocyanin-specific biosynthetic genes in the FlaBS pathway were highly expressed in both flower buds and flowers, suggesting important roles of anthocyanin biosynthesis in flower development (Yao et al. 2016). Interestingly, evolutionary analysis of omega-6 fatty acid desaturase (FAD2) genes among 20 oil-plants unexpectedly suggests that a parallel evolution may occur between *C. oleifera* and *Olea oleifera* (Xia et al. 2014). Additionally transcription factors such as ABI3, FUS3, LEC1, WRI1, TTG2, and DOF4.6 are also found to be associated with oil production in this plant which are involved in oil biosynthesis and fatty acid accumulation, including glycolysis zinc ion binding, positive regulation of fatty acid biosynthetic process, triglyceride biosynthetic process, seed coat development, abscisic acid-mediated signaling pathway, and embryo development (Wu et al. 2019a).

C. nitidissima and *C. chuongtsoensis* are yellow-colored flowers. Transcriptome sequencing was applied for revealing the pigmentation. Furthermore, the expression patterns of secondary metabolism pathway genes were analyzed between yellow- and red-flowered camellias. We showed that the key enzymes involved in glycosylation of flavonoids displayed differential expression patterns, indicating that the direct glycosylation of flavonols rather than anthocyanins was pivotal to coloration and health-improving metabolites in the yellow *Camellia* petals. Finally, the gene family analysis of UDP-glycosyltransferases revealed an expansion of group C members in *C. nitidissima*. Through comparative genomics analysis, we demonstrate that changes of gene expression and gene family members are critical to the variation of natural traits. This work provides valuable insights into the molecular regulation of trait adaptations of floral pigmentation and flowering time (Li et al. 2018c).

8.3.2.3.9 Dormancy

Winter dormancy is an important biological feature for tea plant to survive during cold winters which affects the tea industry due to nonavailability of green leaf production. To discover the bud dormancy regulation mechanism of tea plant in winter, gene expression study of axillary buds at the paradormancy, endodormancy, ectodormancy, and bud flush stages was conducted. Transcriptome study reveals that epigenetic mechanisms, phytohormone signaling pathways, and callose-related cellular communication regulation were found to be associated with this. Furthermore, differentially expressed transcription factors as well as chromatin- and phytohormone-associated genes were identified (Hao et al. 2017). Paul et al. (2014) did RNAseq analysis of dormant as well as actively growing leaf of tea which revealed that (1) operation of mechanisms of winter tolerance, (2) downregulation of genes involved in growth, development, protein synthesis, and cell division, and (3) inhibition of leaf abscission due to modulation of senescence-related processes during winter dormancy in tea. These results validated the relevance of the identified senescence-related processes for leaf abscission and suggested their operation when in need in tea.

8.3.2.3.10 Shade

Shade is a necessary evil for tea plantation. It not only influences the yield, but also it influences the quality of made tea. Therefore in order to understand the effect of shade, a study was undertaken which revealed that shade enhanced chlorophyll accumulation and major catechins, including C, EC, GC, and EGC, decreased significantly in tea buds throughout the whole shading period. RNAseq study also found that the reduction of catechins and flavonols were due to downregulation of biosynthetic genes and TFs associated with flavonoid biosynthesis. Of 16 genes involved in the flavonoid biosynthetic pathway, F3′H and FLS significantly decreased throughout shading, while the others (PAL, CHSs, DFR, ANS, ANR, LAR, etc.) temporally decreased in early or late shading stages. A number of photoreceptors and potential genes involved in UV-B signal transductions (UVR8_L, HY5, COP1, and RUP1/2) showed decreasing expression patterns consistent with structural genes (F3′H, FLS, ANS, ANR, LAR, DFR, and CHSs) and potential TFs (MYB4, MYB12, MYB14, and MYB111) involved in flavonoid biosynthesis, when compared with genes in the UV-A/blue and red/far-red light signal transductions. Further UV-B treatment in the controlled environment confirmed UV-B induction on flavonols and EGCG accumulation in tea leaves (Liu et al. 2018a).

One cultivar, *Baijiguan*, exhibits a yellow leaf phenotype, reduced chlorophyll (Chl) content, and aberrant chloroplast structures under high light intensity. In contrast, under low light intensity, the flush shoot from *Baijiguan* becomes green, the chloroplyll content increases significantly, and the chloroplasts exhibit normal structures. In total, 1993 and 2576 differentially expressed genes (DEGs) were identified in *Baijiguan* plants treated with and days of shade, respectively. Gene Ontology (GO) and pathway enrichment analyses indicated that the DEGs are predominantly involved in the ROS scavenging system, chloroplast development, photosynthetic pigment synthesis, secondary metabolism, and circadian systems. The light-responsive gene *POR* (protochlorophyllide oxidoreductase) and transcription factor *HY5* were identified (Wu et al. 2016a).

8.3.2.3.11 Heterosis Breeding

Heterosis breeding is very important for genetic improvement of polygenic characters such as quality and yield though the molecular mechanisms are not known. In order to find out the mechanism of heterosis of terpenoid volatile and green leaf volatile contents in the F_1 hybrids, transcriptomic study was undertaken. At the molecular level, the comparative transcriptomics analysis revealed that approximately 41% of the genes showed non-additive expression, whereas only 7.83% showed additive expression. Among the non-additive genes, 42.1% showed high parental dominance and 17.6% showed overdominance. In addition, we identified multiple genes (*CsDXS*, *CsAATC2*, *CsSPLA2*, etc.) and transcription factors (*CsMYB1*, *CsbHLH79*, *CsWRKY40*, etc.) that played important roles in tea volatile heterosis. Thus it showed that non-additive action plays a major role in tea volatile heterosis (Zheng et al. 2019b).

During hybridization, combining two genomes results in "genome shock," though molecular mechanism for woody heterozygous species is not known. Here, the gene expression patterns were studied in the F_1 hybrid derived from a cross of *C. azalea* × *C. amplexicaulis* compared with the parents. About 54.5% of all genes were differentially expressed between the F_1 hybrid and at least one of the parents, including 6404 genes with the highest expression level in the F_1 hybrid. A series of genes, related to flower development, essential for RNA-directed DNA methylation and histone methylation, as well as 223 transposable elements, were enriched; and most of them exhibited a higher level of expression in the F_1 hybrid. These results indicated that the genome shock induced by interspecific hybridization in *Camellia* could indeed result in changes of gene expression patterns, potentially through regulating DNA methylation and histone methylation which may be helpful for the maintaining of genome stability and even related to the unique phenotype of the F_1 hybrid (Zhang et al. 2018c).

Genetic divergence of the parent and their F_1 hybrids were used to study the effect of parental genetic divergence on gene expression and regulatory patterns in F_1 hybrids by RNAseq and allelic expression analysis. We found that though the proportion of differentially expressed genes (DEGs) between the hybrids and their parents did not increase, a greater proportion of DEGs would be non-additively (especially transgressively) expressed in the hybrids as genomes between the parents become more divergent. In addition, the proportion of genes with significant evidence of cis-regulatory divergence increased, whereas with trans-regulatory divergence decreased with parental genetic divergence. The discordance within hybrid would intensify as the parents become more divergent, manifesting as more DEGs would be non-additively expressed. Trans-regulatory divergence contributed more to the additively inherited genes than cis; however, its contribution to expression difference would be weakened as cis mutations accumulated over time; and this might be an important reason for that the more divergent the parents are, the greater proportion of DEGs would be non-additively expressed in hybrid (Zhang et al. 2019a).

8.3.2.3.12 Processing

Withering is an important process for green, black, and white teas manufacturing. Optimum withering is essential for correct flavor development for all these tea. Thus to understand the effect of withering, RNAseq study was taken at different time period of withered tea leaf. It was found that genes involved in flavonoid biosynthesis were significantly downregulated, which could be correlated with the reduction of catechins. Enhancement of terpenoids and alpha-linolenic acid metabolism could trigger an increase in the total content and number of volatiles. The increase in free amino acid content could be related to 261 DEGs. Our study suggests that dehydration stress during withering induced significant changes in the gene transcription and content of the tea flavor compounds, which promoted the special flavors in various teas (Wang et al. 2019b).

To elucidate formation mechanism of oolong tea aroma, the released and remaining volatiles during bruising and withering treatment were analyzed using head space solid-phase microextraction/gas chromatography-mass spectrometry. An increase in proportion of the released terpenoid volatiles (TVs) along with a decrease in proportion of the released C6 green leaf volatiles (GLVs) was observed in both cultivars 'Zhejiang-139' and 'Foshou'. Proportion of remaining TVs also fluctuated reversely with GLVs although the level of these volatiles increased remarkably. High ratio of TVs to GLVs was the key chemical foundation of oolong tea characteristic aroma and could be regarded as a good indicator in screening cultivar for suitably producing high-quality oolong tea. Combining with transcriptome analysis, increased TVs and GLVs during the treatment might be largely generated through de novo synthesis and modulated at transcript level through upregulation of genes involved in terpenoid metabolism and enzymatic cleavage of long-chain fatty acids (Hu et al. 2018).

8.3.2.3.13 Seed Oil

C. oleifera is a major tree species for producing edible oil. Its seed oil is well-known for the high level of oleic acids; however, little is known regarding the molecular mechanism of lipid biosynthesis in *C. oleifera*. Here, we measured the oil contents and fatty acid (FA) compositions at four developmental stages and investigated the global gene expression profiles through transcriptome sequencing. We identified differentially expressed genes (DEGs) among the developmental stages and found that the distribution of numbers of DEGs was associated with the accumulation pattern of seed oil. Gene Ontology (GO) enrichment analysis revealed some critical biological processes related to oil accumulation, including lipid metabolism and phosphatidylcholine metabolism. Furthermore, we investigated the expression patterns of lipid biosynthesis genes. We showed that most of the genes were identified with single or multiple copies, and some had correlated profiles along oil accumulation. We proposed that the higher levels of stearoyl-ACP desaturases (SADs) coupled with lower activities of fatty acid desaturase 2 (FAD2) might be responsive to the boost of oleic acid at the late stage of *C. oleifera* seeds' development. This work presents a comprehensive transcriptomics study of *C. oleifera* seeds and uncovers valuable DEGs that are associated with the seed oil accumulation (Lin et al. 2018).

Increasing fruit yield though may not be an important criterion for tea, yet it is necessary for some oil-bearing *Camellia* species. For example, *C. chekiangoleosa* is a novel promising oil tree, but its oil production is limited due to low fruit set. With the availability of high-fruit yield (HY) and low-fruit yield (LY) genotypes, global gene discovery and expression levels of floral buds were taken as fruit count has direct relation with floral bud. Around 2395 differentially expressed genes (DEGs) were identified which were enriched in membrane, energy metabolism, secondary metabolism, fatty acid biosynthesis and metabolism, and 18 other metabolic pathways. Of the DEGs, 12 identified transcription factors, including AP2, are mostly involved in inflorescence and flower development and in hormone networks. Key DEGs in fatty acid biosynthesis (Fab) *FabB*, *FabF*, *FabZ*, and *AccD* were highly expressed in floral buds and associated with high oil yield in fruits. Thus a potential link exists between fruit count and its oil yield (Xie and Wang 2018).

8.3.2.3.14 *Camellia* Species

The golden camellia, *C. nitidissima* Chi., is a well-known ornamental plant that is known as "the queen of camellias" because of its golden yellow flowers. The principal pigments in the flowers are carotenoids and flavonol glycosides. To obtain a comprehensive understanding of flower development in *C. nitidissima*, a number of cDNA libraries were independently constructed during flower development. A differentially expressed genes (DEGs) and co-expression network was constructed to identify unigenes correlated with flower color. The analysis of DEGs and co-expressed network involved in the carotenoid pathway indicated that the biosynthesis of carotenoids is regulated mainly at the transcript level and that *phytoene synthase* (*PSY*), *β-carotene 3-hydroxylase* (*CrtZ*), and *capsanthin synthase* (*CCS1*) exert synergistic effects in carotenoid biosynthesis. The analysis of DEGs and co-expressed network involved in the flavonoid pathway indicated that *chalcone synthase* (*CHS*), *naringenin 3-dioxygenase* (*F3H*), *leucoanthocyanidin dioxygenase* (ANS), and *flavonol synthase* (*FLS*) play critical roles in regulating the formation of flavonols and anthocyanidin. Based on the gene expression analysis of the carotenoid and flavonoid pathways, and determinations of the pigments, it was speculated that high expression of *PSY* and *CrtZ* ensures the production of adequate levels of carotenoids, while the expression of *CHS* and *FLS* ensures the production of flavonols. The golden yellow color is then the result of the accumulation of carotenoids and flavonol glucosides in the petals (Zhou et al. 2017a).

C. ptilophylla, or cocoa tea, is naturally decaffeinated, and its predominant catechins and purine alkaloids are *trans*-catechins and theobromine. Regular tea [*Camellia sinensis* (L.) O. Ktze.] is evolutionarily close to cocoa tea and produces *cis*-catechins and caffeine. Thus to know the catechin biosynthetic pathways, RNAseq of young leaf was done. They found that two *LAR* genes (*CpLAR1* and *CpLAR2*) by *C. ptilophylla* may be advantageous due to the combined effects of this quantitative trait, permitting increased leucoanthocyanidin consumption for the synthesis of *trans*-catechins. In contrast, the only *ANS* gene observed in *C. sinensis* (*CsANS*) shared high identity (99.2%) to one homolog from *C. ptilophylla* (*CpANS1*) but lower identity (~80%) to another (*CpANS2*). Thus it was hypothesized that the diverged *CpANS2* might have lost its ability to synthesize *cis*-catechins. *C. ptilophylla* and *C. sinensis* each contain two copies of ANR, which share high identity and may share the same function (Li et al. 2015b).

8.4 Gene Family Study

Eukaryotic gene expression is complex, controlled by several genes at different levels. Moreover, genes are often duplicated several times in a genome which are then organized in a family, located either in same or in a different chromosome (Ganie et al. 2017), including organelle genome. Among the many reasons, duplication during evolution leads to the occurrence of the gene family. Members of the gene family play important role, many a time all of them are not induced by a particular stress or tissue though all the members have the same domain. With the availability of the genome sequence, it becomes easy to identify the family members based on the domain search, yet it can also be done in limited way from the expressed sequence data such as EST or RNAseq data. For example in tea, Yue et al. (2014) first did the analysis of aquaporin gene family of tea using publicly available EST database (Table 8.2). However, in the post-genomic era, analysis of

Table 8.2 Summary of genome-wide gene family study in tea

Name of the gene family	No. of the member	Validated trait	Reference
R2R3-MYB, bHLH, WD40	73, 49, and 134 genes, respectively	Different developmental stages and treatment conditions, including hormone and wound treatments	Zhao et al. (2013)
Aquaporin (*AQP*) gene family	20	Flower development and opening, bud endodormancy, cold, ABA and salt stress	Yue et al. (2014)
Basic leucine zipper (*bZIP*) TF gene family	18	Cold, high salinity, dehydration stress and ABA treatment	Cao et al. (2015)
Heat shock transcription factors (Hsfs)	16	High and low temperature stress	Liu et al. (2016c)
AP2/ERF TF gene family	89	Cold and heat stress	Wu et al. (2015)
WRKY family	50	Different temperature	Wu et al. (2016b)
Auxin response factors (*ARF*s) gene family	15	IAA, 2,4-D, and NAA treatment	Xu et al. (2016)
DNA binding with one finger (*DoF*) family of TF	29	High and low temperature, salt and drought stress	Li et al. (2016d)
Hexose kinase gene family	11	Cold, salt and drought stress, exogenous ABA treatment	Li et al. (2017f)
Invertase (*INV*) gene family	14	Growth and developmental regulation, cold, drought, ABA and salinity stress	Qian et al. (2016)
UGT genes (*CsUGT*s)	178	Different tissues	Cui et al. (2016)
NAC (NAM-ATAF1/2-CUC) family of TF	45	Growth and development, cold stress	Wang et al. (2016f)
Gibberellin dioxygenase gene family	14	Cold, heat, drought and salinity stress, ABA and GA treatment	Pan et al. (2017)
Teosinte branched1/cycloidea/ proliferating cell factors (*TCP*) family of TF	17	Cell growth and proliferation of leaf tissue	Wu et al. (2017b)
Phenylalanine ammonia-lyases	6	Tissues under different exogenous treatments such as 100 mM methyl jasmonate (MeJA), 100 mM salicylic acid (SA), 100 mM abscisic acid (ABA), 0.74 mM indolebutyric acid (IBA), 50 mM NaCl, 200 mM mannitol, or 90 mM sucrose for 24 h.	Wu et al. (2017c)
Cinnamate 4-hydroxylase	3	Different tissues	Xia et al. (2017b)
bHLH family transcription factors	120	Different abiotic stress	Cui et al. (2018)

(continued)

Table 8.2 (continued)

Name of the gene family	No. of the member	Validated trait	Reference
Heat shock protein (*HSP*) family	47	Heat and drought stress	Chen et al. (2018)
VQ gene family	25	Salt and drought stress	Guo et al. (2018)
R2R3-MYBs	140	Various tissues and organs	Jiang et al. (2018)
HD-Zip family	33	Heat, cold, salinity, drought stress, and ABA treatment	Shen et al. (2019b)
Lateral organ boundaries domain (*LBD*) TF family	31	Hormonal response in plant growth and development	Teng et al. (2018)
WRKY TF family	56	Cold stress and ABA treatment	Wang et al. (2019c)
Late embryogenesis abundant (*LEA*) protein family	33	Cold and dehydration stress	Wang et al. (2018d)
GRAS family proteins	52	GA, salt, drought, high and low temperature	Wang et al. (2018e)
Multidrug and toxic compound extrusion (*MATE*) family proteins	41	Flavonoid accumulation inside vacuole	Chen et al. (2019)
Calcineurin B-like protein (CBL) and CBL-interacting protein kinase (*CIPK*) family genes	CsCBLs—7 CsCIPKs—18	Drought, salt, cold and heat stress	Liu et al. (2019c)
Nuclear factor-Y (*NF-Y*) family of TF	35	Drought and ABA treatment	Wang et al. (2019d)
Nucleotide-binding leucine-rich repeats (*NB-LRR*) family of genes	*C. sinensis* var. *sinensis*—400 *C. sinensis* var. *assamica*—303	A series of biotic and abiotic stress treatments	Wang et al. (2020)
C-repeat-binding factors (*CBF*) gene family	5	Cold treatment	Wang et al. (2019e)
Lox gene family	7	Different growing months	Xu et al. (2019)
Purple acid phosphatase (*PAP*) gene family	19	High level of iron (Fe) stress	Yin et al. (2019)

(continued)

Table 8.2 (continued)

Name of the gene family	No. of the member	Validated trait	Reference
Superoxide dismutase (*SOD*) gene family	10	Cold and drought stress, GA_3 and MeJA treatment	Zhou et al. (2019a, b)
Amino acid permease (*APS*) gene family	19	Different nitrogen (N) treatments	Duan et al. (2020)
Chlorophyll A/B binding (*CAB*) protein family	25	Mechanical injury, NaCl, ABA, mannitol, MeJA and cold stress	Li et al. (2020)
SQUAMOSA promoter binding protein (SBP) box TF family	25	Tolerance to high and low temperature stress	Zhang et al. (2020b)
Pyrabactin resistance-likes (*PYLs*)-Type 2C protein phosphatase (*PP2C*)-SNF-1 related protein kinase 2s (*SnRK2s*) gene family	*PYLs*—14 *PP2C*—84 *SnRK2s*—8	Drought and salinity stress	Xu et al. (2020a, b)
Jasmonate-ZIM domain (*JAZ*) family	12	Low and high temperature, water deficit, MeJA and GA treatment	Shen et al. (2020)
Jasmonate-ZIM domain (*JAZ*) family	13	Cold, drought (PEG) and hormone treatment (ETH, GA, ABA)	Zheng et al. (2020a, b)
Terpenoid synthase (*TPS*) enzyme family	80	Cold acclimation, salt and PEG stress	Zhou et al. (2020b)

TF transcription factor, *ABA* abscisic acid, *PEG* polyethylene glycol, *GA* gibberellic acid, *ETH* ethylene, *MeJA* methyl jasmonate

various gene families has been increased due to the easy identification of the family members. All these studies will serve as a great source to identify the stress-specific members which can be used later either to make a transgenic plant tolerant to a particular stress or it will help to understand the gene regulation under a particular stress.

8.5 miRNA-Mediated Gene Regulation

Non-coding RNAs are considered to be a new player for epigenetic gene regulation in higher eukaryotes in the few recent years (Bhat et al. 2020). They work in the post-transcriptional mode either by methylation or by degrading the mRNAs and hence belong to epigenetics mode of gene expression (Mondal et al. 2014, 2018). Among the various non-coding RNAs, research on miRNAs are dominant in tea as several studies have been conducted (Das et al. 2015; Ganie and Mondal 2015). In tea and its wild species, studies on miRNA regulation have begun during 2010 with in silico analysis of EST data; later it was extended with in silico analysis (Prabu and Mandal 2010; Das and Mondal 2010) as well as the small-scale cloning of miRNAs in a limited scale (Mohanpuria and Yadav 2012). However recently with the advancement of RNAseq technology, several reports of high-throughput sequencing-based discovery of small RNAs have been reported in tea (Zhu and Luo 2013). The predicted miRNAs have been found to be associated with target genes that form a part of a variety of biological mechanisms ranging from stress tolerance, growth and development, biosynthetic pathways, to even some of the target genes that are responsible for encoding products acting as transcription factors and cellular components which are described below.

8.5.1 Low Temperature

Cold is an important stress for tea cultivation as it induces winter dormancy in tea, as a consequence, green leaves are not available for making the made tea. However, molecular mechanisms are not yet known. Thus small RNAseq study was conducted from the cold-treated tea leaves of two contrast genotypes that differ in cold sensitivity which yield several differentially expressed but genotype-specific miRNAs. Several target genes were detected by degradome sequencing. The RLM-5′RACE procedure was successfully used to map the cleavage sites in six target genes of tea which reveal important information about the regulatory mechanism of miRNAs and promote the understanding of miRNA functions during the cold (Zhang et al. 2014). Later, using RNAseq and small RNAseq from the same tissue, regulatory network of tea plants under chilling (4 °C) and freezing (−5 °C) stress was conducted. Differentially expressed miRNAs and target mRNA profiles were obtained based on fold change analysis which showed both coherent and incoherent relationships in the regulatory network. Several important pathways specific to chilling and freezing stress along with the target genes were reported in tea (Zheng et al. 2015).

8.5.2 Growth and Development

Gene regulation for the growth of two leaves and a bud are aways attracted by scientific community in tea as they are the harvestable part of the tea bush. In an attempt to understand the miRNA-mediated pathways, Jeyaraj et al. (2017a) discovered 83 novel and 175 conserved miRNAs in the young leaf of tea plants. Some of the miRNAs were validated through miRNA microarray hybridization and by stem-loop-qRT-PCR. A total of 716 potential target genes of identified miRNAs were predicted. Further, GO and KEGG pathway analysis revealed that most of the target genes were primarily involved in stress response and enzymes related to phenylpropanoid biosynthesis. A negative correlation between expression profiles of three out of four conserved miRNAs (csn-miR160a-5p, csn-miR164a, csn-miR828, and csn-miR858a) and their targets (*ARF17*, *NAC100*, *WER*, and

MYB12 transcription factor) was observed. Further, to identify the miRNAs that regulate bud dormancy/activation transition in tea plant, small RNA libraries from three different periods of bud dormancy–burst transition were studied which yielded 118 differentially expressed miRNAs. The content of caffeine increased continuously from the endodormancy bud to flushing bud, and differentially expressed miRNAs coupling with their targets associated with bud burst were identified. Remarkably, csn-miR319c was downregulated significantly from the quiescent bud to burst bud, while its target gene *CsnTCP2* (teosinte branched1/cycloidea/proliferating cell factor 2) displayed opposite expression patterns indicating that this module is involved in the transition phase (Liu et al. 2019b).

Double flower in ornamental camellias is a desired trait though molecular mechanisms that control this trait are not known in *Camellia.* Hence, comprehensive comparative transcriptomics interrogation of gene expression among floral organs of wild type, "formal double," and "anemone double" was investigated. DEGs between whorls in wild and cultivated *Camellia* were identified which showed that the formation of double flowers tends to demolish gene expression canalization of key functions; the faded whorl specification mechanism was fundamental under the diverse patterns of double flowers. Furthermore, they identified conserved miRNA-targets regulations in the control of double flowers and found that miR172-AP2 and miR156-SPLs were critical regulatory nodes contributing to the diversity of double flower forms and support the roles of "faded ABC model" mechanism and miRNA-targets regulations underlying the double flower domestication (Li et al. 2017e).

Trichomes are unicellular and unbranched structures, considered to be an indicator of quality tea plant. However, trichome formation at the molecular level is poorly understood in plants. In the present study, the trichome-containing and trichomeless tea cultivars Fuding Dabaicha (FDDB) and Rongchunzao (RCZ), respectively, were used to study the trichome formation. In total, 21,599 DEGs were identified between RCZ and FDDB, of which 10,785 DEGs were upregulated and 10,814 DEGs were downregulated. Genes involved in the DNA replication pathway were significantly enriched. Furthermore, between FDDB and RCZ, DEGs related to TFs, phytohormone signals, and cellulose synthesis were identified, suggesting that certain genes involved in these pathways are crucial for trichome initiation in tea plants (Yue et al. 2018).

8.5.3 Catechin Synthesis Regulation

miRNAs also regulate catechin biosynthesis in tea. Catechins determine the quality of made tea. Accumulation of which in tea leaves depends on source of nitrogenous fertilizer given to tea plants. The influences of NH_4^+ or NO_3^- on plant growth and catechin accumulation were studied though tea plant prefers nitrogen source NH_4^+ over NO_3^-. It has been found that total shoot catechins content was significantly higher for NO_3^- treatments than that for NH_4^+ treatment or combined NH_4^+ or NO_3^- treatment, suggesting that, in addition to its influence on plant growth, the nitrogen form regulated the accumulation of catechins in tea. Interestingly, it was found that Cs-miR156 was drastically induced through NH_4^+. Moreover, a potential mechanism of the Cs-miR156 pathway in regulating catechin biosynthesis in tea plants has been suggested, with particular respect to nitrogen forms. Cs-miR156 might repress the expression of the target gene *SPL* to regulate the *DFR* gene, which plays a vital role in catechin biosynthesis (Fan et al. 2015b). Further, it had been found that expression patterns of novel-miR1, novel-miR2, csn-miR160a, csn-miR162a, csn-miR394, and csn-miR396a were negatively correlated with catechin content indicating that they may block the expression of catechin biosynthetic pathway genes (Sun et al. 2017).

Through computational tools, miRNAs that were involved in catechin synthesis were predicted and validated by RNA ligase-mediated rapid amplification of cDNA ends (RLM-RACE) that yielded seven miRNAs cleaving eight catechin synthesis pathway-related genes including chalcone synthase (*CHS*), chalcone isomerase (*CHI*), dihydroflavonol 4-reductase (*DFR*), anthocyanidin reductase (*ANR*), leucoanthocyanidin reductase (*LAR*), and flavanone 3-hydroxylase (*F3H*). An expression analysis of miRNAs and target genes revealed that miR529d and miR156g-3p were negatively correlated with their targets *CHI* and *F3H*, respectively (Ping et al. 2018). These studies collectively suggest that miRNAs are involved in catechin biosynthesis in tea.

Catechin accumulation patterns significantly differ under different nitrogen source that are used for the growth of tea plant. Here, it has been found that flavonoids associated with tea flavor are dominated by different metabolic and transcriptional responses among the four N conditions (N-deficiency, nitrate, ammonia, and nitric oxide). Nitrogen-deficient tea plants accumulate diverse flavonoids, corresponding with higher expression of hub genes including *F3H*, *FNS*, *UFGT*, *bHLH35*, and *bHLH36*. Compared with N-deficiency, N-supply tea plants significantly increased proline, glutamine, and theanine, which are also associated with tea flavor, especially under NH_4^+ supply. As NH_4^+-tolerant species, tea plant exploits the adaptive strategy by substantial accumulation of amino acids including theanine to adapt excess NH_4^+, which attributes to, at least in part, efficient N transport and assimilation and active protein degradation. A distinct divergence of N reallocation in young shoots of tea plant under different N sources contributes to diverse tea flavor (Huang et al. 2018).

In tea current-year sprouts produce better quality of made tea. To understand miRNA-mediated quality formation in made tea, quality-determined metabolites catechins, theanine and caffeine in different tissue including buds, different development stages of leaves as well as stems were quantified and miRNA profiling done from same tissue which identified differentially expressed miRNAs. Interestingly two miRNA-TF-metabolite triplets that participated in both development and quality formation had been identified. Thus conserved miRNAs are more likely to be linked with morphological function in primary metabolism during sprout development. It also holds an important position in secondary metabolism during quality formation in tea plant. Finally, it coordinates with transcription factors in forming regulatory networks in complex multicellular organisms (Zhao et al. 2019).

Terpenoids play important roles not only in tea beverage aroma formation but also in the productivity and quality of tea plantation due to their significant contribution to light-harvesting pigments and phytohormones. Thus to understand the regulation of terpenoid synthase genes by miRNAs, small RNAseq was done which identified four classes of miRNA-TF pairs that might play a central role in the regulation of terpenoid biosynthesis. Thus, mature miRNAs maintained by light regulator at both the transcriptional and post-transcriptional levels negatively regulate the targets to control terpenoid biosynthesis (Zhao et al. 2018).

To understand post-transcriptional regulation of the flavonoid pathway involving miRNAs, 51 out of 200 catechin biosynthesis-related genes/TFs were found to be potentially targeted by 17 conserved and 15 novel miRNAs in tea plants. Several conserved miRNAs (e.g., miR156, miR172, and miR858) are identified in the co-expression network that have been known to play a pivotal role in flavonoid biosynthesis regulation. Thus, the enrichment of miRNA targets indicated an important role for these molecules and revealed an additional layer of control of the expression of "switch" genes involved in the catechin biosynthesis (Zheng et al. 2019c).

8.5.4 Biotic Stress

There are several pests and diseases in tea, and it is known now that small RNA including miRNAs play important role to prevent the diseases (Mukhopadhyay et al. 2016). For example, to identify *E. obliqua*-responsive miRNAs and their target genes in tea plant, small RNA libraries were constructed from leaves challenged with geometrid attack which registered 36 known and 139 novel differentially expressed miRNAs. Several target genes for these miRNAs encode various transcription factors, including ethylene-responsive transcription factors and squamosa promoter-binding-like proteins, which suggest that these miRNAs may regulate stress-responsive transcriptional processes in tea plant (Jeyaraj et al. 2017b).

The role of miRNAs in responses to *C. gloeosporioides* was unexplored in tea plant. It has been found that 239 known and 369 novel miRNAs exhibited significantly differential expression under *C. gloeosporioides* attack. A negative correlation between expression profiles of five miRNAs (PC-5p-80764_22, csn-miR160c, csn-miR828a, csn-miR164a, and csn-miR169e) and their targets (*WRKY*, *ARF*, *MYB75*, *NAC*, and *NFY* transcription factor) were observed. Furthermore, GO and metabolism pathway analysis revealed that most of the target genes were involved in the regulation of auxin pathway, ROS scavenging pathway, salicylic acid mediated pathway, receptor kinases, and transcription factors for plant growth and development as well as stress responses in tea plant against *C. gloeosporioides* attack. This study enriches the resources of stress-responsive miRNAs and their targets in tea plant and thus provides novel insights into the miRNA-mediated regulatory mechanisms, which could contribute to the enhanced susceptibility of *C. gloeosporioides* in tea plant (Jeyaraj et al. 2019).

8.5.5 Low Moisture Stress

Tea is a drought-sensitive plant and requires good amount of water. Thus scientists are always keen to understand the mechanisms for drought tolerance in tea that are modulated by miRNAs. Thus drought stress-responsive miRNA regulation studies were carried out (Guo et al. 2017b). To correlate the miRNA expression, the drought stress effects such as relative water content, leaf chlorophyll concentration, and leaf conductivity were taken into account. A total of 191 conserved and 59 novel miRNAs were detected. The most enriched pathway in KEGG was the D-alanine metabolism. Along with it, differential expression of sulfur metabolism and mineral absorption pathways implied key roles in drought stress response. Three genes related to sulfur metabolism, targeted by drought-related miRNAs, were found to be serine acetyltransferase (targeted by miR5563-5p and miR159a), ATP sulfurylase (targeted by miR395a), and APS kinase (also targeted by miR395a). Five genes associated with mineral absorption pathway such as *DMT1*, three *CTR* genes, and the *ATOX1* gene were found to be targeted by miR854. Regulation of copper transporters and copper metallochaperones indicate copper homeostasis to be an important process during drought stress response in tea. Seven miRNAs (miR166a, miR166g-3p, miR435a, miR894, miR2199, miR24, and miR395a) showed significantly higher expression levels under normal moisture conditions, as compared to drought. KEGG analysis revealed that under severe drought conditions, metabolic pathways related to photosynthesis were significantly affected. miR156, one of the most highly conserved miRNA families, has been found to be upregulated during drought stress. Thus, by controlling the *SPL* expression, it may in turn lead to *DFR* transcription, resulting in activation of catechin biosynthesis. It was hence inferred that tea plants cope up against drought stress by inducing accumulation of secondary metabolites.

MicroRNA (miRNA) regulation under different time point of drought and recovery from drought between a tolerant and sensitive tea plant was studied. It had been found that in total 139 (52.9%) and 96 (36.0%) conserved miRNAs were differentially expressed during the different stages in tolerant and sensitive cultivars, respectively. Around 201 and 218 genes which were specific to the tolerant and sensitive cultivars were identified as targets, respectively, and 395 were common to both cultivars. GO and KEGG analyses revealed the biological roles of these targets and showed that some of the targets responded to drought stress in a stress- and cultivar-dependent manner (Liu et al. 2016b).

8.6 Other Non-coding RNAs

The field of non-coding RNA is expanding very rapidly with the discovery of several novel non-coding RNAs recently (Guha et al. 2019). Among them few have been studied in tea. Long non-coding RNAs are a recent player which controls gene expression by blocking the miRNA and thereby allowing the target protein coding gene to express (Jain et al. 2017). With the progress of sequencing technology coupled with in silico approaches, disclosure of long non-coding RNA (lncRNAs) from RNAseq data has become easy and reliable. Thus, using 170 publicly available RNAseq data of 11 harvestable tissues, 33,400 putative high certainty lncRNAs were discovered, among which 99.76% were found to be novel lncRNAs. The expression analysis of lncRNAs from 53 RNAseq data of seven tissues (axillary bud, a bud and a leaf, a bud and two leaves, apical bud and two leaves, second leaf, fourth leaf, and sixth leaf) demonstrated their tissue-specific expression. Further, around 1654 lncRNAs were found to be endogenous target mimics of miRNAs that interacted with 698 mRNAs of tea genome. Based on their target, i.e., coding-gene partners by GO and KEGG biological pathway showed that 378 lncRNAs are involved in major aroma formation pathway of tea. Thus this study also provides valuable information to uncover the various molecular functions of lncRNAs that regulate aroma formation of tea (Varshney et al. 2019).

Role of lncRNAs during manufacturing of oolong tea was studied to know its involvement for secondary metabolites formation during the withering of oolong tea. By analyzing the transcriptome data from the processed oolong tea, 32,036 lncRNAs were identified. Additionally, 63 differentially expressed lncRNAs (DE-lncRNAs) and 23 target genes involved in secondary metabolite pathways were identified related to the three major pathways, i.e., catechin, theanine, and caffeine biosynthesis. A comparison of the expression profiles of the DE-lncRNAs and their target genes between the solar-withered leaves and indoor-withered leaves revealed several differentially expressed genes related to catechin biosynthetic pathways, terpenoid metabolism, as well as lncRNAs related to JA/MeJA biosynthesis and signal transduction. These results suggested that the expression of DE-lncRNAs and their targets involved in the three pathways may be related to the low abundance of the total polyphenols, flavonoids, and catechins (EGC, CG, GCG, ECG, and EGCG) and the high abundance of terpenoids in the solar-withered leave. Moreover, solar irradiation, high JA and MeJA contents, and the endogenous target mimic (eTM)-related regulatory mechanism in the solar-withered leave were also crucial for increasing the terpenoid levels. These findings provide new insights into the greater contribution of solar-withering to the high-quality flavor of oolong tea compared with the effects of indoor withering (Zhu et al. 2019c).

Salt though may not be very important for tea yet due to the surface irrigation, tea plants are exposed to saline soil. Thus to understand the salinity stress-responsive lncRNAs between salinity treated tea shoots and control tea shoots, a study was performed which yielded 172 differentially expressed lncRNAs (DE-lncRNAs). The results of GO and KEGG enrichment analyses of their cis- and trans-target genes showed that these DE-lncRNAs play important roles in many pathways such as the galactinol synthase and calcium signaling pathway and interact with TFs under salt stress. One lncRNA MSTRG.139242.1 was found to be involved in salinity stress tolerance in tea. In addition, 12 lncRNAs were predicted to be target mimics of 17 known mature miRNAs, such as miR156, that are related to the salt-stress response in tea (Wan et al. 2020).

Another non-coding RNA, circular RNAs (circRNAs), has been recently discovered as a novel class of abundant endogenous stable RNAs produced by circularization with regulatory potential. However, identification of circRNAs in plants, especially in non-model plants with large genomes, is challenging. In this study, Tong et al. (2018) undertook a systematic identification of circRNAs from different stage tissues of tea leaf development using rRNA-depleted circular RNAseq which yielded 342 high-confidence circRNAs. Similar in other plants, tissue-specific expression was also observed for many tea circRNAs. In addition, they found that circRNA abundances were positively correlated with the mRNA transcript abundances of their parental genes. Further the circRNA-miRNA interaction networks were predicted, and 54 of the differentially expressed circRNAs were found to have potential tea plant miRNA binding sites. The gene sets encoding circRNAs were significantly enriched in chloroplast-related GO terms and photosynthesis/metabolite biosynthesis-related KEGG pathways, suggesting the candidate roles of circRNAs in photosynthetic machinery and metabolites biosynthesis during leaf development.

8.7 Housekeeping Gene for Expression Analysis

For several reasons, we need to validate transcript expression which normally is predicted by in silico analysis. It is not only to give reliability of the data generated through informatics but also to identify the expression of a condition-specific member present in a gene family (Chowraisa et al. 2019). These transcripts could be coding RNAs or non-coding RNAs such as miRNA, lncRNA, cirRNA, etc. In tea several condition-specific reference genes have been discovered (Sun et al. 2010; Xie et al. 2015; Liu et al. 2016c). For example, Hao et al. (2014) evaluated 11 genes in 94 experimental samples. The expression stabilities of these 11 genes were ranked using 4 different computation programs and found that 3 commonly used housekeeping genes *CsPTB1*, *CsEF1*, *CsSAND1*, *CsCLATHRIN1*, and *CsUBC1* were in the top 5 appropriate reference genes for qRT-PCR analysis in complex experimental conditions. In a separate study, nine candidate reference genes of tea were examined in five tea leaf developmental stages (i.e., first, second, third, fourth, and older leaves), and normal growth tea leaves subjected to five hormonal stimuli (i.e., ABA, GA, IAA, MeJA, and SA). It was found that *CsTBP* and *CsTIP41* were the most stable genes in tea leaf development and *CsTBP* was the best gene under hormonal stimuli (Wu et al. 2016c). Further, out of the seven candidate reference genes, it was found that *TUA1* (alpha-1 tubulin) has the most stable expression under damage stresses, whereas for drought stresses, *18S rRNA* and *GAPDH* (glyceraldehyde-3-phosphate dehydrogenase) were the most stable genes. Similarly for cold and Al and NaCl stresses, *GAPDH* and *TUA1* were found to be the best (Ma et al. 2016). Later, Zhou et al. (2016) analyzed 15 candidate reference genes of tea in two kinds of post-harvest treatment, turnover and withering, using three algorithms—geNorm, NormFinder, and BestKeeper. They found that *CsACT*, *CsEF-1α*, *CsPPA2*, and *CsTBP* were the top four reference genes in the turnover treatment, while *CsTBP*, *CsPCS1*, *CsPPA2*, *CselF-4α*, and *CsACT* were the five best reference genes in the withering group. The expression level of lipoxygenase genes, which were involved in a number of diverse aspects of plant physiology, including wounding, was evaluated to validate the findings. Wang et al. (2017b) studied 12 candidate reference genes of tea plants, and the stability of their expression was examined systematically in 60 samples exposed to diverse heavy metals (i.e., manganese, aluminum, copper, iron, and zinc). Using three different software, it was found that *PP2AA3* and *18S rRNA* were the most stable expressed genes. Interestingly, it was found that commonly used reference genes such as *GAPDH* and *TBP* were the least appropriate reference genes for most samples. Recently, Xu et al. (2020b) reported that out of the ten reference genes of tea that were examined under five experimental conditions, *CLATHRIN1* and *UBC1*, *TUA1* and *SAND1*, or *SAND1* and *UBC1* were identified as the best combination for normalizing diurnal gene expression in leaves, stems, and roots individually; *CLATHRIN1* and *GAPDH1* were identified as the best combination for jasmonic acid treatment; *ACTIN1* and *UBC1* were identified as the best combination for *Toxoptera aurantii*-infested leaves; *UBC1* and *GAPDH1* were identified as the best combination for *Empoasca onukii*-infested leaves; and *SAND1* and *TBP1* were identified as the best combination for *E. obliqua* regurgitant-treated leaves. Thus collectively it shows that several condition-specific genes have been identified in tea which can be used as internal control for the use of Q-PCR study.

8.8 Proteomics

Proteomics study correlates the protein quantification and potential protein modifications to particular phenotypes through the techniques such as HPLC, SDS-PAGE, two-dimensional gel electrophoresis, in silico protein modeling, matrix-assisted laser desorption/ionization (MALDI) mass spectrophotometer, isobaric tag for relative and absolute quantitation (iTRAQ) and mass spectrometry, such as liquid chromatography-mass spectrometry (LC-MS), tandem mass tag (TMT), etc. In tea, several attempts have been made which are elaborated here.

8.8.1 Growth and Development

Therefore scientists are always interested to reveal the proteins that are responsible for albinism. For an example, 'Anji Baicha' of China in which 3530 lysine succinylation sites were mapped to 2132 proteins which were involved in photosynthesis, carbon fixation, biosynthesis of amino acids porphyrin and chlorophyll metabolism. Lysine succinylation is a novel dynamic and evolutionarily conserved post-translational modification that regulates various biological processes including albinism in tea (Xu et al. 2017b). Further, White leaf No. 1 is another albino tea cultivar color which is divided into three stages: the pre-albinistic stage, the albinistic stage, and the re-greening stage depending on ambient temperature. Proteomics analysis showed that the expression of 61 proteins varied markedly during the three developmental stages. These proteins were involved in metabolism of carbon, nitrogen and sulfur, photosynthesis, protein processing, stress defense, and RNA processing, indicating that these physiological processes may play crucial roles in the periodic albinism. Thus the proteomic analysis revealed that some proteins may have important roles in the molecular events involved in periodic albinism of White leaf No. 1 (Li et al. 2011). Similarly, "Huangjinya" is an albino tea which has chlorophyll-deficient young leaf. It has been found that as high as 173 proteins showed differential accumulations between the chlorotic and normal green leaves which belong to primary carbon metabolism (i.e., carbohydrate synthesis and transport) and were absent in chlorotic tea leaves. Additionally, expression of 4-coumarate-CoA ligase was also reduced in albino leaf, indicating it has major effect on repressing flavonoid metabolism, and abnormal developmental chloroplast inhibited the accumulation of chlorophyll. Additionally, a positive feedback mechanism was found at the protein level (Mg chelatase and chlorophyll b reductase) in the chlorophyll biosynthetic pathway, which might effectively promote the accumulation of chlorophyll b in response to the demand for this pigment in the cells of chlorotic tea leaves in weakened carbon metabolism (Dong et al. 2018). Thus these proteomics study reveal that not only protein but post-translational methods are also involved in albinism of tea.

Proteomic along with transcriptomic profiles of tea leaves with different maturity were studied to understand the proteins that are related to the quality of made tea. A total of 4455 proteins was identified which belong to compositions of flavonoids (catechins and flavonols) in tea leaves, and their expression was found to be higher in young tea leaf (Wu et al. 2019b). Similarly, 233 differentially expressed proteins were identified in buds and the young expanding leaves of tea plants that were involved in energy and carbohydrate metabolism, secondary metabolite metabolism, nucleic acid and protein metabolism, and photosynthesis- and defense-related processes, indicating that polyphenol biosynthesis- and photosynthesis-related proteins regulate the secondary metabolite content of tea plants (Li et al. 2015b, c). Therefore, these works indicate that the protein that belongs to the factors affecting the quality of tea has not been studied.

8.8.2 Abiotic Stress

Most of the abiotic traits are polygenic and controlled at different levels of gene expression such as transcriptomic, proteomics, and even at post-translational levels. Some proteomics study have been made in tea particularly for cold, drought, and aluminum as well as fluorine accumulation in tea. Li et al. (2008) conducted the proteomics analysis to identify the differentially expressed proteins of pollen that were associated with cold stress. Further, changes in protein expression, in the embryo of tea in response to desiccation, had been investigated and found that 23 different proteins related to defense response, metabolism, and redox status were upregulated under desiccation. This finding is particularly important as tea seeds being recalcitrant maintain high rate of metabolism and water content during the seed maturation stage (Chen et al. 2011).

Although Al-induced growth of tea plant is known for quite some time now, the proteomic profiles of tea plants in response to Al are not available. In total, 755 and 1059 differentially expressed proteins were identified in tea roots and leaves that are comprised of 11 and 9 pathways in roots and leaves, respectively. The results indicated that active photosynthesis and glycolysis as well as increased activities of antioxidant enzymes can be considered as a possible reason for the stimulatory effects of Al on the growth of tea plants. Additionally, the downregulation of F5H and the binding of Al and phenolic acids may reduce the accumulation of lignin (Xu et al. 2017c).

The tea plant is also fluoride hyperaccumulator, and fluoride accumulation in its leaves is bad for human health. To dissect molecular mechanisms underlying fluoride accumulation/detoxification, proteomics analyses were done. The results showed that fluoride had no adverse effects on the growth of tea seedlings in spite of high-content fluoride accumulation in their leaves. Several proteins are found that belong to cell wall structure rearrangement, signaling modulation, and the protection of the cells from damages, diverse metabolic reprogramming, energy reallocation, and plant defense. Notably, F treatment to tea seedlings lead to accumulation of several defense-related proteins against fluoride (Liu et al. 2018b). However, hyperaccumulation of these two elements in tea plants is not fully understood.

Winter dormancy is a biological phenomenon, shortening of which for tea plant remains a challenge. Recently a tea genotype, "Dongcha11," was discovered which does not have winter dormancy. Proteomics study of this tea genotype reveals that several biological pathways are active in the leaf of this genotype during winter such as photosynthesis, cell structure, protein synthesis and destination, transporters, metabolism of sugars and polysaccharides, secondary metabolism, disease/defense, and proteins with unknown functions. On the contrary protein of some of the pathways was decreased such as metabolism of polyphenolic flavonoids. Further, it was reported that sugar, amino acid, and polyphenol play important role in winter dormancy of tea (Liu et al. 2017d).

8.8.3 Self-Incompatibility

Tea is a highly cross-pollinated plant. Thus scientists are always keen to understand the mechanisms of self-incompatibility through various omics approaches. Although no systemic study has been done on tea yet, wild species *C. oleifera* was of much interest as it is has late-acting self-incompatibility (LSI) properties. The mechanism of LSI is uncertain, which seriously hinders the research on its genetic characteristics, construction of genetic map, selection of cross-breeding parents, cultivar arrangement, and also bear to low fruit set. Analysis of differentially expressed proteins of self- and cross-pollinated pistils revealed that several candidate proteins are involved in SI of *C. oleifera*, including polygalacturonase inhibitor, UDP-glycosyltransferase A1-like, beta-D-galactosidase, S-adenosylmethionine synthetase, xyloglucan endotransglucosylase/hydrolase, ABC transporter G family member 36-like, and flavonol synthase. Out of these, 11 proteins had been found to be involved in pollen tube growth in *C. oleifera* (He et al. 2020). Functional enrichment analysis revealed that the SI was closely associated with programed cell death (*PCD*)-related genes, mitogen-activated protein kinase (MAPK) signaling pathway, plant hormone signal transduction, ATP-binding cassette (ABC) transporters, and ubiquitin-mediated proteolysis. These particular trends in proteins suggested the involvement of PCD in SI (Zhou et al. 2020a). These two studies provide a solid genetic foundation for elucidating the protein-based regulatory network of PCD-mediated self-incompatibility in *C. oleifera.*

8.8.4 Processing

Aroma of oolong tea is unique which was studied until and unless Gui et al. (2015) reveled that enzymatic hydrolysis of glycosidically bound volatile compounds (GBVs) really contribute to the formation of volatile compounds during the oolong tea manufacturing process. These findings also reveal that, during the enzymatic process of oolong tea, non-disruption of the leaf cell walls resulted in no interaction of GBVs and β-glycosidases. Indole, jasmine lactone, and trans-nerolidol were characteristic volatiles produced from the manufacturing process. Interestingly, the contents of the three volatiles was reduced after the leaf cell disruption, suggesting that mechanical damage with the cell disruption, which is similar to black tea manufacturing, did not induce accumulation of the three volatiles. In addition, 11 volatiles with flavor dilution factor $\geq$44 were identified as relatively potent odorants in the oolong tea suggesting that enzymatic hydrolysis of GBVs was not involved in the formation of volatiles of oolong tea and some characteristic volatiles with potent odorants were produced from the manufacturing process (Gui et al. 2015).

The withering of fresh tea leaves is the first step in tea processing which directly affects tea color, taste, and fragrance. Proteomic analysis of post-harvest tea leaves at four withering stages revealed 863 differentially expressed proteins (DEPs). The results of the functional annotation revealed that the molecular characteristics of tea withering are similar to leaf senescence. The biosynthesis of main tea-specific compounds that constitute tea color, taste, and fragrance of tea is restricted during withering. The substance transformation and degradation may have positive contributions to tea quality in withering technology (Wu et al. 2017a). Altogether these studies basically identified the differentially expressed proteins compared to their control counterparts which at this stage merely indicate the various pathways that are involved in that particular situation. These need to be further validated through various molecular approaches in tea.

8.9 Metabolomics

It is a powerful technique, particularly very relevant in tea as quality is a complex trait and composed of the mixture of several metabolites of various nature. Metabolic characteristics associated with climatic variables were investigated to allow the assessment of quality strategy for green tea production (Lee et al. 2010). Differential metabolites expressed under shade-grown tea plants had been studied which helped to understand the effect of shade on tea growth (Ku et al. 2010). Further, the dependence of global green tea metabolome on plucking positions was found to have positive correlation with the age of plucking shoots and the quality of made tea (Lee et al. 2011). All these studies were done by LC-MS or GC-MS which basically quantified the quality compound with a particular condition of growing tea to optimize better production of tea.

Seasonal accumulation of metabolites that influence the quality of tea was studied. It has been found that flavan-3-ols, theasinensins, procyanidins, quercetin-O-glycosides, apigenin-C-glycosides, and amino acid accumulation depend on seasonal fluctuations as well as on genotypes. An equivalent quantification of tea tastes showed that in summer and autumn teas, the bitterness and astringency were significantly elevated, whereas umami declined. Metabolite content comparisons and partial least-squares analysis suggested that several flavonoids and amino acids are mainly responsible for the seasonal variations in taste quality (Dai et al. 2015).

Glycosylation on small molecular metabolites modulates a series of biological events in plants. In green tea, 120 glucosylated/galactosylated, 38 rhamnosylated, 21 rutinosylated, and 23 primeverosylated metabolites were detected simultaneously. Among them, 61 glycosylated metabolites were putatively identified according to current tea metabolite databases. An additional 40 novel glycosylated metabolites were tentatively elucidated. This work provides a feasible strategy to discover and identify novel glycosylated metabolites in plants, though their biological significance is not known (Dai et al. 2016). White tea is another example, which is a least processed tea that has high medicinal value due to the presence of potent bioactivity compounds. It has been found that there are significant differences in the content of amino acids, catechins, dimeric catechins, flavonol and flavone glycosides, and aroma precursors in white tea compared with green and black teas that were manufactured from the same fresh tea leaves (Dai et al. 2017).

Different species of *Camellia* have different chemical constituents. Comparative chemical analysis of flavan-3-ols, flavan-3-ols glycoside, and procyanidins showed that *C. tachangensis* had a significantly higher content of (-)-epicatechin (EC) and (-)-epigallocatechin (EGC) compared with tea. By contrast, higher levels of galloylated catechins were detected in tea. Main secondary metabolites of *C. tachangensis* were non-galloylated catechins, procyanidin dimmers, trimmers, (−)-epicatechin glucose (EC-glucose), and (-)-epigallocatechin glucose (EGC glucose). Further, the levels of EC-glucose were closely related to the procyanidin dimers content. Thus, it was suggested that EC-glucose might be an important substrate for the biosynthesis of procyanidins (Zhang et al. 2017b).

Organically grown tea fetches better price in the market due to its health benefits. Metabolome analysis showed that epigallocatechin-3-gallate, the major catechin in tea leaves, has a significantly increased accumulation in leaves of tea plants when grown organically compared with those grown traditionally with inorganic fertilizer. The content of L-theanine, the major amino acid in tea leaves, was not significantly changed in tea plants under organic cultivation. Furthermore, aroma compounds were more abundant in tea leaves from the organically grown tea plantation than those from the chemical pesticide-treated tea plantation. Therefore, organically grown teas are more safe than the traditionally grown tea (Li et al. 2019b).

Metabolomics study of albino tea, 'Anji Baicha,' at three different temperature-dependent developmental stages (i.e., the pre-albescent, albescent, and re-greening) indicated several stage-specific metabolites. For example, amino acids (primarily L-theanine, L-glutamate, N2-acetyl-L-ornithine, L-aspartic acid, proline, L-glutamine, L-leucine, and pyroglutamic acid) and 12-OPDA were accumulated significantly higher in the albescent stage, and few compounds such as flavonol and flavonol/flavone glycosides (mainly kaempferol, myricetin, quercetin, cyanidin, and delphinidin glycosides) were detected at the highest levels in the re-greening or pre-albescent stages, indicating that these could be responsible for high quality in this cultivars (Xia et al. 2019a; Zeng et al. 2019).

The quality of made tea depends on the content and their ratio of metabolites. For example, Pu-erh tea, a traditional popular tea, is produced by microbial fermentation. Interestingly in Pu-erh tea, canonical compounds such as epicatechin, epigallocatechin gallate, and theanine get reduced during fermentation, but 33 phenolic acids were found to be accumulated in higher quantity including gallic acid and theogallin. Dynamic profiling revealed the increase of simple phenolic acids and the decrease of most of phenolic acid esters during Pu-erh tea fermentation which is responsible for special aroma (Ge et al. 2019).

Metabolomic regulation of exogenous application of melatonin (MT) and gibberellin (GA) in the tea plant was explored. Results showed that the growth of tea plant was enhanced by MT application. The internode elongation and leaf enlargement were observed by GA treatment. This study along with transcriptomic analysis indicates the upregulation of various biosynthetic pathways such as phenylpropanoid biosynthesis, terpenoid synthesis, and expansions. Therefore, the distinctive effect of MT and GA treatment on tea plant was different. The MT exhibited significant promotion in terpenoid synthesis, especially TPS14 and TPS1. GA was prominent in coordinated regulation of plant hormonal signal transduction (Di et al. 2019).

8.10 Systems Biology

Systems biology is a biology-based interdisciplinary field of study that focuses on complex interactions within biological systems, using a more holistic perspective. Tea quality is a highly complex trait, and tea manufacturing process induces a variety of stresses that affect quality. Using cross-species Affymetrix *Arabidopsis* genome arrays, Marshall et al. (2007) attempted to track transcriptional changes occurring during wounding and withering of the leaves to identify metabolic pathways that could influence tea aroma and flavor. *Arabidopsis* metabolic SBML (Hucka et al. 2003) network data from AraCyc (Mueller et al. 2003), KEGG, and Reactome were collated and merged and then subsequently overlaid with the tea expression data. Sub-networks were constructed by connecting the shortest paths between the differentially expressed genes and the downstream aroma-related compounds, therefore identifying the pathways involved in aroma. However, further research in this line has not been conducted. The resultant co-expression network included 35 co-expression modules, of which 20 modules were significantly associated with the biosynthesis of catechins, theanine, and caffeine among 10 different tissues. Furthermore, they identified several hub genes related to these three metabolic pathways and analyzed their regulatory relationships using RNAseq data. The results showed that these hub genes are regulated by genes involved in all three metabolic pathways, and they regulate the biosynthesis of all three metabolites. It is notable that light was identified as an important regulator for the biosynthesis of catechins. Conclusion: Our integrated omics-level WGCNA analysis provides novel insights into the potential regulatory mechanisms of catechins, theanine, and caffeine metabolism, and the identified hub genes provide an important reference for further research on the molecular biology of tea plants (Tai et al. 2018). Polygenic trait is complex; it is being controlled by several layers of molecular interaction. Thus Zheng et al. (2019a, b, c) constructed a multilayered regulatory network, integrating the gene co-expression relationship with the microRNA-target modules and promoter *cis*-regulatory element information. This allowed them to reveal new uncharacterized TFs and microRNAs that were potentially implicated in different steps of the catechin biosynthesis. Furthermore, they applied metabolic signature-based association method to capture additional key regulators involved in catechin pathway. This provided important clues for the functional characterization of five SCPL1A acyltransferase family members, which might be implicated in the production balance of anthocyanins, galloylated catechins, and proanthocyanidins.

8.11 Bioinformatics and Database Development

In silico analysis is an integral part of the functional genomics today. Initially with the accumulation of large-scale ESTs, later with RNAseq data in the public domain, researchers started to use them for various purposes. The emerging computational approach provides a better alternative process of development of SSRs from these data than the conventional methods of making SSR-enriched libraries. Sahu et al. (2012) mined 12,851 ESTs of tea, for the development of microsatellites. Finally, 6148 (4779 singletons and 1369 contigs) non-redundant ESTs were found using various computational tools with a density of 1 SSR/1.61 kb leading to development of 245 primer pairs. These mined EST-SSR markers will help further in the study of variability, mapping, and evolutionary relationship in tea. In addition, these developed SSRs can also be applied for various studies across species (Sahu et al. 2012).

Using similar approaches, miRNAs of tea were identified along with their stem-loop structure from the ESTs (Das and Mondal 2010; Prabu and Mandal 2010).

Computational approaches were also applied to characterize the genes. Structural analysis of a *CsGS* had been conducted employing computational techniques. This was conducted to compare its structural aspects with other known structures of *CsGS*. The disordered residues and their distribution in *CsGS* were in close comparison to earlier reported *GS*. The 3-D structure of CsGS protein also showed high degree of similarity with the only known crystal structure of GS from maize (Yadav 2009). Similarly, coding sequence of the *ICE1* in tea was analyzed with CodonW, CHIPS (codon heterozygosity in a protein coding sequence), and CUSP (create a codon usage table) programs which was then compared with the genome of tea plant and *ICE1* from seven plant species. The results showed that *ICE1* of tea plant was biased towards the synonymous codons with A and T (Shi et al. 2012).

The stearoyl-acyl carrier protein desaturase (*SAD*) gene is widely present in all kinds of plants. Tea SAD gene (*CsSAD*) sequence was analyzed by various bioinformatics tools such as CodonW, CHIPS, and CUSP programs and compared with publicly available tea sequences, other *SAD* genes from 11 plant species. It had been found that *CsSAD* gene had similar codon usage bias. The *CsSAD* gene had a bias towards the synonymous codons with A and T at the third codon position. Compared with monocotyledons such as *Triticum aestivum* and *Zea mays*, the differences in codon usage frequency between the *CsSAD* gene and dicotyledons such as *Arabidopsis thaliana* and *Nicotiana tabacum* were less. Therefore, *A. thaliana* and *N. tabacum* expression systems may be more suitable for the expression of the *CsSAD* gene. The analysis result of *SAD* genes from the 12 plant species also showed that most of the *SAD* genes were biased towards the synonymous codons with G and C at the third codon position (Pan et al. 2013).

In the last decade with the advancement of next-generation sequencing data, attention has been shifted from individual gene to meta-analysis of RNAseq data. A large number of transcriptomes of tea plants at different developmental stages or under abiotic and/or biotic stresses have been generated to investigate the genetic basis of various traits including secondary metabolites that are related to the quality of made tea. However, these data exhibited large differences, particularly in the total number of reconstructed transcripts and the quality of the assembled transcriptomes. These differences are largely related to optimize the sequencing depth and assembler for transcriptome assembly of structurally complex plant species genomes like tea. Thus Li et al. (2019c) examined five different assemblers (Bridger, BinPacker, SOAPdenovo, Trans-ABySS, and BUSCO) with varying amount of RNAseq data; although the total number of assembled transcripts increased with increasing sequencing data, the proportion of unassembled transcripts became saturated as revealed by plant BUSCO datasets. Among the five representative assemblers, the Bridger package showed the best performance in both assembly completeness and accuracy as evaluated by the BUSCO datasets and genome alignment.

Algorithm development is an important area of bioinformatics. Design principles of gene-regulatory and protein–protein interaction (PPI) networks at subcellular level are largely uncharacterized in plants like tea. Recently, a tea leaf interologous

PPI network (TeaLIPIN) consisting of 11,208 nodes and 197,820 interactions was developed based on a reference transcriptome assembly generated from 44 samples of 6 publicly available leaf transcriptomes dataset. Comparing this network with 10,000 realizations of 2 types of corresponding random networks (Erdős–Rényi and Barabási–Albert models) and examining over 3 network centrality metrics, 2750 bottleneck proteins (having p values <0.01) were predicted. Co-expression analysis of the TeaLIPIN proteins was also performed, and top-ranking modules are elaborated (Singh et al. 2019).

Organizing the data in a user-friendly way is an important aspect of computational biology. Thus in tea, the first integrative and specially designed web-accessible database, Tea Plant Information Archive (TPIA; http://tpia.teaplant.org), was hosted to keep several aspects of recently released genome sequence data of tea. It contained comprehensively annotated tea plant genome as framework and incorporates with abundant well-organized transcriptomes, gene expressions (across species, tissues, and stresses), orthologs, and characteristic metabolites determining tea quality. It also hosts massive transcription factors, polymorphic simple sequence repeats, single nucleotide polymorphisms, correlations, manually curated functional genes, and globally collected germplasm information. A variety of versatile analytic tools (e.g., J Browse, blast, enrichment analysis, etc.) are established helping users to perform further comparative, evolutionary, and functional analysis (Xia et al. 2019b).

Although in tea several SSRs have been generated and validated by various workers across tea-producing countries, there was no organized database for this. Thus, taking the advantage of the recently published tea genomes, large numbers of SSR markers in the tea were mined from various sources. Besides the genomic sequences, other publicly available sequences such as RNAseq, GSS, ESTs, and organelle genomes (chloroplasts and mitochondrial) published literature were also searched to catalog validated set of tea SSR markers. The complete exercise yielded a total of 935,547 SSRs. Using all the information generated from various sources, Dubey et al. (2020) have developed a user-friendly database (TeaMiD; http://indianteagenome.in:8080/teamid/) that hosts SSRs from all the six resources including three nuclear genomes of tea and transcriptome sequences of 17 *Camellia* species.

8.12 Organelle Genome Sequencing

Organelle genomes such as chloroplast and mitochondria are not only important for governing several agronomically important genes, but also they owe useful evolutionary information that are important for taxonomical classification (Borthakur et al. 1998). *Camellia* genus is large enough with more than 320 species and the most important genus in the family Theaceae. However, phylogenetic resolution of the species has proven to be difficult in rapidly evolving new species due to the high outcrossing nature among the species. Consequently, the interspecies relationships of the genus *Camellia* are still not resolved. Thus phylogenomics is an attractive avenue that can be used to reconstruct the tree of life, especially at low taxonomic levels. Till today 17 complete chloroplast (cp) genomes were sequenced using Illumina sequencing technology representing different subdivisions of the genus *Camellia* (Liu and Han 2018). The length of the *Camellia* chloroplast genome was found to be about 157 kb, which contained 123 unique genes. It was found that *Camellia* chloroplast genome was relatively well-conserved but contained enough genetic differences to provide useful phylogenetic information. Phylogenetic relationships were analyzed using complete chloroplast genomes of *Camellia* species. Interestingly, rapidly evolving regions of the cp genome that have the potential to be used for further species identification and phylogenetic resolution were identified. Attempt was made to determine if analyzing completely sequenced chloroplast genomes could help settle the controversies of interspecies relationships in *Camellia*. The results demonstrate that chloroplast genome data are beneficial in resolving species definition because they indicate that organelle-based "barcodes" can be established for a species and then used to unmask interspecies phylogenetic relationships. It reveals that phylogenomics based on chloroplast genomes is an effective approach for achieving phylogenetic resolution between *Camellia* species (Yang et al. 2013). In a separate study, a total of 15 molecular markers were identified with more than 1.5% sequence divergence that may be useful for further phylogenetic analysis and species identification of *Camellia*. These results showed that, rather than functional constrains, it is the regional constraints that strongly affect sequence evolution of the chloroplast genomes. Despite a high degree of conservation between the *Camellia* chloroplast genomes, sequence variation among species could still be detected, representing a wide phylogenetic diversity in the genus. Furthermore, phylogenomic analysis was conducted using 18 complete chloroplast genomes and 5 draft chloroplast genome sequences of *Camellia* species. These results support Chang's taxonomical treatment that *C. pubicosta* may be classified into sect. Thea, and indicate that taxonomical value of the number of ovaries should be reconsidered when classifying the *Camellia* species (Huang et al. 2014). After chloroplast genome, mitochondrial genomes of cv. Yunkang 10 was sequenced which consisted circular contig of 157,100 bp and two complete circular scaffolds (701,719 bp and 177,329 bp), respectively. This generated a total of 141 chloroplast genes and 71 mitochondrial genes. They also detected 478 RNA-editing sites in 42 protein-coding mitochondrial genes, which are ~4.4-fold more than 54 RNA-editing sites detected in 21 protein-coding chloroplast genes (Zhang et al. 2019b).

C. oleifera is one of the four largest woody edible oil plants in the world with high ecological and medicinal values. Due to frequent interspecific hybridization, it was difficult to study its genetics and evolutionary history. The whole chloroplast genome encoded a total of 141 genes (115 different genes), including 88 protein-coding genes, 45 tRNA genes, and 8 rRNA genes. Further, the chloroplast genome structure characteristics of Hainan *C. oleifera* were compared with mainland *C. oleifera* and those of the other eight closely related Theaceae species; it was found that the contractions and expansions of the IR/LSC and IR/SSC regions affected the length of chloroplast genome. It was found that Hainan *C. oleifera* had a similar evolutionary relationship with *C. crapnelliana*, through the use of codons and phylogenetic analysis. This study can provide an effective genomic resource for the evolutionary history of Theaceae family (Zhang et al. 2019c).

Although several chloroplast and mitochondrial genomes were sequenced, no sequence is available for Assamica type of tea, i.e., *C. assamica*. Thus Rawal et al. (2020) successfully assembled and analyzed chloroplast and mitochondrial genome of *C. assamica*. The complete mitochondrial genome of *C. assamica* was sequenced in a single circular contig of 707,441 bp length comprising of a total of 66 annotated genes, including 35 protein-coding genes, 29 tRNAs, and 2 rRNAs. The first-ever chloroplast genome of tea was yielded in a circular contig of 157,353 bp length with a typical quadripartite structure. Phylogenetic analysis based on these organelle genomes showed that *C. assamica* was closely related to *C. sinensis* and *C. leptophylla*. It also supports Caryophyllales as superasterids (Rawal et al. 2020).

8.13 Whole-Genome Sequencing

The progress of genetic improvement of tea is slow for many reasons. Notable among all is the lack of whole genome sequence and high-density linkage map. However, recently, three genomes of tea are decoded de novo simultaneously. The first genome of a tea cultivar Yunkang 10 (CSA) was sequenced in 2018. In this study, basic information of tea genome was generated such as high repeat-rich (80.9%) region. In addition, in a recent whole-genome duplication event, lineage-specific expansions of genes associated with flavonoid metabolic biosynthesis were also discovered. It was also demonstrated that tea genome has an independent and rapid evolution of the caffeine synthesis pathway relative to cacao and coffee. A comparative RNAseq study among 25 *Camellia* species revealed that higher expression levels of most flavonoid and caffeine—but not theanine—related genes that contribute to the increased production of catechins and caffeine and thus enhance tea-processing suitability and tea quality. However, the sequencing was done with only Illumina short reads, which lead to several gaps in the genome (Xia et al. 2017a). Subsequently tea genome with long PacBio read was assembled for the cv. Shuchazao (CSS), a Chinary type genotype which consisted 64% repetitive sequences of the 3.1-Gb genome. The difference of this repeat region with the previously sequenced genome (CSA) could be due to the difference in species of assamica and sinensis. Divergence between two major lineages, CSS and CSA, was calculated to ~0.38 to 1.54 million years ago (MYA). Analysis of genic collinearity reveals that the tea genome is the product of two rounds of whole-genome duplications (WGDs) that occurred ~30 to 40 and ~90 to 100 MYA. Analyses of transcriptome and phytochemistry data showed that amplification and transcriptional divergence of genes encoding a large acyltransferase family and leucoanthocyanidin reductases were associated with the characteristic young leaf accumulation of monomeric galloylated catechins in tea, while functional divergence of a single member of the glutamine synthetase gene family yielded theanine synthetase (Wei et al. 2018). However, this genome could not be assembled to pseudomolecule level due to the absence of long read, yet improved later by adding more long-read data such as PacBio as indicated by improved BUSCO reading (Xia et al. 2019c). However, recently, efforts were made to assemble both the above genomes with Hi-C data to generate 15 pseudomolecules. It was also evidenced that long terminal repeats play critical roles in the genome size expansion of tea and transcriptional diversification of tea genes through preferential gene insertions in promoter regions and introns (Xia et al. 2020). In the meantime, another genome assembly of a tea cv. Biyun using PacBio SMRT technology was developed, though mainly focusing the basic statistic (Zhang et al. 2020a).

All the above tea genome was mainly based on China type as they are the popular cultivated species of China. Interestingly, *Camellia assamica* L. (O). Kuntze. var. TV-1 from Assam, India, is one of the most important tea cultivar of Indian Tea Industry which botanically belongs to Assam type and has distinct morphological as well as chemical constituent in leaf than China type of tea. The genome size of TV-1 was determined to be 3 Gb, and out of that 2.93 Gb size was assembled comprising of 14,824 scaffolds and covering 97.66% of the predicted genome. The N50 value, longest and average contig size for this draft genome, was found to be 538.96 kb, 3.64 Mb, and 197.99 kb, respectively. There were 495,747 SSRs (Di- to Hexa-mers) in the genome, and the repeat contents accounted for 71.87% of the genome. About 94.40% genome completeness was predicted with BUSCO, maximum among the published and reported tea genomes. The clustering of the assembled genome with Hi-C data resulted in 15 clusters ranging from the size of 631.46–14.39 Mb and accounting 99.32% of assembled genome (Mondal et al. 2019). It has also been seen that genome of TV has 85% similarly with CSA and 65% with that of CSS, indicating that this cultivar is distinct from those published genome.

Simultaneously Chen et al. (2020) used HI-C data to generate the 15 pseudomolecules of CSS genome and reported that only one *Camellia* recent tetraploidization event (CRT, 58.9–61.7 MYA) occurred after the core-eudicot common hexaploidization event (146.6–152.7 MYA). Meanwhile, 9243 genes (28.6%) occurred in tandem duplication, and most of these expanded after the CRT event. These gene duplicates increased functionally divergent genes that play important roles in tea-specific biosynthesis or stress response. Sixty-four catechin- and caffeine-related QTLs were anchored to chromosome assembly. Of these, two catechin-related QTL hotspots were derived from the CRT event, which illustrated that polyploidy has played a dramatic role in the diversification of tea germplasms (Chen et al. 2020).

8.14 Conclusion

Functional as well as structural genomics, metabolomics, and proteomics all are new developments of biotechnology. These "omics" approaches along with other advanced techniques of microscopy as well as DNA sequencing are very potential for deciphering several fundamental questions of gene expression. However, in recent years, tea genome including the organelle has been decoded which will help to increase our understanding of genome organization, etc. Although in several aspects of tea molecular biology work can be intended for, priority should be given (1) to develop a platinum standard phased reference genome, (2) to re-sequence popular trait-specific cultivars or extreme genotypes to generate large-scale SNP markers and their utilizations for association mapping, gene introgression, high-density linkage map constructions, etc., (3) effort should be made to identify and introgress the trait-specific QTLs in tea plants, and (4) finally development of technology such as GeneChip and microarray chip will be immensely helpful for tea breeding.

References

Anonymous (2005) L-Theanine. Altern Med Rev 10:136–138

Ashihara H, Kubota H (1987) Biosynthesis of purine alkaloids in *Camellia* plants. Plant Cell Physiol 28:535–539

Ashihara H, Monterio AM, Gillies FM, Crozier A (1996) Biosynthesis of caffeine in leaves of coffee. Plant Physiol 111:747–753

Ashihara H, Gillies FM, Crozier A (1997) Metabolism of caffeine and related purine alkaloids in leaves of tea (*Camellia sinensis* L.). Plant Cell Physiol 38:413–419

Ashihara H, Deng WW, Mullen W, Crozier A (2010) Distribution and biosynthesis of flavan-3-ols in *Camellia sinensis* seedlings and expression of genes encoding biosynthetic enzymes. Phytochemistry 71:559–566

Bharalee R, Gupta S, Bandyopadhyay T, Gohain B, Agarwala N, Kalita CM, Das S (2012) Molecular cloning, expression and computational analysis of a water stress inducible copper-containing amine oxidase gene (*CuAO*) from tea plant [*Camellia sinensis* (L.) O. Kuntze]. Afr J Biotechnol 11:15547–15555

Bhat KV, Mondal TK, Gaikwad AB, Kole PR, Chandel G, Mohapatra T (2020) Genome-wide identification of drought-responsive miRNAs in grass pea (*Lathyrus sativus* L.). Plant Gene 21:100210

Borchetia S, Bora C, Gohain B, Bhagawati P, Agarwala N, Bhattacharya N, Bharalee R, Bhorali P, Bandyopadhyay T, Gupta S, Das SK, Singh HR, Ahmed P, Gogoi M, Das S (2011) Cloning and heterologous expression of a gene encoding lycopene-epsilon-cyclase, a precursor of lutein in tea (*Camellia sinensis* var *assamica*). Afr J Biotechnol 10:5934–5939

Borthakur S, Mondal TK, Parveen SS, Guha A, Sen P, Borthakur AK, Deka PC (1998) Isolation of chloroplast DNA from tea, *Camellia* sp. Indian J Exp Biol 36:1165–1167

Borthakur D, Du YY, Chen H, Lu JL, Lin C, Dong JJ, Ye JH, Zheng XQ, Liang YR (2008) Cloning and characterization of a cDNA encoding phytoene synthase (*PSY*) in tea. Afr J Biotechnol 7:3577–3581

Cao H, Wang L, Yue C, Hao X, Wang X, Yang Y (2015) Isolation and expression analysis of 18 CsbZIP genes implicated in abiotic stress responses in the tea plant (*Camellia sinensis*). Plant Physiol Biochem 97:432–442

Cao D, Liu Y, Ma L, Jin X, Guo G, Tan R, Liu Z, Zheng L, Ye F, Liu W (2018) Transcriptome analysis of differentially expressed genes involved in selenium accumulation in tea plant (*Camellia sinensis*). PLoS One 13:e0197506

Chen L, Zhao LP, Gao QK (2005) Generation and analysis of expressed sequence tags from the tender shoots cDNA library of tea plant (*Camellia sinensis*). Plant Sci 168:359–363

Chen X, Fang WP, Zou Z, Wang YH, Cheng H, Li XH (2009a) Cloning and expression analysis of CBF gene in cold induced tea plant [*Camellia sinensis* (L.) O. Kuntze]. J Tea Sci 29:53–59

Chen H, Tan X, Hu F (2009b) Cloning, bioinformatics analysis and functional identification of a novel small heat shock protein gene from *Camellia oleifera* seed. In: 3rd International Conference on Bioinformatics and Biomedical Engineering, INSPEC Accession Number: 10777459

Chen Q, Yang L, Ahmad P, Wan X, Hu X (2011) Proteomic profiling and redox status alteration of recalcitrant tea (*Camellia sinensis*) seed in response to desiccation. Planta 233:583–592

Chen Y, Fu X, Mei X, Zhou Y, Du B, Tu Y, Yang Z (2016) Characterization of functional proteases from flowers of tea (*Camellia sinensis*) plants. J Funct Foods 25:149–159

Chen J, Yang X, Huang X, Duan S, Long C, Chen J, Rong J (2017) Leaf transcriptome analysis of a subtropical evergreen broadleaf plant, wild oil-tea Camellia (*Camellia oleifera*), revealing candidate genes for cold acclimation. BMC Genomics 18(211):1–14

Chen J, Gao T, Wan S, Zhang Y, Yang J, Yu Y, Wang W (2018) Genome-Wide identification, classification and expression analysis of the *HSP* gene superfamily in tea plant (*Camellia sinensis*). Int J Mol Sci 19(2633):1–19

Chen G, Liang H, Zhao Q, Wu AM, Wang B (2019) Exploiting MATE efflux proteins to improve flavonoid accumulation in *Camellia sinensisin silico*. Int J Biol Macromol 143:732–743

Chen JD, Zheng C, Ma JQ, Jiang CK, Ercisli S, Yao MZ, Chen L (2020) The chromosome-scale genome reveals the evolution and diversification after the recent tetraploidization event in tea plant. Hortic Res 7:63

Choi JY, Mizutani M, Shimizu BI, Kinoshita T, Ogura M, Tokoro K, Lin ML, Sakata K (2007) Chemical profiling and gene expression profiling during manufacturing process of Taiwan oolong tea, "Oriental Beauty". Biosci Biotechnol Biochem 71:1476–1486

Chowraisa S, Kaur H, Mujib A, Mondal TK (2019) Comprehensive evaluation of candidate reference genes for qRT-PCR studies of gene expression in *Oryza coarctata*. Biol Plant 63:496–503

Cui L, Yao S, Dai X, Yin Q, Liu Y, Jiang X, Wu Y, Qian Y, Pang Y, Gao L, Xia T (2016) Identification of UDP-glycosyl transferases involved in the biosynthesis of astringent taste compounds in tea (*Camellia sinensis*). J Exp Bot 67:2285–2297

Cui X, Wang YX, Liu ZW, Wang WL, Li H, Zhuang J (2018) Transcriptome-wide identification and expression profile analysis of the bHLH family genes in *Camellia sinensis*. Funct Integr Genomics 18:489–503

Dai W, Qi D, Yang T, Lv H, Guo L, Zhang Y, Zhu Y, Peng Q, Xie D, Tan J, Lin Z (2015) Nontargeted analysis using ultraperformance liquid chromatography-quadrupole time-of-flight mass spectrometry uncovers the effects of harvest season on the metabolites and taste quality of tea (*Camellia sinensis* L.). J Agric Food Chem 63:9869–9878

Dai W, Tan J, Lu M, Xie D, Li P, Lv H, Lin Z (2016) Nontargeted modification-specific metabolomics investigation of lycosylated secondary metabolites in tea (*Camellia sinensis* L.) based on liquid chromatography–high-resolution mass spectrometry. J Agric Food Chem 64:6783–6790

Dai W, Xie D, Lu M, Li P, Lv H, Yang C, Lin Z (2017) Characterization of white tea metabolome: comparison against green and black tea by a nontargeted metabolomics approach. Food Res Int 96:40–45

Das A, Mondal TK (2010) Computational identification of conserved microRNAs and their targets in tea (*Camellia sinensis*). Am J Plant Sci 1:77–85

Das A, Das S, Mondal TK (2012) Identification of differentially expressed gene profiles in young roots of tea [*Camellia sinensis* (L.) O. Kuntze] subjected to drought stress using suppression subtractive hybridization. Plant Mol Biol Report 30:1088–1101

Das A, Saha D, Mondal TK (2013) An optimized method for extraction of RNA from tea roots for functional genomics analysis. Indian J Biotechnol 12:129–132

Das A, Chaudhury S, Kalita MC, Mondal TK (2015) *In silico* identification, characterization and expression analysis of miRNAs in *Cannabis sativa* L. Plant Gene 2:17–24

Deng WW, Ogita S, Ashihara H (2008) Biosynthesis of theanine (γ-ethylamine-L-Glutamine acid) in seedling of *Camellia sinensis*. Phytochem Lett 1:115–119

Deng W, Wang S, Chen Q, Zhang Z, Hu X (2012) Effect of salt treatment on theanine biosynthesis in *Camellia sinensis* seedlings. Plant Physiol Biotechnol 56:35–40

Deng WW, Zhang M, Wu JQ, Jiang ZZ, Tang L, Li YY, Wei CL, Jiang CJ, Wan XC (2013a) Molecular cloning, functional analysis of three cinnamyl alcohol dehydrogenase (CAD) genes in the leaves of tea plant, *Camellia sinensis*. J Plant Physiol 170:272–282

Deng WW, Fei Y, Wang S, Wan XC, Zhang ZZ, Hu XY (2013b) Effect of shade treatment on theanine biosynthesis in *Camellia sinensis* seedlings. Plant Growth Regul 71:295–299

Deng WW, Wu YL, Li YY, Tan Z, Wei CL (2016) Molecular cloning and characterization of hydroperoxide lyase gene in the leaves of tea plant (*Camellia sinensis*). J Agric Food Chem 64:1770–1776

Deng WW, Wang R, Yang T, Jiang L, Zhang ZZ (2017) Functional characterization of salicylic acid carboxyl methyltransferase from *Camellia sinensis*, providing the aroma compound of methyl salicylate during the withering process of white tea. J Agric Food Chem 65:11036–11045

Di T, Zhao L, Chen H, Qian W, Wang P, Zhang X, Xia T (2019) Transcriptomic and metabolic insights into the distinctive effects of exogenous melatonin and gibberellin on terpenoid synthesis and plant hormone signal transduction pathway in *Camellia sinensis*. J Agric Food Chem 67:4689–4699

Ding ZT, Shen JZ, Pan LL, Wang YU, Li YS, Wang Y, Sun HW (2016) CsSAD: a fatty acid desaturase gene involved in abiotic resistance in *Camellia sinensis* (L.). Genet Mol Res 15:1–12

Dong B, Wu B, Hong W, Li X, Li Z, Xue L, Huang Y (2017) Transcriptome analysis of the tea oil *camellia* (*Camellia oleifera*) reveals candidate drought stress genes. PLoS One 12:e0181835

Dong F, Shi Y, Liu M, Fan K, Zhang Q, Ruan J (2018) iTRAQ-based quantitative proteomics analysis reveals the mechanism underlying the weakening of carbon metabolism in chlorotic tea leaves. Int J Mol Sci 19:3943

Duan Y, Zhu X, Shen J, Xing H, Zou Z, Ma Y, Wang Y, Fang W (2020) Genome-wide identification, characterization and expression analysis of the amino acid permease gene family in tea plants (*Camellia sinensis*). Genomics 112:2866–2874

Dubey H, Rawal HC, Rohilla M, Lama U, Kumar PM, Bandyopadhyay T, Gogoi M, Singh NK, Mondal TK (2020) TeaMiD: a comprehensive database of simple sequence repeat markers of tea. Database:baaa013

Edwards JW, Walker EL, Coruzzi GM (1990) Cell-specific expression in transgenic plants reveals nonoverlapping roles for chloroplast and cytosolic glutamine synthase. Proc Natl Acad Sci 87:3459–3463

Eminoglu A, Dizman AY, Guzel S, Belduz OA (2018) Molecular and *in-silico* cloning, identification, and preharvest period expression analysis of a putative cytochrome P450 monooxygenase gene from *Camellia sinensis* (L.) Kuntze (tea). Turk J Biol 42:1–11

Eungwanichayapant PD, Popluechai S (2009) Accumulation of catechins in tea in relation to accumulation of mRNA from genes involved in catechin biosynthesis. Plant Physiol Biotechnol 47:94–97

Fan Z, Li J, Li X, Wu B, Wang J, Liu Z, Yin H (2015a) Genome-wide transcriptome profiling provides insights into floral bud development of summer-flowering *Camellia azalea*. Sci Rep 5:9729

Fan K, Fan D, Ding Z, Su Y, Wang X (2015b) Cs-miR156 is involved in the nitrogen form regulation of catechins accumulation in tea plant (*Camellia sinensis* L.). Plant Physiol Bioch 97:350–360

Fan K, Wang M, Gao Y, Ning Q, Shi Y (2019a) Transcriptomic and ionomic analysis provides new insight into the beneficial effect of Al on tea roots' growth and nutrient uptake. Plant Cell Rep 38:715–729

Fan K, Zhang Q, Liu M, Ma L, Shi Y, Ruan J (2019b) Metabolomic and transcriptional analyses reveal the mechanism of C, N allocation from source leaf to flower intea plant (*Camellia sinensis* L). J Plant Physiol 232:200–208

Fang WP, Jiang CJ, Yu M, Ye AH, Wan ZX (2006) Differentially expression of Tua 1, a tubulin encoding gene, during flowering of tea plant *Camellia sinensis* (L.) O. Kuntze using cDNA amplified fragment length polymorphism technique. Acta Biochim Biophys Sin 38:653–662

Fang W, Zou Z, Hou X, Zhang D, Duan Y, Yang Y, Li X (2009) Cloning and sequence analysis of cold-induced H1-histone gene from *Camellia sinensis*. Acta Bot Boreali-Occidentalia Sin 8:1514–1519

Fang W, Zhang Y, Zhou L, Wang W, Li X (2013) Isolation and characterization of *Histone1* gene and its promoter from tea plant (*Camellia sinensis*). Mol Biol Rep 40:3641–3648

Feng YF, Liang YR (2001) Cloning and sequencing of S-adenosyl methionine synthetase gene in tea plant. J Tea Sci 21:21–25

Fu J (2012) Molecular cloning and expression analysis of a putative sesquiterpene synthase gene from tea plant (*Camellia sinensis*). Acta Physiol Plant 35:289–293

Fu YJ (2014) Cloning of a new glutathione peroxidase gene from tea plant (*Camellia sinensis*) and expression analysis under biotic and abiotic stresses. Bot Stud 55(1):71–76

Ganie SA, Mondal TK (2015) Genome-wide development of novel miRNA-based microsatellite markers of rice (*Oryza sativa*) for genotyping applications. Mol Breed 35:51–60

Ganie SA, Pani DR, Mondal TK (2017) Genome-wide analysis of DUF221 domain-containing gene family in *Oryza* species and identification of its salinity stress-responsive members in rice. PLoS One 12(8):e0182469

Gao Y, Zhao M, Wu XH, Li D, Borthakur D, Ye JH, Zheng XQ, Lu JL (2019) Analysis of differentially expressed genes in tissues of *Camellia sinensis* during dedifferentiation and root redifferentiation. Sci Rep 9:2935

Ge Y, Bian X, Sun B, Zhao M, Ma Y, Tang Y, Li N, Wu JL (2019) Dynamic profiling of phenolic acids during Pu-erh tea fermentation using derivatization liquid chromatography–mass spectrometry approach. J Agric Food Chem 67:4568–4577

Gohain B, Borchetia S, Bhorali P, Agarwal N, Bhuyan LP, Rahman A, Sakata K, Mizutani M, Shimizu B, Gurusubramaniam G, Ravindranath R, Kalita MC, Hazarika M, Das S (2012) Understanding Darjeeling tea flavour on a molecular basis. Plant Mol Biol 78:577–597

Guha PK, Mazumder A, Das A, Pani DR, Mondal TK (2019) *In silico* identification of long non-coding RNA based simple sequence repeat markers and their application in diversity analysis in rice. Gene Rep 16:100418

Gui J, Fu X, Zhou Y, Katsuno T, Mei X, Deng R, Yang Z (2015) Does enzymatic hydrolysis of glycosidically bound volatile compounds really contribute to the formation of volatile compounds during the oolong tea manufacturing process? J Agric Food Chem 63:6905–6914

Guo F, Guo Y, Wang P, Wang Y, Ni D (2017a) Transcriptional profiling of catechins biosynthesis genes during tea plant leaf development. Planta 246:1139–1152

Guo Y, Zhao S, Zhu C, Chang X, Yue C, Wang Z, Lin Y, Lai Z (2017b) Identification of drought-responsive miRNAs and physiological characterization of tea plant (*Camellia sinensis* L.) under drought stress. BMC Plant Biol 17:211

Guo J, Chen J, Yang J, Yu Y, Yang Y, Wang W (2018) Identification, characterization and expression analysis of the VQ motif containing gene family in tea plant (*Camellia sinensis*). BMC Genomics 19(710):1–12

Guo Y, Zhu C, Zhao S, Zhang S, Wang W, Fu H, Li X, Zhou C, Chen L, Lin Y, Lai Z (2019) *De novo* transcriptome and phytochemical analyses reveal differentially expressed genes and characteristic secondary metabolites in the original oolong tea (*Camellia sinensis*) cultivar 'Tieguanyin' compared with cultivar 'Benshan'. BMC Genomics 20:265

Gupta S, Bharalee R, Bhorali P, Bandyopadhyay T, Gohain B, Agarwal N, Ahmed P, Saikia H, Borchetia S, Kalita MC, Handique K, Das S (2012) Identification of drought tolerant progenies in tea by gene expression analysis. Funct Integr Genomics 12:543–563

Gupta S, Bharalee R, Bhorali P, Das SK, Bhagawati P, Bandyopadhyay T, Gohain B, Agarwal N, Ahmed P, Borchetia S, Kalita MC, Handique AK, Das S (2013) Molecular analysis of drought tolerance in tea by cDNA-AFLP based transcript profiling. Mol Biotechnol 53:237–248

Han Y, Huang Y, Liu Y, Jiao T et al (2017) Functional analysis of two Flavanone-3-hydroxylase genes from *Camellia sinensis*: a critical role in flavonoid accumulation. Gene 300:1–14

Hao X, Horvath DP, Chao WS, Yang Y, Wang X, Xiao B (2014) Identification and evaluation of reliable reference genes for quantitative real-Time PCR analysis in tea plant (*Camellia sinensis* (L.) O. Kuntze). Int J Mol Sci 15:22155–22172

Hao X, Yang Y, Yue C, Wang L, Horvath DP, Wang X (2017) Comprehensive transcriptome analyses reveal differential gene expression profiles of *Camellia sinensis* axillary buds at para-, endo-, ecodormancy, and bud flush stages. Front Plant Sci 8:553

Hao X, Tang H, Wang B, Yue C, Wang L, Zeng J, Yang Y, Wang X (2018) Integrative transcriptional and metabolic analyses provide insights into cold spell response mechanisms in young shoots of the tea plant. Tree Physiol 38:1655–1671

Hattan J, Shindo K, Ito T, Shibuya Y, Watanabe A, Tagaki C, Ohno F, Sasaki T, Ishii J, Kondo A, Misawa N (2016) Identification of a novel hedycaryol synthase gene isolated from *Camellia brevistyla* flowers and floral scent of Camellia cultivars. Planta 243:959–972

He X, Zhao X, Gao L, Shi X, Dai X, Liu Y, Xia T, Wang Y (2018) Isolation and characterization of key genes that promote flavonoid accumulation in purple-leaf tea (*Camellia sinensis* L.). Sci Rep 8:130. p 1–13

He Y, Song Q, Wu Y, Ye S, Chen S, Chen H (2020) TMT-Based quantitative proteomic analysis reveals the crucial biological pathways involved in self-incompatibility responses in *Camellia oleifera*. Int J Mol Sci 21:1987

Hu CJ, Li D, Ma YX, Zhang W, Lin C, Zheng XQ, Liang YR, Lu JL (2018) Formation mechanism of the oolong tea characteristic aroma during bruising and withering treatment. Food Chem 269:202–211

Huang H, Shi C, Liu Y, Mao SY, Gao LZ (2014) Thirteen *Camellia* chloroplast genome sequences determined by high-throughput sequencing: genome structure and phylogenetic relationships. BMC Evol Biol 14:151

Huang H, Yao Q, Xia E, Gao L (2018) Metabolomics and transcriptomics analyses reveal nitrogen influences on the accumulation of flavonoids and amino acids in young shoots of tea plant (*Camellia sinensis* L.) associated with tea flavor. J Agric Food Chem 66:9828–9838

Hucka M, Finney A, Sauro HM, Bolouri H, Doyle JC, Kitano H, Arkin AP, Bornstein BJ, Bray D, Cornish-Bowden A (2003) The Systems Biology Markup Language (SBML): a medium for the representation and exchange of biochemical network models. Bioinformation 19:524–531

Ishida M, Kitao N, Mizuno K, Tanikawa N, Kato M (2009) Occurrence of theobromine synthase genes in purine alkaloid-free species of *Camellia* plants. Planta 229:559–568

Jain P, Sharma V, Dubey H, Singh PK, Kapoor R, Kumari M, Singh J, Pawar DV, Bisht D, Solanke AU, Mondal TK, Sharma TR (2017) Identification of long non-coding RNA in rice lines resistant to Rice blast pathogen *Maganaporthe oryzae*. Bioinformation 13(8):249–255

Jayaswall K, Mahajan P, Singh G, Parmar R, Seth R, Raina A, Swarnkar MK, Singh AK, Shankar R, Sharma RK (2016) Transcriptome analysis reveals candidate genes involved in blister blight defense in tea (*Camellia sinensis* (L) Kuntze). Sci Rep 6:30412

Jeyaraj A, Zhang X, Hou Y, Shangguan M, Gajjeraman P, Li Y, Wei C (2017a) Genome-wide identification of conserved and novel microRNAs in one bud and two tender leaves of tea plant (*Camellia sinensis*) by small RNA sequencing, microarray-based hybridization and genome survey scaffold sequences. BMC Plant Biol 17:212

Jeyaraj A, Liu S, Zhang X, Zhang R, Shangguan M, Wei C (2017b) Genome-wide identification of microRNAs responsive to *Ectropis oblique* feeding in tea plant (*Camellia sinensis* L.). Sci Rep 7:13634

Jeyaraj A, Wang X, Wang S, Liu S, Zhang R, Wu A, Wei C (2019) Identification of regulatory networks of microRNAs and their targets in response to *Colletotrichum gloeosporioides* in tea plant (*Camellia sinensis* L.). Front Plant Sci 10:1096

Jiang C, Li Y, Fang W (2005) cDNA cloning and prokaryotic expression of β-glucosidase in tea plant [*Camellia sinensis* (L.) O. Kuntze]. Chinese J Agric Biotechnol 2:107–111

Jiang C, Wen Q, Chen Y, Xu LA, Huang MR (2011) Efficient extraction of RNA from various *Camellia* species rich in secondary metabolites for deep transcriptome sequencing and gene expression analysis. Afr J Biotechnol 10:16769–16773

Jiang X, Huang K, Zheng G, Hou H, Wang P, Jiang H, Zhao X, Li M, Zhang S, Liu Y, Gao L, Zhao L, Xia T (2018) CsMYB5a and CsMYB5e from *Camellia sinensis* differentially regulate anthocyanin and proanthocyanidin biosynthesis. Plant Sci 270:209–220

Jin X, Liu D, Ma L, Gong Z, Cao D, Liu Y, Li Y, Jiang C (2018) Transcriptome and expression profiling analysis of recalcitrant tea (*Camellia sinensis* L.) seeds sensitive to dehydration. Int J Genomics 2018:5963797

Jun CJ, Hua YL, Ping WF (2005) cDNA cloning and prokaryotic expression of β-glucosidase in tea plant [*Camellia sinensis* (L.) O. Kuntze]. Chinese J Agric Biotech 2:107–111

Kato M, Kanehara T, Shimizu H, Suzuki Gillies FM, Crozier A, Ashihara H (1996) Caffeine biosynthesis in young leaves of *Camellia sinensis*: *in vitro* studies on *N*-methyltransferase activity involved in the conversion of xanthosine to caffeine. Physiol Plant 98:629–636

Kato M, Mizuno K, Fujimura T, Iwama M, Irie M, Crozier A, Ashihara A (1999) Purification and characterization of caffeine synthase from tea leaves. Plant Physiol 120:579–586

Kato M, Mizuno K, Crozier A, Fujimura T, Ashihara A (2000) Caffeine synthase gene from tea leaves. Nature 406:956–957

Kirita M, Honma D, Tanaka Y, Usui S, Shoji T, Sami M, Yokota T, Tagashira M, Muranaka A, Uchiyama M, Kanda T, Maeda-Yamamoto M (2010) Cloning of a novel O-methyltransferase from *Camellia sinensis* and synthesis of o-methylated EGCG and evaluation of their bioactivity. J Agric Food Chem 58:7196–7201

Krishnaraj T, Gajjeraman P, Palanisamy S, Chandrabose SRS, Mandal AKA (2011) Identification of differentially expressed genes in dormant (banjhi) bud of tea (*Camellia sinensis* (L.) O. Kuntze) using subtractive hybridization approach. Plant Physiol Biochem 49:565–571

Ku KM, Choi JN, Kim J, Kim JK, Yoo LG, Lee SJ, Hong YS, Lee CH (2010) Metabolomics analysis reveals the compositional differences of shade grown tea (*Camellia sinensis* L.). J Agric Food Chem 58:418–426

Kumar A, Chawla V, Sharma E, Mahajan P, Shankar R, Yadav SK (2016) Comparative transcriptome analysis of Chinary, Assamica and Cambod tea (*Camellia sinensis*) types during development and seasonal variation using RNA-seq technology. Sci Rep 6:37244

Lee JE, Lee BJ, Chung JO, Hwang JA, Lee SJ, Lee CH, Hong YS (2010) Geographical and climatic dependencies of green tea (*Camellia sinensis*) metabolites: a ^{1}H NMR-based metabolomics study. J Agric Food Chem 58:10582–10589

Lee JE, Lee BJ, Hwang JA, Ko KS, Chung JO, Kim EH, Lee SJ, Hong YS (2011) Metabolic dependence of green tea on plucking positions revisited: a metabolomic study. J Agric Food Chem 59:10579–10585

Li YH, Jiang CJ, Yang SL, Yu YB (2004a) β-glucosidase cDNA cloning in the tea (*Camellia sinensis*) and its prokaryotic expression. J Agric Biotechnol 12:625–629

Li YH, Jiang CJ, Wan XC (2004b) Study on the expression of caffeine synthase gene mRNA in tea plant. J Tea Sci 24:23–28

Li Y, Gu W, Ye S (2007) Expression and location of caffeine synthase in tea plants. Russ J Plant Physiol 54:698–701

Li J, Chen J, Zhang Z, Pan Y (2008) Proteome analysis of tea pollen (*Camellia sinensis*) under different storage conditions. J Agric Food Chem 56:7535–7544

Li XW, Feng ZG, Yang HM, Zhu XP, Liu J, Yuan HY (2010) A novel cold-regulated gene from *Camellia sinensis*, *CsCOR1*, enhances salt- and dehydration-tolerance in tobacco. Biochem Biophy Res Co 394:354–359

Li Q, Huang J, Liu S, Li J, Yang X, Liu Y, Liu Z (2011) Proteomic analysis of young leaves at three developmental stages in an albino tea cultivar. Proteome Sci 9:44

Li XW, Liu HJ, Xie SX, Yuan HY (2013a) Isolation and characterization of two genes of the early light-induced proteins of *Camellia sinensis*. Photosynth Res 51:305–311

Li HC, Yuan XH, Chuan YUE, Lei CMA, Chao XW, Jun YY (2013b) Cloning and expression analysis of Betaine aldehyde dehydragenase gene (*CsBADH1*) from tea plant [*Camellia sinensis* (L.) O. Kuntze]. J Tea Sci 33:99–108

Li HC, Chuan Y, Hua YZ, Lu W, Yuan XH, Jun YY, Chao XW (2014) Molecular cloning and expression of a bZIP transcription factor gene *CsbZIP1* in tea plant (*Camellia sinensis*). Acta Agronomica Sinica 40:17021709

Li CF, Zhu Y, Yu Y, Zhao QY, Wang SJ, Wang XC, Yao MZ, Luo D, Li X, Chen L, Yang YJ (2015a) Global transcriptome and gene regulation network for secondary metabolite biosynthesis of tea plant (*Camellia sinensis*). BMC Genomics 16:560

Li MM, Xue JY, Wen YL, Guo HS, Sun XQ, Zhang YM, Hang YY (2015b) Transcriptomic analysis of *Camellia ptilophylla* and identification of genes associated with flavonoid and caffeine biosynthesis. Genet Mol Res 14:18731–18742

Li Q, Li J, Liu S, Huang J, Lin H, Wang K, Liu Z (2015c) A comparative proteomic analysis of the buds and the young expanding leaves of the tea plant (*Camellia sinensis* L.). Int J Mol Sci 16:14007–14038

Li Q, Lei S, Du K, Li L, Pang X, Wang Z, Wei M, Fu S, Hu L, Xu L (2016a) RNA-seq based transcriptomic analysis uncovers α-linolenic acid and jasmonic acid biosynthesis pathways respond to cold acclimation in *Camellia* japonica. Sci Rep 6:36463

Li CF, Xu YX, Ma JQ, Jin JQ, Huang DJ, Yao MZ, Ma CL, Chen L (2016b) Biochemical and transcriptomic analyses reveal different metabolite biosynthesis profiles among three color and developmental stages in 'Anji Baicha' (*Camellia sinensis*). BMC Plant Biol 16:195

Li H, Huang W, Wang GL, Wu ZJ, Zhuang J (2016c) Expression profile analysis of ascorbic acid-related genes in response to temperature stress in the tea plant, *Camellia sinensis* (L.) O. Kuntze. Genet Mol Res 15:1–10

Li H, Huang W, Liu ZW, Wang YX, Zhuang J (2016d) Transcriptome-based analysis of Dof family transcription factors and their responses to abiotic stress in tea plant (*Camellia sinensis*). Int J Genomics 2016:5614142

Li Y, Huang J, Song X, Zhang Z, Jiang Y, Zhu Y, Zhao H, Ni D (2017a) An RNA-Seq transcriptome analysis revealing novel insights into aluminum tolerance and accumulation in tea plant. Planta 246:91–103

Li W, Xiang F, Zhong M et al (2017b) Transcriptome and metabolite analysis identifies nitrogen utilization genes in tea plant (*Camellia sinensis*). Sci Rep 7:1693

Li QS, Lin XM, Qiao RY, Zheng XQ, Lu JL, Ye JH, Liang YR (2017c) Effect of fluoride treatment on gene expression in tea plant (*Camellia sinensis*). Sci Rep 7:9847

Li H, Huang W, Wang LG, Wang LW (2017d) Transcriptomic analysis of the biosynthesis, recycling, and distribution of ascorbic acid during leaf development in tea plant (*Camellia sinensis* (L.) O. Kuntze). Sci Rep 7:46212

Li X, Li J, Fan Z, Liu Z, Tanaka T, Yin H (2017e) Global gene expression defines faded whorl specification of double flower domestication in *Camellia*. Sci Rep 7:3197

Li NN, Qian WJ, Wang L, Cao HL, Hao XY, Hao XY, Hao XY, Yang YJ, Wang XC (2017f) Isolation and expression features of hexose kinase genes under various abiotic stresses in the tea plant (*Camellia sinensis*). J Plant Physiol 209:95–104

Li M, Li Y, Guo L, Gong N et al (2017g) Functional characterization of tea (*Camellia sinensis*) MYB4a transcription factor using an integrative approach. Front Plant Sci 12:1–17

Li NN, Yue C, Cao HL, Qian WJ, Hao XY, Wang YC, Wang L, Ding CQ, Wang XC, Yang YJ (2018a) Transcriptome sequencing dissection of the mechanisms underlying differential cold sensitivity in young and mature leaves of the tea plant (*Camellia sinensis*). J Plant Physiol 225:144–155

Li QS, Li XM, Qiao RY, Shen EH, Lin XM, Lu JL, Ye JH, Liang YR, Zheng XQ (2018b) *De novo* transcriptome assembly of fluorine accumulator tea plant *Camellia sinensis* with fluoride treatments. Sci Data 5:180194

Li X, Fan Z, Guo H, Ye N, Lyu T, Yang W, Wang J, Wang JT, Wu B, Li J, Yin H (2018c) Comparative genomics analysis reveals gene family expansion and changes of expression patterns associated with natural adaptations of flowering time and secondary metabolism in yellow *Camellia*. Funct Integr Genomics 18:659–671

Li Y, Wang X, Ban Q, Zhu X, Jiang C, Wei C, Bennetzen JL (2019a) Comparative transcriptomic analysis reveals gene expression associated with cold adaptation in the tea plant *Camellia sinensis*. BMC Genomics 20:624

Li J, Zhou Y, Zhou B, Tang H, Chen Y, Qiao X, Tang J (2019b) Habitat management as a safe and effective approach for improving yield and quality of tea (*Camellia sinensis*) leaves. Sci Rep 9:433

Li FD, Tong W, Xia EH, Wei CL (2019c) Optimized sequencing depth and *de novo* assembler for deeply reconstructing the transcriptome of the tea plant, an economically important plant species. BMC Bioinformatics 20:553

Li WX, Zhu LY, Chen YC et al (2020) Cloning and characterization of two chlorophyll A/B binding protein genes and analysis of their gene family in *Camellia sinensis*. Sci Rep 10 (4602):1–9

Lin GZ, Lian YJ, Ryu JH, Sung MK, Park JS, Park HJ, Park BK, Shin JS, Lee MS, Cheon CL (2007) Expression and purification of His-tagged flavonol synthase of *Camellia sinensis* from *Escherichia coli*. Protein Exp Purif 55:287–292

Lin P, Wang K, Zhou C, Xie Y, Yao X, Yin H (2018) Seed transcriptomics analysis in *Camellia oleifera* uncovers genes associated with oil content and fatty acid composition. Int J Mol Sci 19:118

Liu S, Han B (2010) Differential expression pattern of an acidic 9/13-lipoxygenase in flower opening and senescence and in leaf response to phloem feeders in the tea plant. BMC Plant Biol 10:228–238

Liu Y, Han Y (2018) The complete chloroplast genome sequence of endangered camellias (*Camellia pubifurfuracea*). Conserv Genet Resour 10:843–845

Liu Q, Tan X-F, Hu X-Y, Tian X-M (2010) Cloning and sequence analysis of a 14-3-3 protein gene from *Camellia oleifera*. Acta Agric Univ Jiang 5:13–17

Liu SC, Xu YX, Ma JQ, Wang WW, Chen W, Huang DJ, Fang J, Li XJ, Chen L (2016a) Small RNA and degradome profiling reveals important roles for microRNAs and their targets in tea plant response to drought stress. Physiol Plant 158:435–451

Liu ZW, Wu ZJ, Li XH, Huang Y, Li H, Li H, Wang YX, Zhuang J (2016b) Identification, classification, and expression profiles of heat shock transcription factors in tea plant (Camellia sinensis) under temperature stress. Gene 576:52–59

Liu Y et al (2016c) Screening and validation of reference genes for quantitative real-time PCR analysis in tea plant (*Camellia sinensis*) under different nitrogen nutrition. J Tea Sci 36:92–101

Liu ZW, Wu ZJ, Li H, Wang YX, Zhuang J (2017a) L-theanine content and related gene expression: novel insights into theanine biosynthesis and hydrolysis among different tea plant (*Camellia sinensis* L.) tissues and cultivars. Front Plant Sci 8:498

Liu GF, Han ZX, Feng L, Gao LP, Gao MJ, Gruber MY, Zhang ZL, Xia T, Wan XC, Wei S (2017b) Metabolic flux redirection and transcriptomic reprogramming in the albino tea cultivar 'Yu-Jin-Xiang' with an emphasis on catechin production. Sci Rep 7:45062

Liu F, Wang Y, Ding Z, Zhao L, Xiao J, Wang L, Ding S (2017c) Transcriptomic analysis of flower development in tea (*Camellia sinensis* (L.)). Gene 631:39–51

Liu S, Gao J, Chen Z, Qiao X, Huang H, Cui B, Zhu Q, Dai Z, Wu H, Pan Y, Yang C, Liu J (2017d) Comparative proteomics reveals the physiological differences between winter tender shoots and spring tender shoots of a novel tea (*Camellia sinensis* L.) cultivar evergrowing in winter. BMC Plant Biol 17:206

Liu L, Li Y, She G, Zhang X, Jordan B, Chen Q, Zhao J, Wan X (2018a) Metabolite profiling and transcriptomic analyses reveal an essential role of UVR8-mediated signal transduction pathway in regulating flavonoid biosynthesis in tea plants (*Camellia sinensis*) in response to shading. BMC Plant Biol 18:233

Liu Y, Cao D, Ma L, Jin X, Yang P, Ye F, Wei C (2018b) TMT-based quantitative proteomics analysis reveals the response of tea plant (*Camellia sinensis*) to fluoride. J Proteome 176:71–81

Liu GF, Liu JJ, He ZR, Wang FM, Yang H, Yan YF, Gao MJ, Gruber MY, Wan XC, Wei S (2018c) Implementation of CsLIS/NES in linalool biosynthesis involves transcript splicing regulation in *Camellia sinensis*. Plant Cell Environ 41(1):176–186

Liu MY, Tang D, Shi Y, Ma L, Li Y, Zhang Q, Ruan J (2019a) Short-term inhibition of glutamine synthetase leads to reprogramming of amino acid and lipid metabolism in roots and leaves of tea plant (*Camellia sinensis* L.). BMC Plant Biol 19:425

Liu S, Mi X, Zhang R, An Y, Zhou Q, Yang T, Xia X, Guo R, Wang X, Wei C (2019b) Integrated analysis of miRNAs and their targets reveals that miR319c/TCP2 regulates apical bud burst in tea plant (*Camellia sinensis*). Planta 250:1111–1129

Liu H, Wang YX, Li H, Teng RM, Wang Y, Zhuang J (2019c) Genome-wide identification and expression analysis of calcineurin B-like protein and calcineurin B-like protein-interacting protein kinase family genes in tea plant. DNA Cell Biol 38:824–839

Ma C, Qiao X, Chen L (2010) Cloning and expression analysis of leucoanthocyanidin reducase gene of tea plant (*Camellia sinensis*). J Tea Sci 30:27. CNKI:SUN:CYKK.0.2010-01-006

Ma CL, Chen L, Wang XC, Jin JQ, Ma JQ, Yao MZ, Wang ZL (2012) Differential expression analysis of different albescent stages of 'Anji Baicha' (*Camellia sinensis* (L.) O. Kuntze) using cDNA microarray. Sci Hortic 148:246–254

Ma QP, Hao S, Chen X, Li HX (2016) Validation of reliability for reference genes under various abiotic stresses in tea plant. Russ J Plant Physiol 63:423–432

Ma Q, Chen C, Zeng Z, Zou Z, Li H, Zhou Q, Chen X, Sun K, Li X (2018) Transcriptomic analysis between self- and cross-pollinated pistils of tea plants (*Camellia sinensis*). BMC Genomics 19:289

Marshall A, Gollapudi S, de Silva JD, Hodgman C (2007) A systems biology approach to modelling tea (*Camellia sinensis*). BMC Syst Biol 1:13

Matsumoto S, Takeuchi A, Hayatsu M (1994a) Chalcone synthase from *Camellia sinensis*: isolation of the cDNAs and the organ-specific and sugar-responsive expression of the genes. Plant Cell Physiol 35:1011–1018

Matsumoto S, Takeuchi A, Hayatsu M (1994b) Molecular cloning of phenylalanine ammonia-lyase cDNA and classification of varieties and cultivars of tea plants (*Camellia sinensis*) using the tea PAL cDNA probe. Theor Appl Genet 89:671–675

Mingle W, Xujun Z, Weidong W, Xuanqing W, Minglu L, Xinghui L (2015) Molecular cloning and expression analysis of low molecular weight heat shock protein gene *CsHSP17.2* from *Camellia sinensis*. J Nanjing Agric Univ 38:389–394

Mizutani M, Nakanishi H, Ema J, Ma SJ, Noguchi E, Inohara-Ochiai M, Fukuchi-Mizutani M, Nakao M, Sakata K (2002) Cloning of β-primeverosidase from tea leaves, a key enzyme in tea aroma formation. Plant Physiol 130:2164–2176

Mohanpuria P, Yadav SK (2012) Characterization of novel small RNAs from tea (*Camellia sinensis* L.). Mol Biol Rep 39:3977–3986

Mohanpuria P, Kumar V, Joshi R, Gulati A, Ahuja PS, Yadav SK (2009) Caffeine biosynthesis and degradation in tea (*Camellia sinensis* (L.) O. Kuntze) is under developmental and seasonal regulation. Mol Biotechnol 43:104–111

Mondal TK (2001) Camellia biotechnology: a bibliographic search. Int Soc Tea Sci 1:28–37

Mondal TK (2004) Biotechnological improvements of tea. ISB News Lett:24

Mondal TK (2013) Abiotic stress and biotechnology. Regency Publications, New Delhi, pp 57–74

Mondal TK, Rana MK (2013) High throughput DNA sequencing and its implication in plant science research. Everyman Sci 1:15–25

Mondal TK, Sutoh K (2013) Application of next-generation sequencing for abiotic stress tolerance. In: Barh D, Zambare V, Azevedo V (Eds). The world of omics and applications in plant and agricultural sciences CRC Press, Taylor and Francis Group, Boca Raton, FL. pp 383-402.

Mondal TK, Bhattacharya A, Laxmikumaran M, Ahuja PS (2004) Recent advance in tea biotechnology. Plant Cell Tissue Org Cult 75:795–856

Mondal TK, Ganie SA, Neraj RR, Rana MK (2014) Cloning and in silico analysis of a gene encoding a putative *β*-carbonic anhydrase from cowpea (*Vigna unguiculata* L. Walp). J Plant Interact 9(1):504–513

Mondal TK, Panda AK, Rawal HC, Sharma TR (2018) Discovery of microRNA-target modules of African rice (*Oryza glaberrima*) under salinity stress. Sci Rep 8(1):570

Mondal TK, Rawal HC, Bera B, Kumar PM, Choubey M, Saha G, Das B, Bandyopadhyay T, Ilango V, Sharma TR, Barua A, Radhakrishnan B, Singh NK (2019) Draft genome sequence of a popular Indian tea genotype TV-1 [*Camellia assamica* L. (O). Kunze]. BioRxiv:762161

Mueller LA, Zhang P, Rhee SY (2003) AraCyc: a biochemical pathway database for *Arabidopsis*. Plant Physiol 132:453–460

Mukhopadhyay M, Sarkar B, Mondal TK (2013) Omics advances in tea (*Camellia sinensis*). In: Bhar D (ed) Omics applications in crop science. CRC Press, Taylor and Francis Group, Boca Raton, FL, pp 347–366. ISBN:978-1-4665-8582

Mukhopadhyay M, Mondal TK, Chand PK (2016) Biotechnological advances in tea (*Camellia sinensis* [L.] O. Kuntze): a review. Plant Cell Rep 35(2):255–287

Muoki RC, Paul A, Kumar S (2012) A shared response of *thaumatin like protein*, *chitinase*, and *late embryogenesis abundant protein3* to environmental stresses in tea [*Camellia sinensis (L.) O. Kuntze*]. Funct Integr Genomics 12:565–571

Murayama Y, Uematsu K, Koyama H, Morita A (2007) Detection of aluminium responsive genes in tea plant by hybridization with *Arabidopsis thaliana* cDNA. In: The 3rd Int Conf on O-Cha tea culture and science (ICOS) Shizuoka, Japan 29th Feb, 2008

Ohgami S, Ono E, Toyonaga H, Watanabe N, Ohnishi T (2014) Identification and characterization of *Camellia sinensis* glucosyltransferase, UGT73A17: a possible role in flavonolglucosylation. Plant Biotechnol J 31:573–578

Pan LL, Wang Y, Hu JH, Ding ZT, Li C (2013) Analysis of codon use features of stearoyl-acyl carrier protein desaturase gene in *Camellia sinensis*. J Theor Biol 334:80–86

Pan J, Wang W, Li D, Shu Z, Ye X, Chang P, Wang Y (2016) Gene expression profile indicates involvement of NO in *Camellia sinensis* pollen tube growth at low temperature. BMC Genomics 17:809

Pan C, Tian K, Ban Q, Wang L, Sun Q, He Y, Yang Y, Pan Y, Li Y, Jiang J, Jiang C (2017) Genome-wide analysis of the biosynthesis and deactivation of gibberellin-dioxygenases gene family in *Camellia sinensis* (L.) O. Kuntze. Gene 8:235. p 1–20

Pang Y, Abeysinghe ISB, He J, He X, Huhman D, Mewan KM, Sumner LW, Yun J, Dixon RA (2013) Functional characterization of proanthocyanidin pathway enzymes from tea (*Camellia sinensis*) and their application for metabolic engineering. Plant Physiol 161:1103–1116

Park JS, Kim JB, Hahn BS, Kim KH, Ha SH, Kim JB, Kim YH (2004) EST analysis of genes involved in secondary metabolism in *Camellia sinensis* (tea), using suppression subtractive hybridization. Plant Sci 166:953–961

Parmar R, Seth R, Singh P, Singh G, Kumar S, Sharma RK (2019) Transcriptional profiling of contrasting genotypes revealed key candidates and nucleotide variations for drought dissection in *Camellia sinensis* (L.) O. Kuntze. Sci Rep 9:7487

Paul A, Kumar S (2011) Responses to winter dormancy, temperature, and plant hormones share gene networks. Funct Integr Genomics 11:659–664

Paul A, Kumar S (2013) Dehydrin2 is a stress-inducible, whereas Dehydrin1 is constitutively expressed but up-regulated gene under varied cues in tea [*Camellia sinensis* (L.) O. Kuntze]. Mol Biol Rep 40:3859–3863

Paul A, Lal L, Ahuja PS, Kumar S (2012a) Alpha-tubulin (CsTUAa) up-regulated during winter dormancys a low temperature inducible gene in tea [*Camellia sinensis* (L.) O. Kuntze]. Mol Biol Rep 39:3485–3490

Paul A, Muoki RC, Singh K, Kumar S (2012b) CsNAM-like protein encodes a nuclear localized protein and responds to varied cues in tea [*Camellia sinensis (L.) O. Kuntze*]. Gene 502:69–74

Paul A, Jha A, Bhardwaj S, Singh S, Shankar R, Kumar S (2014) RNA-seq-mediated transcriptome analysis of actively growing and winter dormant shoots identifies non-deciduous habit of evergreen tree tea during winters. Sci Rep 4:5932

Phukon M, Namdev R, Deka D, Modi SK, Sen P (2012) Construction of cDNA library and preliminary analysis of expressed sequence tags from tea plant [*Camellia sinensis* (L) O. Kuntze]. Gene 506:202–206

Ping SUN, Zhang ZL, Zhu QF, Zhang GY, Xiang P, Lin YL, Lai ZX, Lin JK (2018) Identification of miRNAs and target genes regulating catechin biosynthesis in tea (*Camellia sinensis*). J Integr 17:1154–1164

Prabu GR, Mandal AKA (2010) Computational identification of miRNAs and their target genes from expressed sequence tags of tea (*Camellia sinensis*). Genomics Proteomics Bioinf 8:113–121

Prabu GR, Thirugnanasambantham K, Mandal AKA, Saravanan A (2012) Molecular cloning and characterization of nucleoside *diphosphate kinase* 1 cDNA in tea. Biol Plant 56:140–144

Punyasiri PA, Abeysinghe IS, Kumar V, Treutter D, Duy D, Gosch C, Martens S, Forkmann G, Fischer TC (2004) Flavonoid biosynthesis in the tea plant *Camellia sinensis*: properties of enzymes of the prominent epicatechin and catechin pathways. Arch Biochem Biophys 431:22–30

Qian WJ, Yue C, Wang YC, Cao HL, Li NN, Wang L, Hao XY, Wang XC, Xiao B, Yang YJ (2016) Identification of the invertase gene family (INVs) in tea plant and their expression analysis under abiotic stress. Plant Cell Rep 35:2269–2283

Qian Y, Zhang S, Yao S, Xia J, Li Y, Dai X, Wang W, Jiang X, Liu Y, Li M, Gao L, Xia T (2018) Effects of *in vitro* sucrose on quality components of tea plants (*Camellia sinensis*) based on transcriptomic and metabolic analysis. BMC Plant Biol 18:121

Qiao J, Zhang Y, Zhu X (2011) Identification of genes up-regulated by *Ectropicobliqua* feeding in tea seedlings. Acta Hort Sinica 38:783–789

Qiao D, Yang C, Chen J, Guo Y, Li Y, Niu S, Cao K, Chen Z (2019) Comprehensive identification of the full-length transcripts and alternative splicing related to the secondary metabolism pathways in the tea plant (*Camellia sinensis*). Sci Rep 9:2709

Rana NK, Mohanpuria P, Yadav SK (2008a) Cloning and characterization of a cytosolic glutamine synthetase from *Camellia sinensis* (L.) O. Kuntze that is upregulated by ABA, SA, and H_2O_2. Mol Biotechnol 39:49–56

Rana NK, Mohanpuria P, Yadav SK (2008b) Expression of tea cytosolic glutamine synthetase is tissue specific and induced by cadmium and salt stress. Biol Plant 52:361–364

Rana NK, Mohanpuria P, Kumar V, Yadav SK (2010) A CsGS is regulated at transcriptional level during developmental stages and nitrogen utilization in *Camellia sinensis* (L.) O. Kuntze. Mol Biol Rep 37:703–710

Rani A, Singh K, Sood P, Kumar S, Ahuja PS (2009) p-Coumarate: CoA ligase as a key gene in the yield of catechins in tea [*Camellia sinensis* (L.) O. Kuntze]. Funct Integr Genomics 9:271–275

Rani A, Singh K, Ahuja PS, Kumar S (2012) Molecular regulation of catechins biosynthesis in tea [*Camellia sinensis* (L.) O. Kuntze]. Gene 495:205–210

Rawal HC, Kumar PM, Bera B, Singh NK, Mondal TK (2020) Decoding and analysis of organelle genomes of Indian tea (*Camellia assamica*) for phylogenetic confirmation. Genomics 112:659–668

Rothenberg DO, Yang H, Chen M, Zhang W, Zhang L (2019) Metabolome and transcriptome sequencing analysis reveals anthocyanin metabolism in pink flowers of anthocyanin-rich tea (*Camellia sinensis*). Molecules 24:E1064

Roy SC, Chakraborty BN (2009) Cloning and sequencing of chitinase gene specific PCR amplified DNA fragment from tea plant (*Camellia sinensis*) and analyzed the nucleotide sequence using bioinformatics algorithms. Can J Pure Appl Sci 3:795–800

Sahu J, Sarmah R, Dehury B, Sarma K, Sahoo S, Sahu M, Barooah M, Modi MK, Sen P (2012) Mining for SSRs and FDMs from expressed sequence tags of *Camellia sinensis*. J Bioinformatics 8:260–266

Samynathan R, Palanisamy PC, Gandhi S, Mandal AA et al (2015) Isolation and characterization of Polyphenol oxidase from UPASI selected clone of *Camellia sinensis* (L.) O. Kuntze. Indo Am J Pharm Res 5:241–252

Senthilkumar P, Thirugnanasambantham K, Mandal AKA (2012) Suppressive subtractive hybridization approach revealed differential expression of hypersensitive response and reactive oxygen species production genes in tea (*Camellia sinensis (L.) O. Kuntze*) leaves during *Pestalotiopsisthea* Infection. Appl Biochem Biotechnol 168:1917–1927

Seth R, Bhandawat A, Parmar R, Singh P, Kumar S, Sharma RK (2019) Global transcriptional insights of pollen-pistil interactions commencing self-incompatibility and fertilization in tea [*Camellia sinensis* (L.) O. Kuntze]. Int J Mol Sci 20:539

Sharma P, Kumar S (2005) Differential display-mediated identification of three drought-responsive expressed sequence tags in tea [*Camellia sinensis* (L.) O. Kuntze]. J Biosci 30:231–235

Shen J, Zhang D, Zhou L, Zhang X, Liao J, Duan Y, Wen B, Ma Y, Wang Y, Fang W, Zhu X (2019a) Transcriptomic and metabolomic profiling of *Camellia sinensis* L. cv. 'Suchazao' exposed to temperature stresses reveals modification in protein synthesis and photosynthetic and anthocyanin biosynthetic pathways. Tree Physiol 39:1583–1599

Shen W, Li H, Teng R, Wang Y, Wang W, Zhuang J (2019b) Genomic and transcriptomic analyses of HD-Zip family transcription factors and their responses to abiotic stress in tea plant (*Camellia sinensis*). Genomics 111:1142–1151

Shen J, Zou Z, Xing H, Duan Y, Zhu X, Ma Y, Wang Y, Fang W (2020) Genome-wide analysis reveals stress and hormone responsive patterns of JAZ Family genes in *Camellia sinensis*. Int J Mol Sci 21(7):E2433

Shi CY, Yang H, Wei CL, Yu O, Zhang ZZ, Jiang CJ, Sun J, Li YY, Chen Q, Xia T, Wan XC (2011) Deep sequencing of the *Camellia sinensis* transcriptome revealed candidate genes for major metabolic pathways of tea-specific compounds. BMC Genomics 12:131–150

Shi H, Wang Y, Yang L-C, Ding Z-T (2012) Analysis of codon bias of the cold regulated transcription factor *ICE1* in tea plant. Acta Hort Sin 39:1341–1352

Shi J, Ma C, Qi D, Lv H, Yang T, Peng Q, Chen Z, Lin Z (2015) Transcriptional responses and flavor volatiles biosynthesis in methyl jasmonate-treated tealeaves. BMC Plant Biol 15:233

Shi YL, Sheng YY, Cai ZY, Yang R, Li QS, Li XM, Li D, Guo XY, Lu JL, Ye JH, Wang KR, Zhang LJ, Liang YR, Zheng XQ (2019) involvement of salicylic acid in anthracnose infection in tea plants revealed by transcriptome profiling. Int J Mol Sci 20:E2439

Singh K, Raizada J, Bhardwaj P, Ghawana S, Rani A, Singh H, Kaul K, Kumar S (2004) 26S rRNA-based internal control gene primer pair for reverse transcription-polymerase chain reaction-based quantitative expression studies in diverse plant species. Anal Biochem 335:330–333

Singh K, Rani A, Kumar S, Sood P, Mahajan M, Yadav SK, Singh B, Ahuja PS (2008) An early gene of the flavonoid pathway, flavanone 3-hydroxylase, exhibits a positive relationship with the concentration of catechins in tea (*Camellia sinensis*). Tree Physiol 28:1349–1356

Singh K, Kumar S, Yadav SK, Ahuja PS (2009a) Characterization of dihydroflavonol 4-reductase cDNA in tea [*Camellia sinensis* (L.) O. Kuntze]. Plant Biotechnol Rep 3:95–101

Singh K, Rani A, Paul A, Dutt S, Joshi R, Gulati A, Ahuja PS, Kumar S (2009b) Differential display mediated cloning of anthocyanidin reductase gene from tea (*Camellia sinensis*) and its relationship with the concentration of epicatechins. Tree Physiol 29:837–846

Singh K, Kumar S, Rani A, Gulati A, Ahuja PS (2009c) Phenylalanine ammonia-lyase (PAL) and cinnamate 4-hydroxylase (C4H) and catechins (flavan-3-ols) accumulation in tea. Funct Integr Genomics 9:125–134

Singh K, Kumar S, Ahuja PS (2009d) Differential expression of Histone H3 gene in tea (*Camellia sinensis* (L.) O. Kuntze) suggests its role in growing tissue. Mol Biol Rep 36:537–542

Singh K, Paul A, Kumar S, Ahuja PS (2009e) Cloning and differential expression of *QM* like protein homologue from tea [*Camellia sinensis* (L.) O. Kuntze]. Mol Biol Rep 36:921–927

Singh G, Singh V, Singh V (2019) Construction and analysis of an interologous protein-protein interaction network of *Camellia sinensis* leaf (TeaLIPIN) from RNA-Seq data sets. Plant Cell Rep 38:1249–1262

Song L, Ma Q, Zou Z, Sun K, Yao Y, Tao J, Kaleri NA, Li X (2017) Molecular link between leaf coloration and gene expression of flavonoid and carotenoid biosynthesis in *Camellia sinensis* cultivar 'Huangjinya'. Front Plant Sci 8:803

Su Xia X, Yuan Z, Kai Lin G, Lei W, Hong-Yu Y (2009) Cloning and expression analysis of a flavonol synthase gene from *Camellia sinensis*. Plant Physiol Commun 45:1093–1097

Sugiyama T, Sadzuka Y (2003) Theanine and glutamate transporter inhibitors enhance the antitumor efficacy of chemotherapeutic agents. Biochem Biophys Acta 1653:47–59

Sugiyama T, Sadzuka Y (2004) Theanine, a specific glutamate derivative in green tea, reduces the adverse reactions of doxorubicin by changing the glutathione level. Cancer Lett 212:177–184

Sun M et al (2010) Reference genes for real-time fluorescence quantitative PCR in *Camellia sinensis*. Chin Bull Bot 45:579–587

Sun P, Cheng C, Lin Y, Zhu Q, Lin J, Lai Z (2017) Combined small RNA and degradome sequencing reveals complex microRNA regulation of catechin biosynthesis in tea (*Camellia sinensis*). PLoS One 12:0171173

Sun M, Zhang C, Lu M, Gan N, Chen Z, Deng WW, Zhang ZZ (2018) Metabolic flux enhancement and transcriptomic analysis displayed the changes of catechins following long-term pruning in tea trees (*Camellia sinensis*). J Agric Food Chem 66:8566–8573

Suzuki T (1972) The participation of *S*-adenosylmethionine in the biosynthesis of caffeine in the tea plants. FEBS Lett 24:18–20

Tai Y, Wei C, Yang H, Zhang L, Chen Q, Deng W, Wei S, Zhang J, Fang C, Ho C, Wan X (2015) Transcriptomic and phytochemical analysis of the biosynthesis of characteristic constituents in tea (*Camellia sinensis*) compared with oil tea (*Camellia oleifera*). BMC Plant Biol 15:190

Tai Y, Liu C, Yu S, Yang H, Sun J, Guo C, Huang B, Liu Z, Yuan Y, Xia E, Wei C, Wan X (2018) Gene co-expression network analysis reveals coordinated regulation of three characteristic secondary biosynthetic pathways in tea plant (*Camellia sinensis*). BMC Genomics 19:616

Takechi A, Matsumoto S (2003) Effect of light on gene expression of catechin biosynthetic expression in tea (*Camellia sinensis*) leaves. Tea Res J 96:27–32

Takeo T (1981) Production of linalool and geraniol by hydrolytic breakdown of bound forms in disrupted tea shoots. Phytochemistry 120:2145–2147

Takeuchi A, Matsumoto S, Hayatsu M (1994a) Chalcone synthase from *Camellia sinensis* isolation of the cDNAs and the organ-specific and sugar-responsive expression of the genes. Plant Cell Physiol 35:1011–1018

Takeuchi A, Matsumoto S, Hayatsu M (1994b) Amplification of β-tubulin cDNA from *Camellia sinensis* by PCR. Bull Natl Res Inst Veg Ornam Plants Tea 7:13–20

Tan X, Chen H, Zhang D, Zeng Y, Li W et al (2008) Cloning of full-length cDNA of *FAD2* gene from *Camellia oleifera*. Scientia Silvae Sinicae 44:70–75

Tang L, Jiang ZZ, Zhang RJ, Zhang RJ, Wei LC (2013) Cloning and functional analysis of a NAC-like gene *Csnac* in tea plant (*Camellia sinensis*). J Tea 39:279–286

Taniguchi F, Tanaka J (2004) Characterization of genes for ammonium assimilation in *Camellia sinensis*. In: Proc of 2004 Int Con on O-Cha culture and Sci. Shizuaka, Japan, pp 217–218

Taniguchi F, Fukuoka H, Tanaka J (2012) Expressed sequence tags from organ-specific cDNA libraries of tea (*Camellia sinensis*) and polymorphisms and transferability of EST-SSRs across *Camellia* species. Breed Sci 62:186–195

Tateishi N, Ozaki Y, Okubo H (2010) Molecular cloning of the genes involved in anthocyanin biosynthesis in *Camellia japonica*. J Facul Agric Kyushu Univ 55:21–28

Teng MR, Wang XY, Wang LW, Li H, Shen W, Zhuang J (2018) Genome-wide identification, classification and expression pattern of LBD gene family in *Camellia sinensis*. Biotechnol Biotechnol Equip 32(6):138797

Thirugnanasambantham K, Prabu G, Palanisamy S, Subhas Chandrabose SR, Mandal AKA (2013) Analysis of dormant bud (Banjhi) specific transcriptome of tea (*Camellia sinensis (L.) O. Kuntze*) from cDNA library revealed dormancy-related genes. Appl Biochem Biotechnol 169:1405–1417

Tomimoto Y, Ikehashi I, Kakeda K, Kowyama Y (1999) A pistil specific PR-1 like protein in Camellia, its expression, sequence and genealogical position. Breed Sci 49:97–104

Tong W, Yu J, Hou Y et al (2018) Circular RNA architecture and differentiation during leaf bud to young leaf development in tea (*Camellia sinensis*). Planta 248:1417–1429

Varshney D, Rawal HC, Dubey H, Bandyopadhyay T, Bera B, Mohan Kumar P, Singh NK, Mondal TK (2019) Tissue specific long non-coding RNA landscape and their association with aroma formation of tea. Ind Crop Prod 133:79–89

Venkatesh P, Jaiprakash M, Prasad P, Pillai B, Sadhale PP, Sinkar VP (2006) Flavonoid biosynthesis in tea (*Camellia sinensis*). p 11–17. http://www.teascience.org/pdf/chapter-2-6-2.pdf

Wan S, Zhang Y, Duan M, Huang L, Wang W, Xu Q, Yu Y (2020) Integrated analysis of long non-coding RNAs (lncRNAs) and mRNAs reveals the regulatory role of lncRNAs associated with salt resistance in *Camellia sinensis*. Front Plant Sci 11:218

Wang CX, Li YY, Jiang CJ, Yu YB (2005) Molecular cloning and sequence analysis on cDNA of cystatin gene from tea leaves. J Tea Sci 25:177–182

Wang L, Li X, Zhao Q, Jing S, Chen S, Yuan H (2009) Identification of genes induced in response to low-temperature treatment in tea leaves. Plant Mol Biol Report 27:257–265

Wang XC, Yang YJ, Liang C, Chun-Lei MA, Yao MZ (2010) Construction and preliminary analysis of the suppression subtractive hybridization cDNA libraries between dormant and sprouting buds of tea plant (*Camelliasinensis*). J Tea Sci 30:129–135

Wang P, Su R, Zheng J, Cheng S, Zhu G (2011a) Molecular cloning and phylogenetic analysis of a novel BURP domain-containing gene from *Camellia sinensis*. Afr J Biotechnol 10:15470–15476

Wang X, Ma C, Yang C, Yao M, Jin J (2011b) Cloning and expression analysis of auxin-repressed protein gene CsARP1 in tea plant (*Camellia sinensis*). J Nucl Agric Sci 25:910. CNKI:SUN: HNXB.0.2011-05-014

Wang X, Yang Y, Ma C, Jin J, Ma J, Cao H (2011c) Cloning and expression analysis of cyclin gene (*CsCYC1*) of tea plant. Acta Bot Boreali-Occidentalia Sin 31:2365–2372

Wang YS, Gao LP, Wang ZR, Liu YJ, Sun M, Zeng W, Yuan H (2012a) Light induced expression of genes involved in phenylpropanoid biosynthetic pathways in callus of tea (*Camellia sinensis* (L.) O. Kuntze). Sci Hortic 133:72–83

Wang YS, Gao LP, Shan Y, Liu YJ, Tian YW, Xia T (2012b) Influence of shade on flavonoid biosynthesis in tea (*Camellia sinensis (L.) O.* Kuntze). Sci Hortic 141:7–16

Wang X, Ma C, Yang Y, Jin J, Ma J, Cao H (2012c) cDNA cloning and expression analysis of cyclin dependent kinase (CsCDK) gene in tea plant. Acta Hort Sinica 39:333–342

Wang Y, Jiang C, Li Y, Wei C, Deng W (2012d) CsICE1 and CsCBF1: two transcription factors involved in cold responses in *Camellia sinensis*. Plant Cell Rep 31:27–34

Wang P, Song P, Li X, Su R, Wang H, Zhu G (2012e) Study on soluble expression of glutamate dehydrogenase from tea plant in *Escherichia coli* using fusion tags. Afr J Biotechnol 11:6241–6250

Wang XC, Zhao QY, Ma CL, Zhang ZH, Cao HL, Kong YM, Yue C, Hao XY, Chen L, Ma JQ, Jin JQ, Li X, Yang YJ (2013) Global transcriptome profiles of *Camellia sinensis* during cold acclimation. BMC Genomics 14:415

Wang ZW, Jiang C, Wen Q, Wang N, Tao YY, Xu LA (2014a) Deep sequencing of the *Camellia chekiangoleosa* transcriptome revealed candidate genes for anthocyanin biosynthesis. Gene 538:1–7

Wang L, Yue C, Cao H, Zhou Y, Zeng J, Yang Y, Wang X (2014b) Biochemical and transcriptome analyses of a novel chlorophyll-deficient chlorina tea plant cultivar. BMC Plant Biol 14:352

Wang D, Li CF, Ma CL, Chen L (2015a) Novel insights into the molecular mechanisms underlying the resistance of *Camellia sinensis* to *Ectropis oblique* provided by strategic transcriptomic comparisons. Sci Hortic 192:429–440

Wang L, Cao H, Chen C, Yue C (2015b) Complementary transcriptomic and proteomic analyses of a chlorophyll-deficient tea plant cultivar reveal multiple metabolic pathway changes. J Proteome 130:160–169

Wang Y, Shu Z, Wang W, Li D, Pan J, Li X (2015c) CsWRKY2, a novel WRKY gene from *Camellia sinensis*, is involved in cold and drought stress responses. Biol Planta 60:443–451

Wang L, Cao H, Chen C, Yue C, Hao X, Yang Y, Wang X (2016a) Complementary transcriptomic and proteomic analyses of a chlorophyll-deficient tea plant cultivar reveal multiple metabolic pathway changes. J Proteome 130:160–169

Wang L, Wang Y, Cao H, Hao X, Zeng J, Yang Y, Wang X (2016b) Transcriptome analysis of an anthracnose-resistant tea plant cultivar reveals genes associated with resistance to *Colletotrichum camelliae*. PLoS One 11(e0148535):1–26

Wang YN, Tang L, Hou Y, Wang P, Yang H, Wei CL (2016c) Differential transcriptome analysis of leaves of tea plant (*Camellia sinensis*) provides comprehensive insights into the defense responses to Ectropis oblique attack using RNA-Seq. Funct Integr Genomics 16:383–398

Wang W, Xin H, Wang M, Ma Q, Wang L, Kaleri NA, Wang Y, Li X (2016d) Transcriptomic analysis reveals the molecular mechanisms of drought-stress-induced decreases in *Camellia sinensis* leaf quality. Front Plant Sci 7:385

Wang W, Sheng X, Shu Z, Li D, Pan J, Ye X, Chang P, Li X, Wang Y (2016e) Combined cytological and transcriptomic analysis reveals a nitric oxide signaling pathway involved in cold-inhibited *Camellia sinensis* pollen tube growth. Front Plant Sci 7:456

Wang YX, Liu ZW, Wu ZJ, Zhuang J (2016f) Transcriptome-wide identification and expression analysis of the NAC gene family in tea plant [*Camellia sinensis (L.) O. Kuntze*]. PLoS One 11: e0166727

Wang MX, Ma QP, Han BY, Li XH (2016g) Molecular cloning and expression of a jasmonate biosynthetic gene allene oxide cyclase from *Camellia sinensis*. Can J Plant Sci 96:109–116

Wang WW, Jun ZW, Wei ZL, Xin C, Jing Z (2017a) Cloning and expression profiles analysis of development related gene *CsARGOS* in tea plant (*Camellia sinensis*). Plant Physiol J 53:38–44

Wang ML, Li QH, Xin HH, Chen X, Zhu XJ, Li XH (2017b) Reliable reference genes for normalization of gene expression data in tea plants (*Camellia sinensis*) exposed to metal stresses. PLoS One 12:e0175863

Wang M, Zhuang J, Zou Z, Li Q, Xin H, Li X (2017c) Overexpression of a *Camellia sinensis* DREB transcription factor gene (*CsDREB*) increases salt and drought tolerance in transgenic *Arabidopsis thaliana*. J Plant Biol 60:452–461

Wang L, Cao H, Qian W, Yao L, Hao X, Li N, Yang Y, Wang X (2017d) Identification of a novel bZIP transcription factor in *Camellia sinensis* as a negative regulator of freezing tolerance in transgenic Arabidopsis. Ann Bot 119:1195–1209

Wang W, Zhou Y, Wu Y, Dai X, Liu Y, Qian Y, Li M, Jiang X, Wang Y, Gao L, Xia T (2018a) Insight into catechins metabolic pathways of Camellia sinensis based on genome and transcriptome analysis. J Agric Food Chem 66:4281–4293

Wang Y, Hao X, Lu Q, Wang L, Qian W, Li N, Ding C, Wang X, Yang Y (2018b) Transcriptional analysis and histochemistry reveal that hypersensitive cell death and H_2O_2 have crucial roles in the resistance of tea plant (*Camellia sinensis* (L.) O. Kuntze) to anthracnose. Hortic Res 5:18

Wang M, Zhang X, Li Q, Chen X, Li X (2018c) Comparative transcriptome analysis to elucidate the enhanced thermotolerance of tea plants (*Camellia sinensis*) treated with exogenous calcium. Planta 249:775–786

Wang W, Gao T, Chen J, Yang J, Huang H, Yu Y (2018d) The late embryogenesis abundant gene family in tea plant (*Camellia sinensis*): genome-wide characterization and expression analysis in response to cold and dehydration stress. Plant Physiol Bioch 135:1–19

Wang YX, Liu ZW, Wu ZJ, Li H, Wang WL, Cui X, Zhuang J (2018e) Genome-wide identification and expression analysis of GRAS family transcription factors in tea plant (*Camellia sinensis*). Sci Rep 8(3949):1–19

Wang J, Jiyuan WB, Yin H, Fan Z, Li X, He L (2018f) Overexpression and silent expression of *CrGA20ox1* from *Camellia reticulata* 'Hentiangao' and its effect on morphological alterations in transgenic tobacco plants. Plant Breed 137:903–911

Wang P, Zhang L, Jiang X, Dai X et al (2018g) Evolutionary and functional characterization of leucoanthocyanidin reductases from *Camellia sinensis*. Planta 247:139–154

Wang B, Tan X, Jiang J, Zhang L (2018h) Molecular cloning and expression of two genes encoding accase subunits of *Camellia Oleifera* (Theaceae). Pak J Bot 50:103–110

Wang X, Liu BY, Zhao Q, Sun X, Li Y, Duan Z, Miao X, Luo S, Li J (2019a) Genomic variance and transcriptional comparisons reveal the mechanisms of leaf color affecting palatability and stressed defense in tea Plant. Genes (Basel) 10:E929

Wang Y, Zheng PC, Liu PP, Song XW, Guo F, Li YY, Ni DJ, Jiang CJ (2019b) Novel insight into the role of withering process in characteristic flavor formation of teas using transcriptome analysis and metabolite profiling. Food Chem 272:313–322

Wang P, Yue C, Chen D, Zheng Y, Zhang Q, Yang J, Ye N (2019c) Genome-wide identification of WRKY family genes and their response to abiotic stress in tea plant (*Camelia sinensis*). Genes Genomics 41:17–33

Wang P, Zheng Y, Guo Y, Chen X, Sun Y, Yang J, Ye N (2019d) Identification, expression, and putative target gene analysis of nuclear factor-Y (NF-Y) transcription factors in tea plant (*Camellia sinensis*). Planta 250:1671–1686

Wang P, Chen X, Guo Y, Zheng Y, Yue C, Yang J, Ye N (2019e) Identification of CBF transcription factors in tea plants and a survey of potential CBF target genes under low temperature. Int J Mol Sci 20:5137

Wang Y, Lu Q, Xiong F, Hao X, Wang L, Zheng M, Li N, Ding C, Wang X, Yang Y (2020) Genome-wide identification, characterization, and expression analysis of nucleotide-binding leucine-rich repeats gene family under environmental stresses in tea (*Camellia sinensis*). Genomics 112:1351–1362

Wei CL, Jiang CJ, Tao HZ, Wan XC (2003) Cloning and bioinformatics analysis of sequence signature of violaxanthin de-epoxidase cDNA in tea plant (*Camellia sinensis* (L.) O. Kuntze). J Nanjing Agric Univ 26:14–19

Wei CL, Jiang CJ, Tao HZ, Wan XC (2004) Site-directed mutation of violaxanthin de-epoxidase from tea plant (*Camellia sinensis*) *in vitro* and expression of bioactivity assay of the mutants. Chinese J Biochem Mol Biol 20:73–78

Wei K, Wang L, Cheng H, Zhang C, Ma C, Zhang L, Gong W, Wu L (2013) Identification of genes involved in indole-3-butyric acid-induced adventitious root formation in nodal cuttings of *Camellia sinensis* (L.) by suppression subtractive hybridization. Genes 514:91–98

Wei K, Wang LY, Wu LY, Zhang CC, Li HL, Tan LQ, Cao HL, Cheng H (2014) Transcriptome analysis of indole-3-butyric acid-induced adventitious root formation in nodal cuttings of *Camellia sinensis* (L.). PLoS One 9:e107201

Wei K, Wang L, Zhang C, Wu L, Li H, Zhang F, Cheng H (2015) Transcriptome analysis reveals key flavonoid 3′-Hydroxylase and flavonoid 3′,5′-Hydroxylase genes in affecting the ratio of dihydroxylated to trihydroxylated catechins in *Camellia sinensis*. PLoS One 10:e0137925

Wei K, Zhang Y, Wu L, Li H, Ruan L, Bai P, Zhang C, Zhang F, Xu L, Wang L, Cheng H (2016) Gene expression analysis of bud and leaf color in tea. Plant Physiol Bioch 107:310–318

Wei C, Yang H, Wang S, Zhao J, Liu C, Gao L, Xia E, Lu Y, Tai Y, She G, Sun J, Cao H, Tong W, Gao Q, Li Y, Deng W, Jiang X, Wang W, Chen Q, Zhang S, Li H, Wu J, Wang P, Li P, Shi C, Zheng F, Jian J, Huang B, Shan D, Shi M, Fang C, Yue Y, Li F, Li D, Wei S, Han B, Jiang C, Yin Y, Xia T, Zhang Z, Bennetzen JL, Zhao S, Wan X (2018) Draft genome sequence of *Camellia sinensis* var. *sinensis* provides insights into the evolution of the tea genome and tea quality. Proc Natl Acad Sci U S A A115:E4151–E4158

Wu YL, Pan LP, Yu SL, Li HH (2010) Cloning, microbial expression and structure-activity relationship of polyphenol oxidases from *Camellia sinensis*. J Biotechnol 145:66–72

Wu H, Chen D, Li J, Yu B, Qiao X, Huang H, He Y (2013) *De novo* characterization of leaf transcriptome using 454 sequencing and development of EST-SSR markers in tea (*Camellia sinensis*). Plant Mol Biol Report 31:524–538

Wu ZJ, Li XH, Liu ZW, Xu ZS, Zhuang J (2014a) De novo assembly and transcriptome characterization: novel insights into catechins biosynthesis in *Camellia sinensis*. BMC Plant Biol 14:277

Wu Z, Lu L, Li X, Fang W, Zhou L, Tan G, Zhuang J (2014b) Isolation and expression profiles analysis of AP2/ERF-B3 group transcription factor from *Camellia sinensis*. J Nanjing Agric Univ 37:67–75

Wu ZJ, Li XH, Liu ZW, Li H, Wang YX, Zhuang J (2015) Transcriptome-based discovery of AP2/ERF transcription factors related to temperature stress in tea plant (*Camellia sinensis*). Funct Integr Genomics 15:741–752

Wu Q, Chen Z, Sun W, Deng T, Chen M (2016a) *De novo* sequencing of the leaf transcriptome reveals complex light-responsive regulatory networks in *Camellia sinensis* cv. Baijiguan. Front Plant Sci 7:332

Wu ZJ, Li XH, Liu ZW, Li H, Wang YX, Zhuang J (2016b) Transcriptome-wide identification of *Camellia sinensis* WRKY transcription factors in response to temperature stress. Mol Genet Genomics 291:255–269

Wu Z, Tian C, Jiang Q, Li X-H, Zhuang J (2016c) Selection of suitable reference genes for qRT-PCR normalization during leaf development and hormonal stimuli in tea plant (*Camellia sinensis*). Sci Rep 6:19748

Wu ZJ, Ma HY, Zhuang J (2017a) iTRAQ-based proteomics monitors the withering dynamics in postharvest leaves of tea plant (*Camellia sinensis*). Mol Genet Genom 293:45–59

Wu ZJ, Wang WL, Zhuang J (2017b) *TCP* family genes control leaf development and its responses to hormonal stimuli in tea plant [*Camellia sinensis* (L.) O. Kuntze]. J Plant Growth Regul 83:43–53

Wu Y, Wang W, Li Y, Dai X, Ma G, Xing D, Zhu M, Gao L, Xia T (2017c) Six phenylalanine ammonia-lyases from *Camellia sinensis*: evolution, expression, and kinetics. Plant Physiol Biochem 118:413–421

Wu B, Ruan C, Han P, Ruan D, Xiong C, Ding J, Liu S (2019a) Comparative transcriptomic analysis of high- and low-oil *Camellia oleifera* reveals a coordinated mechanism for the regulation of upstream and downstream multigenes for high oleic acid accumulation. 3 Biotech 9:257

Wu LY, Fang ZT, Lin JK, Sun Y, Du ZZ, Guo YL, Liu JH, Liang YR, Ye JH (2019b) Complementary iTRAQ proteomic and transcriptomic analyses of leaves in tea plant (*Camellia sinensis* L.) with different maturity and regulatory network of flavonoid biosynthesis. J Proteome Res 18:252–264

Xia EH, Jiang JJ, Huang H, Zhang LP, Zhang HB, Gao LZ (2014) Transcriptome analysis of the oil-rich tea plant, *Camellia oleifera*, reveals candidate genes related to lipid metabolism. PLoS One 9:e104150

Xia EH, Zhang HB, Sheng J, Li K, Zhang QJ, Kim C, Zhang Y, Liu Y, Zhu T, Li W (2017a) The tea tree genome provides insights into tea flavor and independent evolution of caffeine biosynthesis. Mol Plant 882(10):866–877

Xia J, Liu Y, Yao S, Li M, Zhu M, Huang K, Gao L, Xia T (2017b) Characterization and expression profiling of *Camellia sinensis* cinnamate 4-hydroxylase genes in phenylpropanoid pathways. Genes (Basel) 8(8):193

Xia E, Tong W, Hou Y, An Y, Chen L, Wu Q, Liu Y, Yu J, Li F, Li R, Li P, Zhao H, Ge R, Huang J, Mallano AI, Zhang Y, Liu S, Deng W, Song C, Zhang Z, Zhao J, Wei S, Zeng C, Lin H, Liu Z, Liu Z (2019a) Analysis of young shoots of 'Anji Baicha' (*Camellia sinensis*) at three developmental stages using nontargeted LC-MS-based metabolomics. J Food Sci 84:1746–1757

Xia EH, Li FD, Tong W, Li PH, Wu Q, Zhao HJ, Ge RH, Li RP, Li YY, Zhang ZZ, Wei CL, Wan XC (2019b) Tea Plant Information Archive: a comprehensive genomics and bioinformatics platform for tea plant. Plant Biotechnol J 17:1938–1953

Xia E, Li F, Tong W et al (2019c) The tea plant reference genome and improved gene annotation using long-read and paired-end sequencing data. Sci Data 6:122

Xia E, Tong W, Hou Y, An Y et al (2020) The reference genome of tea plant and resequencing of 81 diverse accessions provide insights into its genome evolution and adaptation. Mol Plant 13:1013–1026

Xiao Z, Fu R, Li J, Fan Z, Yin H (2015) Overexpression of the Gibberellin 2-Oxidase gene from *Camellia lipoensis* induces dwarfism and smaller flowers in *Nicotiana tabacum*. Plant Mol Biol Report 34:182–191

Xie Y, Wang X (2018) Comparative transcriptomic analysis identifies genes responsible for fruit count and oil yield in the oil tea plant *Camellia chekiangoleosa*. Sci Rep 8:6637

Xie XF, Tian XF, Jiang CJ, Li YY (2015) Screening of microRNA reference genes for real-time fluorescence quantitative PCR under cold stress in *Camellia sinensis*. J Tea Sci 35:596–604

Xin Z, Zhang J, Ge L, Lei S, Han J, Zhang X, Li X, Sun X (2017) A putative 12-oxophytodienoate reductase gene *CsOPR3* from *Camellia sinensis*, is involved in wound and herbivore infestation responses. Gene 615:18–24

Xu YX, Mao J, Chen W, Qian TT, Liu SC, Hao WJ, Li CF, Chen L (2016) Identification and expression profiling of the auxin response factors (ARFs) in the tea plant (*Camellia sinensis* (L.) O. Kuntze) under various abiotic stresses. Plant Physiol Biochem 98:46–56

Xu Q, Zhu J, Zhao S, Hou Y, Li F, Tai Y, Wan X, Wei C (2017a) Transcriptome profiling using single-molecule direct RNA sequencing approach for *in-depth* understanding of genes in secondary metabolism pathways of *Camellia sinensis*. Front Plant Sci 8:1205

Xu YX, Shen CJ, Ma JQ, Chen W, Mao J, Zhou YY, Chen L (2017b) Quantitative succinyl-proteome profiling of *Camellia sinensis* cv. "Anji Baicha" during periodic Albinism. Sci Rep 7:1873

Xu Q, Wang Y, Ding Z, Fan K, Ma D, Zhang Y, Yin Q (2017c) Aluminum induced physiological and proteomic responses in tea (*Camellia sinensis*) roots and leaves. Plant Physiol Biochecm 115:141–151

Xu Q, Cheng L, Mei Y, Huang L, Zhu J, Mi X, Yu Y, Wei C (2019) Alternative splicing of key genes in LOX pathway involves biosynthesis of volatile fatty acid derivatives in tea plant (*Camellia sinensis*). J Agric Food Chem 67:13021–13032

Xu P, Zhang X, Su H, Liu X, Wang Y, Hong G (2020a) Genome-wide analysis of *PYL-PP2C-SnRK2s* family in *Camellia sinensis*. J Bioeng 11:103–115

Xu W, Dong Y, Yu Y, Zhang X, Hou X, Sun X (2020b) Identification and evaluation of reliable reference genes for quantitative real-time PCR analysis in tea plants under differential biotic stresses. Sci Rep 10:2429

Yadav SK (2009) Computational structural analysis and kinetic studies of a cytosolic glutamine synthetase from *Camellia sinensis* (L.) O. Kuntze. Protein J 28:428–434

Yang H, Xie S, Wang L, Jing S, Zhu X, Li X, Zeng W, Yuan H (2011) Identification of up-regulated genes in tea leaves under mild infestation of green leaf hopper. Sci Hortic 130:476–448

Yang D, Liu Y, Suna M, Zhao L, Wang Y, Chen X, Wei C, Gao L, Xia T (2012a) Differential gene expression in tea (*Camellia sinensis L.*) calli with different morphologies and catechin contents. J Plant Physiol 169:163–175

Yang YJ, Wang XC, Ma CL (2012b) Cloning and bioinformatics analysis of full-length cDNA of actin gene (*CsActin*1) from tea plant (*Camellia sinensis* (L.) O. Kuntze). Bull Bot Res 32:69–76

Yang JB, Yang SX, Li HT, Yang J, Li DZ (2013) Comparative chloroplast genomes of *Camellia* species. PLoS One 8:e73053

Yang H, Wang Y, Li L, Li F, He Y, Wu J, Wei C (2019) Transcriptomic and phytochemical analyses reveal root-mediated resource-based defense response to leaf herbivory by *Ectropisoblique* in tea plant (*Camellia sinensis*). J Agric Food Chem 67(19):5465–5476

Yao QY, Huang H, Tong Y, Xia EH, Gao LZ (2016) Transcriptome analysis identifies candidate genes related to triacylglycerol and pigment biosynthesis and photoperiodic flowering in the ornamental and oil-producing plant, *Camellia reticulata* (Theaceae). Front Plant Sci 7:163

Ye Y, Jiang CJ, Zhu L, Yu M, Wang ZX, Deng WW, Wei CL (2009) Cloning and sequencing of a full length cDNA encoding the RuBPCase small subunit (RbcS) in tea (*Camellia sinensis*). Agric Sci China 8:161–166

Yin C, Wang F, Fan H, Fang Y, Li W (2019) Identification of tea plant purple acid phosphatase genes and their expression responses to excess Iron. Int J Mol Sci 20:1954

Ying CMA, Peng H, Zhi L, Yue Z, Guo Y, Tan L, Feng J (2013) Cloning and prokaryotic expression of Flavonoid O-methyltransferase from *Camellia sinensis*. Scien Agria Sinica 46:325–333

Ying Z, Ling Z, Jiadong G, Fang D et al (2015) Molecular cloning and characterization of a short-chain dehydrogenase showing activity with volatile compounds isolated from *Camellia sinensis*. Plant Mol Biol Report 33:253–263

Yoshida K, Homma T (2005) Isolation of wound/pathogen inducible cDNA from tea by mRNA differential display. In: Inter Tea Symp-2005, Hanzou, China, pp 62–65

Yu LY, Juan YC, Xu W, Rong HT (2017) Cloning and function analysis of metal tolerance gene (*CsMTP11*) in tea plant (*Camellia sinensis* L. O. Kuntze). Acta Agronomica Sinica 43:708–717

Yue C, Cao H, Wang LU, Zhou Y, Hao X, Zeng J, Wang X, Yang Y (2014) Molecular cloning and expression analysis of tea plant aquaporin (AQP) gene family. Plant Physiol Biochem 83:65–76
Yue C, Cao HL, Chen D, Lin HZ, Wang Z, Hu J, Yang GY, Guo YQ, Ye NX, Hao XY (2018) Comparative transcriptome study of hairy and hairless tea plant (*Camellia sinensis*) shoots. J Plant Physiol 229:41–52
Ze L, Jianyong W, Hongxu L, Yanling Z et al (2017) Cloning and expression analysis of an aldehyde dehydrogenase gene from *Camellia oleifera*. Nanosci Nanotechnol Lett 9:364–373
Zeng Y, Tan X, Zhang L, Jiang N, Cao H (2014) Identification and expression of Fructose-1,6-bisphosphate aldolase genes and their relations to oil content in developing seeds of tea oil tree (*Camellia oleifera*). PLoS One 9:1–9
Zeng C, Lin H, Liu Z, Liu Z (2019) Analysis of young shoots of 'Anji Baicha' (*Camellia sinensis*) at three developmental stages using nontargeted LC-MS-based metabolomics. J Food Sci 84:1746–1757
Zhang YL, Zhao LP, Chun-Lei MA, Chen L (2007) Molecular identification, bioinformatic analysis and prokaryotic expression of the cyclophilin gene full length cDNA from tea plant (*Camellia sinensis*). J Tea Sci 27:120–126
Zhang YL, Qiao XY, Chen L (2008a) Full-length cDNA cloning and bioinformatic analysis of ACC synthase gene from the tea plant (*Camellia sinensis*). J Tea Sci 28:235–241
Zhang YL, Qiao XY, Chen L (2008b) Molecular cloning and expression analysis of ACC oxidase gene full-length cDNA from the tea plant (*Camellia sinensis*). J Tea Sci 28:459–466
Zhang X, Liu Y, Gao K, Zhao L, Liu L, Wang Y, Sun M, Gao L, Xia T (2012) Characterization of anthocyanidin reductase from Shuchazao green tea. J Sci Food Agric 92:1533–1539
Zhang Y, Zhu X, Chen X, Song C, Zou Z, Wang Y, Wang M, Fang W, Li X (2014) Identification and characterization of cold-responsive microRNAs in tea plant (*Camellia sinensis*) and their targets using high-throughput sequencing and degradome analysis. BMC Plant Biol 14:271
Zhang HB, Xia EH, Huang H, Jiang JJ, Liu BY, Gao LZ (2015) *De novo* transcriptome assembly of the wild relative of tea tree (*Camellia taliensis*) and comparative analysis with tea transcriptome identified putative genes associated with tea quality and stress response. BMC Genomics 16:298
Zhang CC, Wang LY, Wei K, Wu LY, Li HL, Zhang F, Cheng H, Ni DJ (2016a) Transcriptome analysis reveals self-incompatibility in the tea plant (*Camellia sinensis*) might be under gametophytic control. BMC Genomics 17(359):1–15
Zhang CC, Tan QL, Wang YL, Wei K, Wu YL, Zhang F, Cheng H, Ni JD (2016b) Cloning and characterization of an S-RNase gene in *Camellia sinensis*. Sci Hort 207:218–224
Zhang LQ, Wei K, Cheng H, Wang YL, Zhang CC (2016c) Accumulation of catechins and expression of catechin synthetic genes in *Camelliasinensis* at different developmental stages. Bot Stud 57(31):1–8
Zhang Q, Liu M, Ruan J (2017a) Integrated transcriptome and metabolic analyses reveals novel insights into free amino acid metabolism in Huangjinya tea cultivar. Front Plant Sci 8:291
Zhang L, Tai Y, Wang Y, Meng Q, Yang Y, Zhang S, Wan X (2017b) The proposed biosynthesis of procyanidins by the comparative chemical analysis of five *Camellia* species using LC-MS. Sci Rep 7:46131
Zhang S, Zhang L, Tai Y, Wang X, Ho CT, Wan X (2018a) Gene discovery of characteristic metabolic pathways in the tea plant (*Camellia sinensis*) using 'Omics'-based network approaches: a future perspective. Front Plant Sci 9:480
Zhang Y, Wei K, Li H, Wang L, Ruan L, Pang D, Cheng H (2018b) Identification of key genes involved in catechin metabolism in tea seedlings based on transcriptomic and HPLC analysis. Plant Physiol Bioch 133:107–115
Zhang M, Liu XK, Fan W, Yan DF, Zhong NS, Gao JY, Zhang WJ (2018c) Transcriptome analysis reveals hybridization-induced genome shock in an interspecific F_1 hybrid from *Camellia*. Genome 61:477–485
Zhang M, Ttang YW, Qi J, Liu XK, Yan DF, Zhong NS, Tao NQ, Gao JY, Wang YJ, Song ZP, Yang J, Zhang WZ (2019a) Effects of parental genetic divergence on gene expression patterns in interspecific hybrids of Camellia. BMC Genomics 20:828

Zhang F, Li W, Gao CW, Zhang D, Gao LZ (2019b) Deciphering tea tree chloroplast and mitochondrial genomes of *Camellia sinensis* var. assamica. Sci Data 6:209

Zhang W, Zhao Y, Yang G, Peng J, Chen S, Xu Z (2019c) Determination of the evolutionary pressure on *Camellia oleifera* on Hainan island using the complete chloroplast genome sequence. PeerJ 7:e7210

Zhang QJ, Li W, Li K, Nan H, Shi C, Zhang Y, Dai ZY, Lin YL, Yang XL, Tong Y, Zhang D, Lu C, Feng LY, Wang CF, Liu XX, Huang JA, Jiang WK, Wang XH, Zhang XC, Eichler EE, Liu ZH, Gao LZ (2020a) The Chromosome-level reference genome of tea tree unveils recent bursts of non-autonomous LTR retrotransposons to drive genome size evolution. Mol Plant 13 (7):935–938

Zhang D, Han Z, Li J, Qin H, Zhou L, Wang Y, Zhu X, Ma Y, Fang W (2020b) Genome-wide analysis of the *SBP-box* gene family transcription factors and their responses to abiotic stresses in tea (*Camellia sinensis*). Genome 1112(3):2194–2202

Zhao D, Liu ZS, Xi B (2001) Cloning and alignment of polyphenol oxidase cDNA of tea plant. J Tea Sci 21:94–98

Zhao LP, Chen L, Wang XC, Yao MZ (2006a) Quantitative detection of β-glucosidase and β-primeveroside gene expressions in different leaves of tea plant (*Camellia sinensis*) by real-time PCR analysis. J Tea Sci 26:11–16

Zhao LP, Gao QK, Chen L, Wang XC, Yao MZ (2006b) Development and preliminary application of cDNA microarray of tea plant (*Camellia sinensis*). J Tea Sci 5:3–7

Zhao L, Gao L, Wang H, Chen X, Wang Y, Yang H, Wei C, Wan X, Xia T (2012) The R2R3-MYB, bHLH, WD40, and related transcription factors in flavonoid biosynthesis. Funct Integr Genomics 13:75–98

Zhao L, Gao L, Wang H, Chen X, Wang Y, Yang H, Wei C, Wan X, Xia T (2013) The R2R3-MYB, bHLH, WD40, and related transcription factors in flavonoid biosynthesis. Funct Integr Genomics 13:75–98

Zhao S, Wang X, Yan X, Guo L, Mi X, Xu Q, Zhu J, Wu A, Liu L, Wei C (2018) Revealing of the microRNA involved regulatory gene networks on terpenoid biosynthesis in *Camellia sinensis* in different growing time points. J Agric Food Chem 66:12604–12616

Zhao L, Chen C, Wang Y, Shen J, Ding Z (2019) Conserved microRNA act boldly during sprout development and quality formation in PingyangTezaocha (*Camellia sinensis*). Front Genet 10:237

Zheng C, Zhao L, Wang Y, Shen J, Zhang Y, Jia S, Li Y, Ding Z (2015) Integrated RNA-Seq and sRNA-Seq analysis identifies chilling and freezing responsive key molecular players and pathways in tea plant (*Camellia sinensis*). PLoS One 10:e0125031

Zheng C, Wang Y, Ding Z, Zhao L (2016) Global transcriptional analysis reveals the complex relationship between tea quality, leaf senescence and the responses to cold-drought combined stress in *Camellia sinensis*. Front Plant Sci 7:1858

Zheng C, Ma JQ, Ma CL, Shen SY, Liu YF, Chen L (2019a) Regulation of growth and flavonoid formation of tea plants (*Camellia sinensis*) by blue and green light. J Agric Food Chem 67:2408–2419

Zheng Y, Wang P, Chen X, Sun Y, Yue C, Ye N (2019b) Transcriptome and metabolite profiling reveal novel insights into volatile heterosis in the tea plant (*Camellia sinensis*). Molecules 24:3380

Zheng C, Ma JQ, Chen JD, Ma CL, Chen W, Yao MZ, Chen L (2019c) Gene coexpression networks reveal key drivers of flavonoid variation in eleven tea cultivars (*Camellia sinensis*). J Agric Food Chem 67:9967–9978

Zheng Y, Chen X, Wang P, Sun Y, Yue C, Ye N (2020a) Genome-wide and expression pattern analysis of JAZ family involved in stress responses and postharvest processing treatments in *Camellia sinensis*. Sci Rep 10(2792):1–2713

Zheng C, Ma JQ, Chen JD, Ma CL, Chen W, Yao MZ, Chen LJ (2020b) Gene coexpression networks reveal key drivers of flavonoid variation in eleven tea cultivars (*Camellia sinensis*). J Agric Food Chem 67:9967–9978

Zhou WX, Fan QZ, Chen Y, Zhu LY, Li YJ, Yin FH (2013) Functional analyses of a Flavonol synthase–like gene from *Camellia nitidissima* reveal its roles in flavonoid metabolism during floral pigmentation. J Biosci 38:593–604

Zhou TS, Zhou R, Yu YB, Xiao Y, Li DH, Xiao B, Yu O, Yang YJ (2016) Cloning and characterization of a Flavonoid 3′-hydroxylase gene from tea plant (*Camellia sinensis*). Int J Mol Sci 17:1–17

Zhou X, Li J, Zhu Y, Ni S, Chen J, Feng X, Zhang Y, Li S, Zhu H, Wen Y (2017a) *De novo* assembly of the *Camellia nitidissima* transcriptome reveals key genes of flower pigment biosynthesis. Front Plant Sci 8:1545

Zhou Y, Liu Y, Wang S, Shi C, Zhang R, Rao J, Wang X, Gu X, Wang Y, Li D, Wei C (2017b) Molecular cloning and characterization of Galactinol synthases in *Camellia sinensis* with different responses to biotic and abiotic stressors. J Agric Food Chem 65:2751–2759

Zhou C, Zhu C, Fu H, Li X, Chen L, Lin Y, Guo Y (2019a) Genome-wide investigation of superoxide dismutase (SOD) gene family and their regulatory miRNAs reveal the involvement in abiotic stress and hormone response in tea plant (*Camellia sinensis*). PLoS One 14 (e0223609):1–23

Zhou Z, Deng H, Wu Q, Liu B, Yue C, Deng T, Lai Z, Sun Y (2019b) Validation of reference genes for gene expression studies in post-harvest leaves of tea plant (*Camellia sinensis*). PeerJ 7:e6385

Zhou J, Lu M, Yu S, Liu Y, Yang J, Tan X (2020a) In-depth understanding of *Camellia oleifera* self-incompatibility by comparative transcriptome, proteome and metabolome. Int J Mol Sci 21:1600

Zhou CH, Shamala FL, Yi KX, Yan Z, Wei S (2020b) Analysis of terpene synthase family genes in *Camellia sinensis* with an emphasis on abiotic stress conditions. Sci Rep 10(933):1–13

Zhu QW, Luo YP (2013) Identification of miRNAs and their targets in tea (*Camellia sinensis*). J Zhejiang Univ Sci B 14:916–923

Zhu L, Deng WW, Ye AH, Yu M, Wang ZX, Jiang CJ (2008a) Cloning of two cDNAs encoding a family of ATP sulfurylase from *Camellia sinensis* related to selenium or sulfur metabolism and functional expression in *Escherichia coli*. Plant Physiol Biochem 46:731–738

Zhu L, Jiang CJ, Deng WW, Gao X, Wang RJ, Wan XC (2008b) Cloning and expression of selenocysteine methyltransferase cDNA from *Camellia sinensis*. Acta Physiol Plant 30:167–174

Zhu B, Chen LB, Lu M, Zhang J, Han J, Deng WW, Zhang ZZ (2019a) Caffeine content and related gene expression: novel insight into caffeine metabolism in *Camellia* plants containing low, normal, and high caffeine concentrations. J Agric Food Chem 67:3400–3411

Zhu J, Pan J, Nong S, Ma Y, Xing A, Zhu X, Wen B, Fang W, Wang Y (2019b) Transcriptome analysis reveals the mechanism of fluoride treatment affecting biochemical components in *Camellia sinensis*. Int J Mol Sci 20:237

Zhu C, Zhang S, Fu H, Zhou C, Chen L, Li X, Guo Y (2019c) Transcriptome and phytochemical analyses provide new insights into long non-coding RNAs modulating characteristic secondary metabolites of oolong tea (*Camellia sinensis*) in solar-withering. Front Plant Sci 10:1638

Index

T. K. Mondal, *Tea: Genome and Genetics*,
https://doi.org/10.1007/978-981-15-8868-6

The manufacturer's authorised representative in the EU is Springer Nature Customer Service Centre GmbH, Europaplatz 3, 69115 Heidelberg, Germany. If you have any concerns regarding our products, please contact ProductSafety@springernature.com

Printed and bound by CPI Group (UK) Ltd, Croydon, CR0 4YY
28/07/2026
02179234-0001